THE PLACE OF ASTRONOMY
IN THE ANCIENT WORLD

THE PLACE OF ASTRONOMY
IN THE ANCIENT WORLD

A JOINT SYMPOSIUM OF
THE ROYAL SOCIETY
AND
THE BRITISH ACADEMY

ORGANIZED BY
D. G. KENDALL, F.R.S.
S. PIGGOTT, F.B.A.
D. G. KING-HELE, F.R.S.
AND I. E. S. EDWARDS, F.B.A.

EDITED BY F. R. HODSON

PUBLISHED FOR
THE BRITISH ACADEMY
BY
OXFORD UNIVERSITY PRESS, LONDON
1974

Oxford University Press, Ely House, London W. 1

GLASGOW NEW YORK TORONTO MELBOURNE WELLINGTON
CAPE TOWN IBADAN NAIROBI DAR ES SALAAM LUSAKA ADDIS ABABA
DELHI BOMBAY CALCUTTA MADRAS KARACHI LAHORE DACCA
KUALA LUMPUR SINGAPORE HONGKONG TOKYO

Hardbound edition ISBN 0 19 725944 8

This hardbound edition is published for the British Academy
by Oxford University Press London.

This report is also published, in a paper-bound edition, by
the Royal Society in its series *Philosophical Transactions A* vol.
276, no. 1257

Printed in Great Britain
at the University Printing House, Cambridge
(Brooke Crutchley, University Printer)

CONTENTS

PREFACE

This volume contains the proceedings of the second joint symposium sponsored by the British Academy and the Royal Society. The first of these, on 'The impact of the natural sciences on archaeology', took place on 11 and 12 December 1969, and the Royal Society–Romanian Academy binational conference on 'Mathematics in the archaeological and historical sciences' (Mamaia, 1970) might be regarded as further evidence of the strong desire of the British Academy and the Royal Society to act vigorously in concert whenever the occasion offers, for the Mamaia meeting owed a great deal to the whole-hearted support of the then Secretary of the Academy, Sir Mortimer Wheeler, F.R.S., F.B.A.

The present symposium, organized for the Academy by I. E. S. Edwards, F.B.A., and S. Piggott, F.B.A., and for the Royal Society by D. G. Kendall, F.R.S., and D. G. King-Hele, F.R.S., depended for its realization on the work of an enlarged committee including R. J. C. Atkinson, G. E. Daniel, F. R. Hodson and J. E. S. Thompson, F.B.A. An attempt has been made to examine 'the place of astronomy' in all parts of the ancient world, and the proceedings will be found to have been arranged so as first to present the evidence relating to literate societies, treated 'in geographical order', and then to present the evidence relating to non-literate societies, and to societies not known to have been literate. The actual Discussion Meeting was spread over the two days 7 and 8 December 1972, but it proved impossible to record more than a fraction of the very lively informal discussions which followed the presentation of the papers collected here.

Professor Hodson generously undertook the laborious tasks of acting both as Secretary to the Planning Committee and as Editor of the present work. To him most especially, and to all who assisted him, we wish to record our gratitude and appreciation. We also wish to thank the then Secretary of the Academy, Mr D. F. Allen, F.B.A., and the Deputy Executive Secretary of the Royal Society, Dr R. W. J. Keay and their colleagues, who encouraged us to act with the freedom such an operation requires, and who dealt with all difficulties as soon as they arose.

D. G. KENDALL

S. PIGGOTT

Phil. Trans. R. Soc. Lond. A. **276**, 5–20 (1974)   [5]
Printed in Great Britain

I. ASTRONOMY IN ANCIENT LITERATE SOCIETIES

Introduction to some basic astronomical concepts

By R. R. NEWTON

Applied Physics Laboratory, Johns Hopkins University
Silver Spring, Maryland 20910, *U.S.A.*

In modern celestial mechanics, the goal is to construct a coordinate system in the heavens, and then to develop theories that will allow us to calculate the coordinates of the Sun, Moon and planets as functions of time. The goal of astronomy among ancient peoples was probably quite different; it was probably to develop methods of dealing with certain appearances in a coordinate system based upon the observer's horizon. These appearances include eclipses, the positions at which celestial objects rise and set, and methods of telling both the time of day and the time of the year by observations of risings and settings. This paper is concerned with discussions of these appearances. It deals with theories of motion only to the extent needed in discussing the appearances.

1. INTRODUCTION

In studying the place of astronomy in the ancient world, it is necessary to have some knowledge of the basic astronomical phenomena that ancient peoples observed. The purpose of this paper is to outline some basic elements for the sake of those who have not had a previous opportunity to study them. Much if not all of the paper will be familiar to many readers.

The first task is to establish coordinate systems which can be used in describing the important phenomena that concern us. Some coordinate systems are fixed with respect to an observer on the Earth's surface. Others are fixed in the heavens; that is, the stars appear to be at constant positions in these systems except for some very slow motions. After we establish the necessary coordinate systems and the relations between them, we turn our attention to three important classes of problem.

The first class concerns the risings and settings of celestial bodies. We shall be concerned with the places on the horizon at which rising and setting occur. We shall also be concerned with the times of rising and setting, both with regard to time of day and time of year.

The second class concerns the motions of the Sun, Moon, and planets with respect to the stars. Only the most basic elements of motion will be discussed here, in modern terms. The terms in which planetary motion was discussed by ancient peoples will be left as a matter for the specialists in those subjects.

Finally, we shall be concerned with the theory of eclipses, since eclipses often played a dramatic part in the thinking of ancient peoples.

In all the discussions, I shall treat the Earth as if it were an exact sphere. I shall also neglect various other small effects such as refraction.

2. EARTH-FIXED AND STAR-FIXED COORDINATES

The coordinate system that was most obvious to an ancient observer was probably the one based upon his own horizon, which we can assume to be level for present purposes. Suppose that an observer is looking north from the point marked O in figure 1, so that he sees half of his horizon along the arc WHNE. Z is his zenith and N is the direction of north. Thus the plane ZON is part of his meridian plane. Other vertical planes are drawn in lightly.

Suppose that the observer sees an object at the point Q. He locates it by first drawing the vertical plane HOZQ through the point and by measuring the angle HON in his horizon plane. HON is called the azimuth of the point Q. HON is usually considered positive when it measured clockwise from N, thus the azimuth of Q is a negative number as figure 1 is drawn. Various conventions about the starting-point and the direction of measuring azimuth have been used in the past. Azimuth will be denoted by the symbol A.

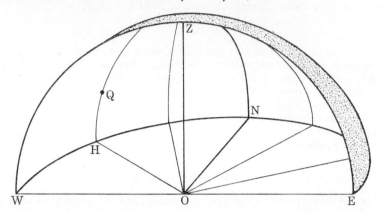

FIGURE 1. The coordinate system of azimuth and elevation.

The observer completes the location of Q by measuring the length of the arc QH, which is called the elevation a of the point Q.

I shall use the term anthropocentric for the coordinate system of figure 1, in order to emphasize that it is based upon the perception of a specific individual at a specific spot. The term topocentric is probably more common, but it removes attention from the individual.

We need the distance from the observer to the object in order to complete the anthropocentric system. For astronomical bodies, the distance is usually so large that its value is unimportant, and the distance coordinate will receive little attention in this paper.

Let us now turn to star-fixed coordinates. I shall assume that the reader knows that the Earth spins on an axis. The points at which the axis pierces the surface of the earth are called the north and south poles. The points at which the axis pierces the celestial sphere, that is, the spherical surface upon which the stars and planets seem to lie, are called the north and south celestial poles. With the aid of these poles, we can set up a coordinate system in the heavens that is analogous to the longitude–latitude system used on earth.

This coordinate system is illustrated in figure 2. In this figure, we are looking at a hemisphere as seen from inside. The figure shows circles that look much like the lines of longitude and latitude shown on terrestrial globes. The quantity analogous to longitude is called the right ascension α, and the quantity analogous to latitude is called the declination δ. Declination, like latitude, is positive north of the equator and negative south of it. Right ascension, like longitude, increases as we go eastwards. There is one main difference between figure 2 and the corresponding figure of a terrestrial hemisphere. In representations of the Earth, we show how it looks from the outside, but in figure 2 we are looking from the inside. Thus, with north at the top, east is to the left rather than to the right in figure 2.

Except for small effects that can be neglected here, a star has a fixed position in the coordinate

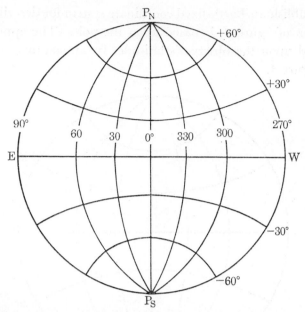

FIGURE 2. The coordinate system of right ascension and declination.

system of figure 2. Obviously a star does not have a fixed position in the system of figure 1. Our task is now to relate positions in the two systems.

In establishing the connexion between the coordinate systems, I am going to assume that the observer's horizon plane passes through the centre of the Earth instead of being tangent to the Earth. The effect of this assumption is shown in figure 3. In this figure, we see the observer's actual horizon drawn tangent to the Earth's surface. The parallel plane through the centre of the Earth can be called the translated horizon. It is clear that the translation of the horizon plane does not change the azimuth of any point. The figure shows how the elevation angle changes from one horizon to the other. The difference between the two elevation angles is called the parallax.

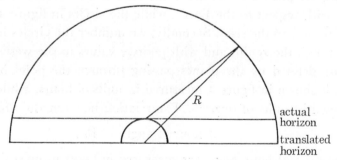

FIGURE 3. The phenomenon of parallax, shown as a function of distance R and of elevation angle.

The parallax depends upon the elevation of the point considered, and it is obviously greatest for a point on the horizon. The parallax for a point on the horizon is called the horizontal parallax, which clearly depends upon the distance R of the point from the centre of the Earth. For the Moon, the horizontal parallax is about 1°. For every other body that will concern us here, the horizontal parallax is much less than 1'. I shall neglect parallax in the remaining discussion of coordinate systems. This is equivalent to taking the translated horizon as the actual horizon.

An observer can establish an Earth-fixed coordinate system for describing the heavens that, like the celestial system of figure 2, is based upon the poles. The appearance of this system will necessarily depend upon the observer's latitude. Its appearance for an observer on the equator is shown in figure 4.

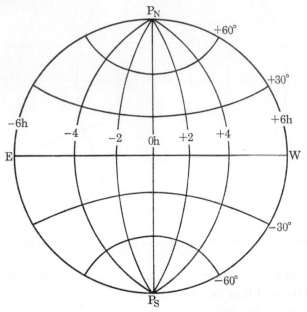

FIGURE 4. The coordinate system of hour angle and declination, for an observer on the Equator.

Suppose that an observer is lying on his back at the equator, with his head pointing toward the north. The celestial poles P_N and P_S are on the horizon. Planes parallel to the equator cut the celestial sphere in circles that are identical to the declination circles in figure 2, and we can continue to use the term declination for the corresponding coordinate. Planes passing through the poles cut the celestial sphere in circles that look like the circles of right ascension in figure 2. However, there are two main differences. The circles in figure 2 are fixed in the stars and rotate with respect to the Earth, while the circles in figure 4 are fixed to the Earth and rotate with respect to the stars. Secondly, we number the circles in figure 4 with the zero circle passing through the zenith and with positive values to the west.

The coordinate defined by the planes passing through the poles in figure 4 is called the hour angle h. It is shown in figure 4 measured in units of hours, while the right ascension in figure 2 was shown in units of degrees. The relation between the units is

$$1 \text{ hour of angle} = 15°. \tag{2.1}$$

Both right ascension and hour angle are measured in hours in most literature, but the use of degrees is probably increasing.

Now suppose that the observer is in the northern hemisphere at latitude ϕ, and let us suppose that he is standing looking toward the north as he was in figure 1. Then the elevation angle of the north celestial pole, namely the arc NP in figure 5, equals ϕ. The declination circles for $\delta \geqslant 90° - \phi$ are above the horizon for their full circuit. If the observer turns toward the south, he finds that the south celestial pole is below the horizon by the angle ϕ; the declination circles that lie within the angle ϕ from the south pole never rise above the horizon. Intermediate declination circles are visible in part but not in full.

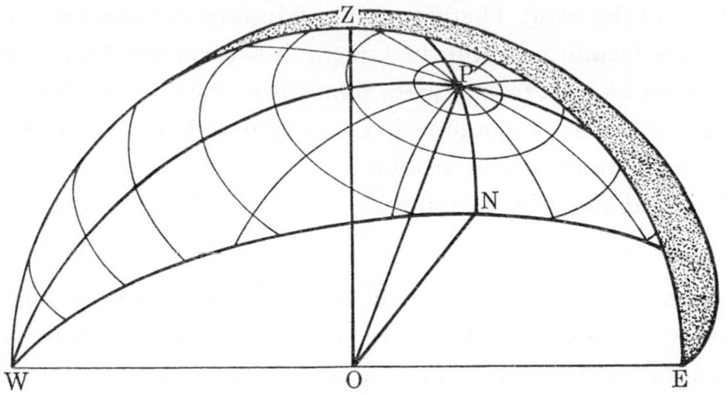

FIGURE 5. The coordinate system of hour angle and declination, for an observer in the northern hemisphere.

The hour-angle arcs radiate from the pole in figure 5. The hour angle is zero along the arc PZ. It is −6h along the arc PE and +6h along the arc PW. Along the arc PN, it can be called either −12h or +12h, whichever is convenient. Some arcs for other values of the hour angle are drawn in lightly in figure 5.

A star or other celestial object with a declination δ seems to travel around the Earth on the corresponding declination circle, like a bead sliding on a string. Thus the stars near the north pole never set for the observer in figure 5, although they may become invisible in the daytime because of the brightness of the sky. Stars near the south pole never rise. All stars in between rise, remain above the horizon for some time, and then set. Since the declination circles in figure 5 do not intersect the horizon at right angles, an object does not rise straight from the horizon. Instead, it travels southward as it rises and, similarly, it travels northward as it sets. Only at the equator do objects rise vertically.

I shall take up the formal relations between the various coordinate systems in the next section. After the formal relations have been established, I shall take up rising and setting problems.

3. Transformations of coordinates; time

The azimuth-elevation system in figure 1 and the declination-hour angle system in figures 4 and 5 are fixed with respect to the Earth. Hence the transformation between the systems does not involve the time, but it does involve the latitude ϕ of the observer. The transformation from (a, A) to (δ, h) is

$$\left.\begin{aligned} \delta &= \arcsin \{\sin a \sin \phi + \cos a \cos A \cos \phi\}, \\ h &= \arctan \{-\cos a \sin A/(\sin a \cos \phi - \cos a \cos A \sin \phi)\}. \end{aligned}\right\} \tag{3.1}$$

The inverse transformation from (δ, h) to (a, A) is

$$\left.\begin{aligned} a &= \arcsin \{\sin \delta \sin \phi + \cos \delta \cos h \cos \phi\}, \\ A &= \arctan \{-\cos \delta \sin h/(\sin \delta \cos \phi - \cos \delta \cos h \sin \phi)\}. \end{aligned}\right\} \tag{3.2}$$

Since δ and a are restricted to lie between -90 and $+90°$, there is no ambiguity in the inverse sine function.

The transformation to the right ascension–declination system does involve the time. In order to derive this transformation, we note that right ascension α increases to the east while hour angle h increases to the west. Therefore, at any instant of time, the sum $\alpha + h$ is the same for all bodies. At a later instant, α is still the same as it was earlier, for a fixed star, but all stars have moved to the west as seen from Earth. Thus h increases with time for a given star. In other words, the sum $\alpha + h$ is a monotone in creasing function of time, but the sum has the same value for all stars at any specific instant.

The function of time defined by the sum $\alpha + h$ is called local sidereal time, which I shall abbreviate as L.S.T. Thus:

$$\text{L.S.T.} = \alpha + h. \tag{3.3}$$

Although the value of L.S.T. does not depend upon the star used, it does depend upon the terrestrial longitude of the observer. The value of L.S.T. for $0°$ longitude is called Greenwich sidereal time, often abbreviated G.M.S.T. The 'M' in G.M.S.T. stands for 'mean'. In very accurate work, it is necessary to distinguish between 'mean' and 'apparent' sidereal time, but this distinction need not concern us in the study of ancient astronomy.

The adjective 'sidereal' refers to the stars and in conversation we often say that sidereal time is the time measured by the fixed stars. This is not quite correct. From equation (3.3) we see that local sidereal time is equal to the hour angle of the point whose right ascension is zero. This point is called the vernal equinox, or the first line of Aries. The vernal equinox moves slowly westward with respect to the stars and makes one complete circuit in about 26 000 years. Thus, in about 26 000 years, the time that would be measured by the stars will be 1 day greater than sidereal time as defined by equation (3.3). This fact also means that it is not quite correct to call the coordinate system (α, δ) of figure 2 a star-fixed system. However, I shall ignore this and continue to call figure 2 a star-fixed system.

Since δ is the same in both the star-fixed and Earth-fixed systems of figures 2, 4 and 5, the transformation between the systems involves only α and h. Equation (3.3) is thus the desired transformation.

Equation (3.3) also furnishes a practical means of measuring time. We have only to measure the hour angle of an object whose right ascension α is known. It is possible to mount a telescope or other sighting tube on gimbals so that one axis is parallel to the Earth's axis, making one plane of rotation parallel to the equator. With such an 'equatorial' mount, the hour angle can be measured directly; however, ancient and medieval astronomers do not seem to have used this method much.

The method of measuring time astronomically that was most used, at least if we judge by the surviving data, was that of measuring the elevation of a star. Since δ is known for the star, we find h from a by eliminating A from equations (3.1). This is in fact just what we did in deriving the first of equations (3.2), so

$$\cos h = (\sin a - \sin \delta \sin \phi)/\cos \delta \cos \phi. \tag{3.4}$$

This is not useful when h is near 0, because $\cos h$ changes too slowly there. However, the astronomers seem to have applied this method only to stars fairly near the horizon. We should note that $\cos h$ has the same value for positive or negative h, so that it is necessary to know whether the star used was in the east or the west.

It would work just as well to measure the azimuth, although this does not seem to have been done as often as measuring the elevation. In order to find h from A, we solve the second of equations (3.2) for h. This requires converting the equation into a quadratic equation for $\sin h$ (or $\cos h$), solving the quadratic, and then finding the inverse function. The cumbersomeness of this method may have inhibited its use.

The time commonly used in daily life is not sidereal time but solar time, based upon the apparent daily rotation of the Sun. Until 1925, astronomers for many centuries and perhaps millennia had defined solar time to be the hour angle of the Sun. Since the hour angle of the Sun is zero at midday, this meant that astronomers began their day at midday, which put them at variance with most of the population. Various peoples have had various conventions for beginning the day, such as sunrise, sunset, or midnight, but few if any (except astronomers) have ever begun their day at the time that we call noon. Accordingly, the definition of solar time used by astronomers was changed in 1925, and it is now defined as 12h plus the hour angle of the sun.

Solar time as defined depends upon the longitude of the observer, and its value for the meridian of Greenwich is called G.M.T. (Greenwich mean time). As we did with sidereal time, we shall ignore the significance of the adjective 'mean' inserted into this definition. G.M.T. is almost synonymous with the quantity called 'universal time'.

G.M.T. and G.M.S.T. can be related as soon as we know the right ascension of the Sun at any time during the year. We can also calculate one from the other simply but tediously by noting that the times are equal by the old definition when the Sun is at the vernal equinox and that they are equal by the new definition when the Sun is at the autumnal equinox. During a tropical year, the stars make one more apparent revolution than the Sun; that is, if there are D solar days in a tropical year, there are $D+1$ sidereal days. Since D is approximately 365.2422,

$$\text{sidereal time/solar time} = 366.2422/365.2422 \tag{3.5}$$
$$= 1.002738,$$

if we reckon both times from a common epoch. Ephemeris publications contain tables that aid in applying equation (3.5) at any time during the year.

In sum, one way that time appears in astronomy is through the transformation between Earth-fixed and star-fixed coordinates. If an astronomer wishes to measure time during daylight, he does so by finding the hour angle of the Sun in his anthropocentric coordinate system. If he wishes to measure time during the dark hours, he can do so by finding the hour angle of a star and thence finding sidereal time from equation (3.3). He usually converts this time to solar time by using equation (3.5) or the equivalent.

One of the oldest ways of using this idea was to use the special case in which $a = 0$, that is, to use celestial objects on the horizon. This brings us to the subject of risings and settings.

4. RISINGS AND SETTINGS

From figure 5, we see that an object never sets if $\delta > 90° - \phi$, and that it never rises if $\delta < \phi - 90°$, for an observer in the northern hemisphere. Symmetrical relations apply in the southern hemisphere. Objects with intermediate declinations do rise and set, and we see from figure 5 that the azimuths of the rising and setting points depend only upon the declination. Thus a given star always rises and sets at the same points on the horizon.

This fact furnishes a simple way of measuring declination if we know our latitude, and conversely. We measure the azimuth A of an object either at rising or setting. Then, in order to relate declination and latitude, we set $a = 0$ in the first of equations (3.1), obtaining simply

$$\sin \delta = \cos A \cos \phi. \tag{4.1}$$

If the horizon is not level, we must use the correct value of a at the rising or setting point, which gives us a slightly more complex relation.

The hour angle at rising or setting is also a unique function of declination in a given anthropocentric system. The relation between declination and hour angle, when $a = 0$, is a special case of equation (3.4):

$$\cos h = -\tan \delta \tan \phi. \tag{4.2}$$

Risings and settings furnish a convenient way of measuring time. If we observe that a certain star, say, is about to set, we use its known declination to find h from equation (4.2). We then use its known right ascension to find sidereal time from equation (3.3), and we can then convert to solar time if we wish.

For a given star, the value of h at rise, say, is the same every sidereal day. That is, a star always rises at the same sidereal hour. The relation between the sidereal hour and the solar hour changes steadily throughout the year, however, so the solar time of rising also changes steadily with the time of year.

Conversely, if we note what star has just risen or is about to set at some solar time, sunset say, we can tell the season. This fact was important to farmers and others whose activities depend upon the season. A number of ancient works tell how to judge the correct times for planting by observing what stars are rising, say, at sunset.

We can use equation (4.2) to find how the length of the day varies with the seasons. In order to do so, we must use a fact that will be discussed in more detail below, namely that the declination of the Sun varies from about $-23°$ in midwinter to about $+23°$ in midsummer. Let δ_S denote the declination of the Sun at any time. We note that $\cos h$ has the same value whether h is positive or negative. Hence equation (4.2) furnishes two equal and opposite values of h; the negative value applies at rise and the positive value at set. Thus the length of the day is twice the value of h at sunset. Hence

$$\text{length of day} = 2 \arccos (-\tan \delta_S \tan \phi). \tag{4.3}$$

Let us evaluate this for a latitude of $50°$. In midwinter, $\delta_S = -23°$, and $\cos h = +0.5$ rather closely. Hence h at set is 4h and the day is 8h long. At the equinoxes, $\delta_S = 0$, $\cos h = 0$, h at set is 6h, and the day is 12h long. In midsummer, $\delta_S = +23°$, $\cos h = -0.5$, h at set is 8h, and the day is 16h long. These figures are for the length of time the centre of the Sun is above the horizon, in the absence of refraction; other definitions of sunrise and sunset give slightly different values.

5. The ecliptic and the zodiac; unequal hours

Ancient observers noted that most of the objects in the heavens are fixed relative to each other, but that a few move in the heavens in a regular fashion. Thus they spoke of the fixed stars and the wandering stars or planets. In their terminology, there were seven planets: Sun, Moon, Mercury, Venus, Mars, Jupiter, and Saturn. Uranus, Neptune, and Pluto had not been discovered yet.

Within the accuracy that is possible for observations with the naked eye, the Sun moves in a plane that is fixed with respect to the stars. This plane is called the ecliptic. The ecliptic plane cuts the celestial equator in two points called the vernal equinox and the autumnal equinox; these points are also called the first line of Aries and the first line of Libra. The angle between the planes is called the obliquity ϵ. At noon on 1899 December 31, ϵ equalled 23° 27' 08.26" according to the best estimate. Because of perturbations of other planets upon the apparent motion of the Sun (that is, upon the orbital motion of the Earth in modern terms), ϵ varies with time. On the scale of geologic time, the variation is probably oscillatory. Within the relatively short scale of historic time, however, ϵ has been decreasing steadily at a rate of 46.845" per century, to high accuracy.

At the present time, ϵ is closer to 23° than to any other integral value, and I used 23° earlier for illustrative purposes. In ancient historical times, however, ϵ was close to 24°. This fact is important in studies of ancient astronomy.

Since the ecliptic makes an angle with the equator, the Sun is north of the equator ($\delta_S > 0$) about half the time and it is south ($\delta_S < 0$) about half the time. The vernal equinox is the point at which δ_S changes from negative to positive and the autumnal equinox is the point at which δ_S changes in the opposite direction. The terms 'vernal equinox' and 'autumnal equinox' are used to mean either the points in the sky that were defined above or the instants in time when the Sun passes these points. The context usually makes it clear which meaning is intended.

After the Sun passes the vernal equinox, its declination continues to increase until it reaches the value ϵ. The instant and position of this event are called the summer solstice. The declination of the Sun then begins to decrease. When it is 0, we are at the autumnal equinox and when it is at its minimum $-\epsilon$, we are at the winter solstice.

I have already mentioned the fact that the equinoxes move with respect to the stars. This motion is caused by planetary and lunar perturbations, as is the change in the obliquity. Since the plane of the equator and the plane of the ecliptic both move, there is no simple way to describe the path of the equinoxes. However, the ecliptic moves much less than the equator and, for simplicity, we can say that the equinox moves westward along the (fixed) plane of the ecliptic; it makes a complete circle around the ecliptic in about 26 000 years. This motion is called the precession of the equinoxes.

The planets other than the Sun (including the Moon) move in a narrow band in the heavens that is centred on the ecliptic and that lies within a few degrees of it. This band is called the zodiac. The zodiac is divided into 12 equal parts called signs. The names of the signs, in order going eastward, are: Aries, Taurus, Gemini, Cancer, Leo, Virgo, Libra, Scorpio, Sagittarius, Capricorn, Aquarius and Pisces. These names furnish part of the common terminology of astrology, which was not carefully distinguished from astronomy until modern times.

After the precession of the equinoxes was discovered in about the year 150 B.C., these names had to serve double duty. They were originally, and still are, the names of constellations, and the divisions of the zodiac were taken from these prominent constellations that lay within them. The vernal equinox, as we have said, is still often called the first line of Aries. Precession has carried the vernal equinox westward from the constellation Aries and it now lies within the constellation Pisces. However, by the time that the precession was discovered, the names were firmly established in the terminology used for specifying the positions of stars. Hellenistic astronomers such as Hipparchus and Ptolemy decided to retain the names to indicate equal sections of the zodiac, beginning with Aries at the vernal equinox and continuing eastward. That is, Aries, when considered as a sign of the zodiac rather than as a constellation, begins at the vernal equinox and includes 30° of the ecliptic circle going eastward. Taurus begins 30° east of the vernal equinox and ends 60° east, and so on.

The signs of the zodiac have thus become merely ways of helping to specify the astronomical coordinate called celestial longitude, which will be defined in the next section. Before we turn to that definition, it is desirable to take up the matter of unequal or seasonal hours, which are closely related to the ecliptic.

The ecliptic cuts a great circle on the celestial sphere, and the horizon of any observer cuts another great circle which, for observers in temperate or tropical latitudes, never coincides with the ecliptic circle. Therefore half of the ecliptic is always above the horizon, if we restrict the discussion to observers outside the frigid zones.

Consider the Sun at the instant of sunset. It is in the ecliptic, at the point B, say. Let C denote the point on the ecliptic diametrically opposite to B. Since B is on the western horizon, by definition, C is on the eastern horizon. Since point B is the point of the next sunrise, except for the small angle that the sun moves along the ecliptic between sunset and the next sunrise, all the points on the ecliptic from C eastward 180° around to B rise during the night.

Suppose that we divide the ecliptic into 24 equal parts. Then 12 parts will rise during any night, regardless of the season and regardless of the observer's latitude. That is, if we keep time by observing the point of the zodiac that is on the horizon, instead of using the right ascension of a point on the horizon, we divide each night into 12 parts, called 'hours of the night', 'seasonal hours', or 'unequal hours'. We can also arrange for a sundial or other means of marking time to divide daylight into 12 parts, called 'hours of the day', or seasonal or unequal hours.

When the word 'hour' occurs in European literature through perhaps the fourteenth century, it commonly means a seasonal hour and not an hour as we know it. The hours were usually counted from sunrise or sunset, not from midnight and midday.

6. Celestial latitude and longitude

Celestial latitude and longitude are much like declination and right ascension except that they are based upon the plane of the ecliptic instead of the plane of the celestial equator. In order to find the celestial latitude and longitude of any point Q in the heavens, we start by passing the plane through Q that is normal to the ecliptic. The arc that this plane cuts on the celestial sphere from Q to the ecliptic is the celestial latitude β. The plane cuts the ecliptic in a point E, and the angle from the vernal equinox to E is the celestial longitude λ.

Celestial longitude is taken as positive if it is measured in an easterly direction from the vernal equinox. Celestial latitude is taken as positive for points that lie in a northerly direction from the ecliptic.

The transformation from α, δ to β, λ involves only a rotation about the direction of the equinoxes and is thus rather simple. It is

$$\left. \begin{aligned} \lambda &= \arctan \{(\cos \delta \sin \alpha \cos \epsilon + \sin \delta \sin \epsilon)/\cos \delta \cos \alpha\}, \\ \beta &= \arcsin \{-\cos \delta \sin \alpha \sin \epsilon + \sin \delta \cos \epsilon\}. \end{aligned} \right\} \qquad (6.1)$$

When the terms latitude and longitude are used, the context usually makes it clear whether the terrestrial or the celestial quantities are meant. Hence writers commonly use latitude and longitude without modifiers unless it is necessary to prevent ambiguity.

The classical planets (including the Sun and Moon) move nearly in the plane of the ecliptic. Further, the plane of the ecliptic changes its orientation in the heavens much less rapidly than the plane of the celestial equator. For these reasons, latitude and longitude are often more convenient coordinates than right ascension and declination. However, it is simpler and more precise to observe the latter pair of coordinates than it is to observe the former pair. Thus latitude and longitude are often used in theories of motion, but modern observations are almost always given in terms of right ascension and declination.

7. PLANETARY MOTIONS

I am still using the term planet to include the Sun and Moon. It is simplest to describe the motions of the Sun and Moon in geocentric terms and to describe the other motions in heliocentric terms. The descriptions, whether geocentric or heliocentric, are most simply referred to the plane of the ecliptic, with the vernal equinox as the basic reference line in that plane.

To high accuracy, the orbit of the Sun about the Earth is an ellipse, and this ellipse lies in the plane of the ecliptic. We need several quantities in order to describe the elliptic motion fully.

Two quantities specify the size and shape of the ellipse. In astronomy, the semi-major axis and the eccentricity are commonly used for this purpose. The semi-major axis is half the length of the long axis of the ellipse, while the eccentricity is related to the difference between the axes. The eccentricity is zero for a circle.

A third quantity specifies the orientation of the ellipse in the plane of the ecliptic. In order to specify the orientation, we usually start by finding the point of perigee, that is, the point at which the Sun comes closest to the Earth. We then give the angle Γ from the vernal equinox to the position of perigee. Γ is called the longitude of perigee.

While these three quantities tell us completely the path along which the Sun moves in space, they do not give enough information to let us calculate the point which the Sun occupies at a given time. Even when the path is known, calculation of position along the path is a complex matter that I shall not try to treat in this paper. I shall point out only that we need two additional quantities. One of them is the position at some conventional epoch. The conventional epoch used most often in current writing is noon on 1899 December 31, and the longitude at this time is called the longitude at the epoch.† The second additional quantity needed is the period of the motion, that is, the time needed to complete one circuit of the elliptical path. In the case of the solar orbit, the period is called the year.

To high accuracy, the orbit of the Moon about the Earth and the orbits of the other planets about the Sun are also ellipses. For each body, then, we must give the period, the position at some epoch, the orientation of perigee (in a heliocentric description, we use the term perihelion instead of perigee), the semi-major axis, and the eccentricity. However, we have an additional problem that we did not have with the solar orbit. The orbits of the Moon and of the other planets lie in planes that differ from each other and from the plane of the ecliptic. For each body, we must specify the plane of the orbit, and this takes two additional quantities. These quantities are illustrated in figure 6.

In figure 6, Υ is the vernal equinox and the plane containing Υ and N is the plane of the ecliptic. The plane NP is the plane of the orbit being considered. As the body moves along its orbit, it will clearly pass through the plane of the ecliptic at two points during each period.

† More accurately, the quantity usually given is the longitude of the mean Sun at the epoch. I shall not take the space to define the mean Sun here. If the reader makes a guess at what the mean Sun means, he will probably be close to being right.

These points are called the nodes of the orbit. At one node, the body passes from the north side of the ecliptic to the south side; this node is called the descending node. The other node, called the ascending node, is the point at which the body passes from the south side of the ecliptic to the north side. N represents the ascending node in figure 6.

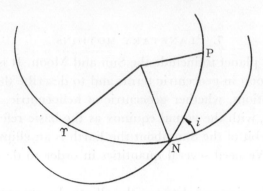

FIGURE 6. The quantities needed to specify the plane of an orbit for a body other than the Sun.

The angle from ♈ to N, measured in an eastward direction, is called the longitude of the ascending node, and it is frequently denoted by Ω. Ω is one of the two additional quantities needed to specify the plane of the orbit. The other is the inclination i, which is the dihedral angle between the planes, as shown in figure 6.

There are two methods in common use for giving the orientation of perigee (or perihelion). One is to give the angle ω measured in the direction of motion from N around to P in figure 6. This angle is called the argument of perigee or perihelion. The other method is to give a quantity Γ' defined by

$$\Gamma' = \Omega + \omega. \tag{7.1}$$

Γ' is often called the longitude of perigee or perihelion, although it is not the same as the celestial longitude of the point P. However, Γ' approaches the celestial longitude of P as the inclination i approaches 0.

Mercury has the most eccentric orbit of any major body in the solar system. Even for it, the axes of the orbit have the ratio 0.98; for other bodies, the ratio is even closer to 1. If a person saw the orbit of any major body drawn to scale, he would probably not recognize, with the unaided eye, the departure of the orbit from a circle. However, he would probably realize from the drawing that the centre of gravitation is not at the geometrical centre of the orbit.

The quantities that specify an elliptic orbit are often called the Kepler parameters of an orbit. If the orbits actually were ellipses, the Kepler parameters would be constants for a given planet. However, no real orbit is exactly an ellipse. The departures from elliptic motion can be described in a variety of ways. The way that is probably most valuable for intuitive thinking, as opposed to quantitative calculation, is to let the Kepler parameters vary with time.

We have already mentioned that the obliquity ϵ has been decreasing slowly throughout the historical period, and that the position of the vernal equinox is moving slowly westward through the fixed stars. The only other changes in parameters that need to concern us here are the changes in Ω and i for the lunar orbit.

The node of the lunar orbit rotates in a westerly direction around the plane of the ecliptic, making a complete revolution in about 18.61 years. This motion, and this time interval, are important in eclipse theory, as we shall discuss in the next section. This motion results mostly from the torque exerted on the Moon by the Earth's equatorial bulge. As such, it is a reaction to the precession of the equinoxes, which results mostly from the torque exerted on the bulge by the Moon's gravitation.

There are several types of variation in the lunar inclination. The largest one is produced by the Sun's gravitation, which causes an oscillation in the inclination. The amplitude of this oscillation is slightly less than 9′ of angle. The period of the oscillation is half of the period of revolution of the node, that is, it is about 9.3 years.

8. Eclipses

A celestial object B which is smaller than the Sun S casts a shadow whose geometry is shown in figure 7. First, there is the part shown solid between B and the point C; I shall call this part the inner umbra. Next there is the part shown solid but lying on the other side of C; I shall call this the outer umbra. Finally there is the part shown cross-hatched that surrounds the umbra; I shall call it the penumbra. We have an eclipse when another body enters, wholly or in part, either the umbra or the penumbra. It is clear that a body cannot enter the umbra without first entering the penumbra. Therefore the conditions for the occurrence of eclipses are governed by the penumbra.

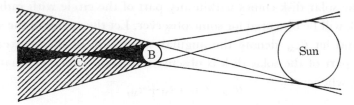

FIGURE 7. The geometry of the shadow cast by a body B. The part shown solid between B and C is the inner umbra, the part shown solid beyond C is the outer umbra, and the cross-hatched part is the penumbra.

When B represents the Moon, and the Earth enters the Moon's shadow, we have an eclipse of the Sun, as seen by people on Earth. People on the Moon, however, would see an eclipse of the Earth. When B represents the Earth, and the Moon enters the Earth's shadow, we have a lunar eclipse for Earth inhabitants and a solar eclipse for lunar ones. In the rest of the discussion, I shall consider only eclipses as seen by an inhabitant of the Earth.

Because the distances from the Sun to the Earth and moon are constantly changing, the dimensions of the shadows are continually changing. For our purposes, we can ignore this fact

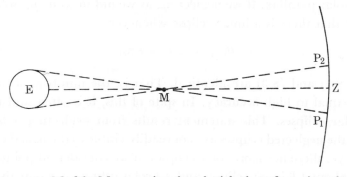

FIGURE 8. Positions of the centre M of the Moon, against the celestial sphere, for various observers on th e Earth E Any point between P_1 and P_2 is the apparent centre of the Moon for some observer on Earth.

and assume that the shadows have constant dimensions. We can also assume that the Sun has the same direction in the heavens regardless of the point on Earth from which it is viewed; that is, we can neglect the solar parallax. However, we cannot neglect the lunar parallax.

Consider where the centre of the Moon seems to be for an observer anywhere on Earth. Let M in figure 8 denote the centre of the lunar disk, and let E denote the Earth. For some observer, the Moon is in the zenith and its centre is at the point Z. There are many observers for whom the Moon is on the horizon. For one of these, the centre of the Moon is at the Point P_1 on the celestial sphere; for the diametrically opposite observer, the centre of the Moon is at the point P_2. Now draw the circle on the celestial sphere whose radius is P_1Z (which equals P_2Z) and whose centre is Z, and consider any point within this circle. There is some observer on Earth for whom the centre of the Moon lies at this point.

The radius $P_1Z = P_2Z = \pi_M$, say, is the quantity that was called the horizontal parallax in § 2.

Consider again the observer who sees the centre of the Moon at the point P_1. For him, the Moon blocks out a circle whose centre is P_1 and whose radius is s_M, say; s_M is called the semi-diameter of the Moon. For the diametrically opposite observer, the Moon blocks a circle whose centre is P_2 and whose radius is s_M. Clearly, if we draw a circle of radius $s_M + \pi_M$ centred on Z, any point within this circle is blocked from the view of some observer.

If any point of the solar disk comes within any part of the circle with radius $s_M + \pi_M$, some part of the solar disk will be obscured for some observer. Let the radius of the solar disk subtend an angle s_S, say, and let θ_{SM} denote the angular separation of the centres of the Sun and Moon. Then some part of the solar disk is obscured for some observer, that is, there is some sort of solar eclipse, whenever

$$\theta_{SM} < s_S + s_M + \pi_M, \tag{8.1}$$

approximately. The reader should remember that this relation and other relations to be obtained must be modified slightly to account for the solar parallax and other small effects.

In order to find a condition for lunar eclipses, let us imagine the Earth's shadow to be cast upon a sphere whose radius equals the distance to the Moon. Let the angular radius of the penumbral shadow be denoted by s_{ES}. Further, let U denote the centre of the shadow (this is the point on the celestial sphere diametrically opposite to the Sun), and let θ_{UM} be the angle from U to the centre of the Moon. Then there will be a lunar eclipse whenever

$$\theta_{UM} < s_M + s_{ES}. \tag{8.2}$$

The reader can readily convince himself, with the aid of a sketch, that $s_{ES} = \pi_M + s_S + \pi_S$, in which π_S is the solar parallax. If we neglect π_S, as we did in studying solar eclipses, we find from relation (8.2) that there is a lunar eclipse whenever

$$\theta_{UM} < s_S + s_M + \pi_M, \tag{8.3}$$

approximately.

The criteria in (8.1) and (8.3) are identical. That is, the numbers of solar eclipses and of lunar eclipses are equal to high accuracy. In spite of this, it is often said that there are considerably more solar eclipses. This statement results from neglecting a large class of lunar eclipses. However, the neglected eclipses are not readily visible to the naked eye. For observation with the unaided eye, there are more solar eclipses, if we consider all points on Earth. On the average, there are about 1.6 lunar eclipses and 2.4 solar eclipses per year that can be seen with the unaided eye. For a particular observer, however, there are many more lunar eclipses.

The quantity that appears on the right of (8.1) and (8.3) is slightly less than 1.5°. Thus there can be an eclipse only when the centre of the Moon is within about 1.5° of the ecliptic. If the inclination of the lunar orbit were less than 1.5°, there would be a solar eclipse at each new Moon and a lunar eclipse at each full Moon. However, the inclination is about 5.1°, so that most new and full Moons occur without an eclipse.

If the Moon is to be close enough to the ecliptic to make an eclipse possible, it must be within about 17° of one of the points where it crosses the ecliptic; that is, the Moon must be within about 17° of one of its nodes at new or full Moon for an eclipse to happen. Suppose, for example, that one of the nodes is at longitude 0°, so that the other is at 180°, and suppose that the critical angle is exactly 17°. Then, in order to have an eclipse, the longitude of the Moon must either be between −17° and +17°, or it must be between 163° and 197°.

At new moon, the longitudes of the Moon and Sun are the same and at full Moon they differ by exactly 180°. Therefore the Sun must also lie within the ranges of longitude stated above in order for eclipses to occur. Suppose that the Sun has longitude 0° at the epoch E, say. Then it will be within one of the ranges of longitude from about E − 17 days to about E + 17 days. This interval of about 34 days is called an eclipse season. At all times within the season, the Sun is in a position that makes an eclipse possible. During one lunar revolution, the Moon takes on all longitudes. Since 34 days is greater than the orbital period of the Moon, the Moon is necessarily in the correct position for eclipses at some time during a season.

Thus there is at least one solar eclipse in each season, and there is at least one lunar eclipse if we consider all eclipses. If we consider only lunar eclipses that can be seen easily with the naked eye, however, it is possible for a season to pass without a visible lunar eclipse. In fact, it is possible for an entire year to pass without one.

If the nodes of the lunar orbit stayed in the same place, the eclipse seasons would come at the same time each year. However, as I said earlier, a node of the lunar orbit moves westward and makes a complete circuit in 18.61 years. This interval is therefore an important one in studying the recurrence of eclipses.

If the sun is at the position of one of the nodes at epoch E, it will be at the same node at the epoch E + 346.6 days, approximately. This interval of about 346.6 days is called an eclipse year; it is less than an ordinary year because of the westward movement of the node. There are two eclipse seasons in each eclipse year. On the average, there are slightly more than two seasons in an ordinary year. A given eclipse season moves through all the times of the ordinary year in an interval that averages about 18.61 years.

Concluding remarks

The purpose of this paper has been to supply some of the basic information about astronomy that is needed in order to study the role of astronomy in ancient civilizations. It is clearly not possible to supply all the information needed in a short paper, but I hope that this paper gives enough information to start the reader off on the subject.

The paper is odd in that it has given no references or citations. The reason is that the material in it has long since passed from the realm of research into the realm of standard knowledge. When so many references are available, to single out a few of them for citation would be invidious. Further, it is difficult to know the background of the reader in mathematics and the physical sciences, and thus it is difficult to know what references would be useful to him.

There is one reference, however, that should be mentioned. This is the *Explanatory supplement to The Astronomical Ephemeris and The American Ephemeris and Nautical Almanac* (London: Her Majesty's Stationery Office, 1961). This work explains how the tables in the ephemeris works of the United Kingdom and the United States are calculated, and it is invaluable for anyone who needs to do independent calculations about the behaviour of the solar system.

I thank my colleagues H. D. Black, B. B. Holland and R. E. Jenkins for valuable advice on the preparation of figures 1 to 5. I also thank Mr Jenkins for a critical review of the original typescript.

The preparation of this paper was supported by the Department of the Navy, under Contract N 00017-62-C-0604 with the Johns Hopkins University.

Phil. Trans. R. Soc. Lond. A. **276**, 21–42 (1974)   [21]

Printed in Great Britain

Scientific astronomy in antiquity

By A. Aaboe

Yale University, New Haven, Connecticut, U.S.A.

[Plates 1 and 2]

The character and content of Babylonian scientific or mathematical astronomy, as we know it from texts of the last half millennium B.C., are sketched. This late-Babylonian astronomy is set in contrast to earlier Babylonian astronomy as well as to the kinds of astronomy found in other ancient cultures, and an attempt is made at a very broad classification of such pre-scientific astronomies.

The lateness and uniqueness of Babylonian mathematical astronomy is emphasized, and it is shown that its creation depended upon the availability of a peculiar set of ingredients, e.g., a particular type of mathematics, and a tradition of making and recording observations of certain astronomical phenomena.

It is finally argued that all subsequent varieties of scientific astronomy, in the Hellenistic world, in India, in Islam, and in the West – if not indeed all subsequent endeavour in the exact sciences – depend upon Babylonian astronomy in decisive and fundamental ways.

In the present paper I shall describe to you, of necessity in barest and crudest outline, the character and content of Babylonian mathematical astronomy which was the first and highly successful attempt at giving a refined mathematical description of astronomical phenomena. Further, I shall try to give expression to my conviction that subsequent scientific astronomy – in Hellenistic Greece, in India, in Islam, and in the Latin West – depended in decisive ways upon Babylonian astronomy.

My choice of a title for this lecture was, I fear, somewhat clumsy – mathematical or theoretical astronomy in antiquity might better describe my topic – and so to avoid misunderstanding, I should like to make clear what I mean by scientific astronomy; and let me begin by giving some examples of what I do not wish to include in that category. To talk about what I shall not talk about may not be as idle as it sounds, for it will serve to set my principal topic in relief, and also to provide a first swift glimpse of subjects which will be dealt with more properly by others. I shall exclude astronomical observations from my discussion; not that I consider them irrelevant to my topic, on the contrary, but a treatment of observational techniques and records would lead me too far from my main concern. I shall, none the less, have something to say later about the connexion between Babylonian observations and theories.

I find it useful to distinguish between two levels of prescientific astronomy. To the less advanced level I assign astronomical achievements such as the naming of prominent stars and constellations; drawing the distinction between fixed stars and planets; the awareness of the morning star and the evening star being just different appearances of one and the same celestial body, namely Venus; the realization that a fixed star which is not circumpolar always rises and sets at the same two places on the horizon, while the Sun, Moon, and planets do not; the discovery that the first appearance of a fixed star after its interval of invisibility happens at the same time of year and may be used as a seasonal indicator.

As examples of this last practice I need only remind you of the part of Hesiod's *Works and days*, beginning with line 383, in which certain rustic tasks are tied to the first or last visibility of certain fixed stars, and of the ancient Egyptian use of Sirius (or Sothis) at its first visibility as a herald of the time of flooding of the Nile. This sort of primitive astronomical knowledge, a farmer's or shepherd's astronomy, if you will, I should not be at all surprised at finding in any settled community, whether literate or not.

The more advanced level of prescientific astronomy you will find in some, but not all, of the ancient literate cultures, namely, in China, in the pre-Hellenistic and early Hellenistic Greek world, in early Mesopotamia, and in Mesoamerica. You will hear about these things in some detail from others; here let me say that however different such early astronomies may be in their aims and approach, they have as a common characteristic that they employ various cycles concerning the motion of the Sun, Moon and the five classical planets, that is, those visible to the naked eye: Mercury, Venus, Mars, Jupiter and Saturn.

Let me give some examples, and as the first the so-called 'Metonic' cycle which sets 19 years equal to 235 synodic months. This cycle which happens to be very excellent was the basis of the calendar introduced by Meton in Athens. We find it in Babylonia and China as well.

My second example is the Venus cycle of 8 years during which Venus runs its synodic course five times; that is, it becomes a morning star five times, an evening star five times, retrograde five times, and so on, and at the end of which it returns very nearly to the same place with respect to the stars and with respect to the Sun. This cycle was known in Babylonia and remained in use, because of its excellence, with but slight modifications, until the end of cuneiform literature, side by side with much more sophisticated schemes for the other planets. It was also known in China and it plays an important role in Mayan astronomy.

Thirdly, there are various so-called eclipse cycles. Though too crude to yield reliable predictions of eclipses, such cycles could be used, as we now know, for issuing eclipse warnings in advance, that is, for pointing to syzygies (i.e. conjunctions or oppositions of Sun and Moon) at which there was a danger of either a solar or a lunar eclipse, while guaranteeing safety or freedom from eclipses elsewhere (Aaboe 1972). One such cycle consists of 135 lunar months in the course of which one issues 23 eclipse warnings according to a certain pattern. We have evidence of this cycle from Babylon and it was known and used in China in the Han period; the Mayas used a cycle of three times 135 months.†

It is really not surprising that such cycles are discovered in literate civilizations that have made a habit of recording important events, among them celestial phenomena. Thus I would say that the occurrence of the same period of this character in two civilizations is not by itself evidence of contact between them. Indeed, it would be absurd to suggest that a connexion existed between the Babylonian and the Mayan civilizations just because the Venus cycle and a certain eclipse cycle are known in both. (It is perhaps foolish to make even such a negative statement, for it might lead some intrepid enthusiast to sail on inflated goat skins from the Euphrates to the Panama canal.)

I shall here mention three uses of such periods. The first is for calendric purposes, where various cycles are combined to form larger cycles which assure repetition of several kinds of phenomena. The Metonic cycle which I already mentioned is a very simple example. In China and in Mesoamerica much more complicated periods are built up, as you will doubtless hear.

Secondly, such periods may be used to furnish parameters for astronomical models of various kinds, as I shall mention later in my discussion of some Greek geometrical models.

† The factor 3 is easily explained because 135 months happen to be two-thirds of a day more than a whole number of days, so that thrice this period is very nearly a whole number of days as required by Mayan astronomy in which one day, and not one lunation, was the basic unit.

Thirdly, they can be made to yield predictions of astronomical events, particularly when used in conjunction with a series of observational records. A perfect example of this approach is the class of late Babylonian texts which Sachs has called Goal Year Texts (Sachs 1948). A goal year text written for this year, A.D. 1972, would contain information about the planets and the Moon in the following fashion: under the heading of Jupiter it would tell about Jupiter in 1901, for what Jupiter did 71 years ago it will do again this year on the same dates, 71 years being a very good period of Jupiter; for Venus it would tell what happened in 1964, for 8 years is a good Venus period; for Mercury it would have what happened 46 years ago; for Saturn what happened 59 years ago, and so on for Mars and the Moon. The information presented would be taken from observational records for the relevant years. This turns out to be a very efficient way of giving astronomical predictions, but it still falls short of what I mean by a truly scientific astronomy.

I do not wish to call an astronomical theory scientific until it gives us control over the irregularities within each period and thus frees us from constant consultation of observational records.†

What I mean by a scientific astronomical theory is then a mathematical description of celestial phenomena capable of yielding numerical predictions that can be tested against observations. I think this distinction will become clear when I introduce you to a truly scientific astronomical text.

My first illustration (figure 1, plate 1) shows a photograph of the reverse of a Babylonian clay tablet, pieced together from many fragments, which was published most recently by Neugebauer (1955) as A.C.T. no. 122. The scribe tells us in the colophon that it was written in year 209 of the Seleucid Era, month IX, day 18, and it gives us information about new moons for the three years 208, 209 and 210 of the same era, so that the text must, at least in part, be a forecast. I chose this text not because it is typical of the material we have to work with, which, alas, it is not, but rather because it is one of the most complete examples of a lunar text that has reached us.

The tablet has the oblong shape characteristic of lunar ephemerides, as we are accustomed to call these texts, even though they do not proceed day by day, but month by month.

Figure 2a, b, c shows Neugebauer's transcription and restoration of the text, and one's first impression, I dare say, is of quantities and quantities of numbers arranged very neatly in columns that continue from the obverse over the bottom edge onto the reverse. Horizontal alinement is strictly observed in this text, as in all similar texts, and indeed all the numbers in one line running across from the left edge to the right concern one particular month or rather, in this case, the situation around the new moon, that happens at the end of one month, and the first visibility of which signals the beginning of the next. The first column, which is not preserved, but which can be reconstructed with confidence, gives the year number and the month.

The first line of the text concerns month XII of the year 207 of the Seleucid Era, and whatever is told there has to do with the conjunction that took place at the end of March in 104 B.C., to transform to a date in our era (Parker & Dubberstein 1956). As you read down the first column, we simply have years and the corresponding months listed in order. You will

---

† A modern parallel to the Goal Year texts is the kind of meteorology where a replica of the current pattern of temperatures, pressures, and wind velocities is found in past records, and predictions are given according to what happened then; this methodology would, in my usage, be prescientific.

obverse [O]			I	II	III	IV	V		VI	[VII]	
1		XII	[29, 8, 3]9, 18	2, 2, 6, 20	[ḫ]un	2, 56	1, 32	6, 5, 30	sig	[11, 30]	3, 59, 52, 30	1
	3, 28	I	[28, 50, 39,] 18	[5]2, 45, 38	múl	3, 14	1, 23	9, 46, 30	sig	[11, 16, 10]	4, 22, 22, 30	
		II	[28, 3]2, 39, 18	29, 25, 24, 56	múl	3, 26	1, 17	5, 54	sig	11, [52, 10]	4, 14, 1, 40	
		III	[28,] 14, 39, 18	27, 40, 4, 14	maš	3, 34	1, 13	2, 1, 30	sig	12, 2[8, 10]	3, 51, 31, 40	
5		IV	[2]8, 24, 40, 2	26, 4, 44, 16	kušú	3, 32	1, 14	1, 51	bar	13, 4, 10	3, 29, 1, 40	5
		V	[2]8, 42, 40, 2	24, 47, 24, 18	a	3, 24	1, 18	2, 43, 30	nim	[13,] 40, 10	3, 6, 31, 40	
		VI	29, . ., 40, 2	23, 48, 4, 20	absin	3, 9	1, 25	6, 36	nim	14, 16, 10	2, 44, 1, 40	
		VI₂	29, 18, 40, 2	23, 6, 44, 22	rín	2, 51	1, 34	9, 16	nim	14, 52, 10	2, 21, 31, 40	
		VII	[2]9, 36, 40, 2	22, 43, 24, 24	gír-tab	2, 36	í, 42	5, 23, 30	nim	15, 4	1, 59, 1, 40	
10		VIII	29, 54, 40, 2	[22, 38,] 4, 26	pa	2, 27	1, 46	1, 31	nim	14, 28	2, 8, 37, 30	10
		IX	[29,] 51, 17, 5[8]	[22, 29, 22,] 24	máš	2, 27	1, 46	2, 21, 30	bar	13, 5[2]	2, 31, 7, 30	
		X	[29,] 33, 17, 58	[22, 2, 40, 22	g]u	2, [3]6	1, 42	[3, 14	sig]	[13, 16]	2, 53, 37, 30	
		XI	[2]9, 15, 17, 58	21, 17, 58, 20	zib-me	2, [50]	1, [35]	[7, 6, 30	sig]	[12, 40]	3, 16, 7, 30	
		XII	[2]8, 57, 17, 58	20, 15, 16, 18	ḫun	3, 8	1, 26	[8, 45,] 30	sig	12, [4]	3, 38, 37, 30	
15	3, 29	I	[2]8, 39, 17, 58	18, 54, 34, 16	múl	3, 22	1, 19	[4,] 53	sig	11, 28	4, 1, 7, 30	15
		II	[28,] 21, 17, 58	17, 15, 52, 14	maš	3, 32	1, 14	[1,] . ., 30	sig	11, 18, 10	4, 23, 37, 30	
		III	[28, 1]8, 1, 22	15, 33, 53, 36	kušú	3, 35	1, 12	[2,] 52	bar	11, 5[4], 10	4, 12, 46, 40	
		IV	[28, 3]6, 1, 22	14, 9, 54, 58	a	3, 28	1, 16	[3,] 44, 30	nim	12, 30, 10	3, 50, 16, 40	
		V	[28, 54, 1,] 22	13, 3, 56, 20	absin	3, 15	1, 22	[7, 37	ni]m	[13, 6, 10]	3, 27, 46, 40	
20		VI	[29, 12, 1, 2]2	12, 15, 57, 42	rín	2, 5[8]	[1, 31]	[8, 15	nim]	[13, 42, 10]	3, 5, 16, 40	20

reverse [O]			I	II	III	IV	V		VI	VII	
1		VII	[29, 30,] 1, 22	11, 45, 59, 4	gír-tab	2, 40	1, 40	4, [22, 30	nim]	[14, 18, 10]	[2, 42, 46, 40]	1
		VIII	[29,] 48, 1, 22	11, 34, . ., 26	pa	2, 29	1, 45	. ., [30	nim]	[14, 5]4, [10]	2, [20, 16, 40]	
		IX	[2]9, 57, 56, 38	11, 31, 57, 4	máš	2, 25	1, 47	. ., [22,] 30	bar	15, 2	1, 57, 46, 40	
		X	29, 39, 56, 38	11, 11, 53, 42	gu	2, 31	1, 44	4, [1]5	sig	14, 26	2, 9, 52, 30	
5		XI	[2]9, 21, 56, 38	10, 33, 50, 20	zib-me	2, 43	1, 38	8, 7, 30	sig	13, 50	2, 32, 22, 30	5
		XII	[2]9, 3, 56, 38	9, 37, 46, 58	ḫun	3, 1-	1, 29	7, 44, 30	sig	13, 14	2, 54, 52, 30	
	3, 30	I	[28,] 45, 56, 38	8, 23, 43, 36	múl	3, 18	1, 21	3, 52	sig	12, 38	3, 17, 22, 30	
		II	[28, 2]7, 56, 38	6, 51, 40, 14	maš	3, 29	1, 15	. ., . ., 30	bar	12, 2	3, 39, 52, 30	
		III	[28, 1]1, 22, 42	5, 3, 2, 56	kušú	3, 35	1, 12	. ., 53	nim	11, 26	4, 2, 22, 30	
10		IV	[28, 2]9, 22, 42	3, 32, 25, 38	a	3, 31	1, 14	4, 45, 30	nim	11, 20, 10	4, 24, 52, 30	10
		V	[28,] 47, 22, 42	2, 19, 48, 20	absin	3, 20	1, 20	8, 38	nim	11, 56, 10	4, 11, 31, 40	
		VI	[29,] 5, 22, 42	1, 25, 11, 2	rín	3, 4	1, 28	7, 14	nim	12, 32, 10	3, 49, 1, 40	
		VII	[2]9, 23, 22, 42	48, 33, 44	gír-tab	2, 46	1, 37	3, 21, 30	[n]im	13, 8, 10	3, 26, 31, 40	
		VIII	[2]9, 41, 22, 42	29, 56, 26	pa	2, 33	1, 43	31	bar	[1]3, 44, 10	3, [4, 1, 40]	
15		IX	[29,] 59, 22, 42	29, 19, 8	máš	2, 26	1, 47	1, 23, 30	sig	14, 20, 10	2, 41, 31, 40	15
		X	[2]9, 46, 35, 18	15, 54, 26	gu	2, 28	1, 46	5, 16	sig	14, 56, 10	2, 19, 1, 40	
		XI	[29,] 28, 35, 18	29, 44, 29, 44	gu	2, 39	1, 40	9, 8, 30	sig	15	1, 56, 31, 40	
		XII	[29, 10, 3]5, 18	28, 55, 5, 2	zib-me	2, 54	1, 33	6, 43, 30	sig	14, 24	2, 11, 7, 30	
		XII₂	[28, 52, 35,] 18	27, 47, 40, 20	ḫun	3, 12	1, 24	2, 5[1	sig]	[13, 4]8	2, 33, 37, 30	

FIGURE 2a

	VIII	IX	X	XI				XII			
1	[20, 20]	[7, 19 lal]	[3, 52, 33, 30]	[še	29	1, 2, 43, 50	šú]	[29	., 30	kur]	→ 1
	[14, 52, 30]	[22, 11, 30 lal]	[4, ., 11]	[bar	28	5, 2, 54, 50	šú]	[29	., 26	d]u	
	[8, 5]	[30, 16, 30 lal]	[3, 43, 45, 10]	[gu₄	28	2, 46,] 40	šú	28	1, 2[9	n]im	
	[1, 17,] 30	31, 34 lal	3, 19, 57, 40	sig	29	., 6, 37, 40	šú	29	1, [7	kur]	
5	[5,] 30	27, 5[2 lal]	3, 1, 9, 40	š[u	28	3, 7, 47, 20	šú]	28	1, [38	šú]	5
	[12,] 17, 30	[15, 34, 30 lal]	[2, 50, 57, 10]	[izi	28	5, 58, 44, 30	šú]	[29	1, 16	d]u	
	[19, 5]	[3, 30, 3]0 tab	2, 47, 32, 10	[ki]n	28	2, 46, 16, 40	šú	[28	1, 21	ni]m	
	[16, 7, 30]	[19, 3]8 tab	2, 41, 9, 40	ki[n 2-kam	2]8	5, 27, 26, 20	šú	2[9	1, 1	d]u	
	[9, 20]	[28,] 58 tab	2, 27, 59, 40	[du₆]	29	1, 55, 26	šú	2[9	., 13	ni]m	
0	[2, 32, 30]	[31,] 30, 30 tab	2, 40, 8	[api]n	28	4, 35, 34	šú	2[9	., 22	d]u	10
	[4, 15]	[2]9, 10, 30 tab	3, ., 18	gan	29	1, 35, 52	šú	2[9	., 10	ku]r	
	[11, 2, 30]	18, 8 tab	3, 11, 45, 30	ab	28	4, 47, 37, 30	šú	2[9	., 29	d]u	
	[17, 5]0	18 tab	3, 16, 25, 30	zíz	29	2, 4, 3	šú	[29	., 29	nim]	
5	[17,] 22, 30	17, 4, 30 lal	3, 21, 33	še	28	5, 25, 36	[šú]	[29	., 51	du]	
	[10, 35]	[27, 39, 30 lal]	[3,] 33, 28	b[ar]	28	[2,] 59, 4	[šú]	[28	1, 40	nim]	15
	[3, 47, 30]	[31, 27 lal]	[3,] 52, 10, 30	gu₄	29	51, 14, 30	[šú]	[29	., 23	kur]	
	[3]	30, 29 lal	3, 42, 17, 40	sig	28	4, 33, 32, [10	šú]	[28	., 14	šú]	
	[9, 47, 3]0	20, 41, 30 lal	3, 29, 35, 10	šu	28	2, 3, 7, 20	[šú]	[28	., 47	nim]	
	[16, 3]5	4, 6, 30 lal	3, 23, 40, 10	izi	28	5, 26, [47, 30	šú]	[29	., 49	du]	
0	[18,] 37, 30	14, 31 tab	3, 19, 47, 40	kin	29	[2, 46, 35, 10	šú]	[29	1, 15	nim]	20 →

	VIII	IX	X	XI				XII			
1	[11, 50]	[26, 21 tab]	[3, 9, 7, 40]	[du₆	28	5, 55, 42, 50	šú]	[2]9	1, 35	du	→ 1
	[5, 2, 30]	[31, 23,] 30 tab	2, 51, 40, 10	apin	28	2, 47, 23	šú	28	1, 2	nim	
	1, 45	31, 47, 30 tab	2, 29, 34, 10	gan	28	5, 16, 57, 10	šú	29	1, [4	d]u	
	8, 32, 30	23, 15 tab	2, 33, 7, 30	ab	29	1, 50, 4, 40	šú	29	., [6]	nim	
5	15, 20	7, 55 tab	2, 40, 17, 30	zíz	28	4, 30, 22, 10	šú	29	., 9	du	5
	19, 52, 30	11, 57, 30 lal	2, 42, 55	še	29	1, 13, 17, 10	šú	29	15	kur	
	13, 5	25, 2, 30 lal	2, 52, 20	bar	28	4, 5, 37, 10	šú	28	34	šú	
	6, 17, 30	31, 20 lal	3, 8, 32, 30	gu₄	29	1, 14, 9, 40	šú	29	., 1	kur	
	., 30	31, 50 lal	3, 30, 32, 30	sig	28	4, 44, 42, 10	šú	28	[.,] 3	[šú]	
0	7, 17, 30	25, 48, 30 lal	3, 59, 4	šu	28	2, 43, 46, 10	šú	28	1, 29	[nim]	10
	14, 5	11, 43, 30 lal	3, 59, 48, 10	izi	29	43, 34, 20	šú	29	3[7	kur]	
	20, 52, 30	9, 9 tab	3, 58, 10, 40	kin	28	4, 41, 45	šú	29	9	[du]	
	14, 20	23, 29 tab	3, 50, ., 40	du₆	29	2, 31, 45, 40	šú	29	5[5	nim]	
	7, 32, 30	31, [1,] 30 tab	3, 35, 3, 10	[a]pin	29	6, 48, 50	šú	29	[1, 36	kur]	
5	[., 45]	31, 4[6, 30 tab]	3, 13, [18,] 10	[gan	2]8	[3,] 20, 7	[šú]	[28	53	šú]	15
	6, 2, 30	27, [7] tab	2, 46, 8, 40	ab	29	6, 15, 40	šú	2[9	1, 4]0	kur	
	12, 50	[14,] 17 tab	2, 10, 48, 40	zíz	28	2, 17, 4, 20	šú	2[8	37	nim]	
	19, 37, 30	5, 20, 30 lal	2, 5, 47	še	28	4, 22, 51, 20	šú	28	., 4	šú	
	15, 35	20, 55, 30 lal	2, 12, 42	dir-še	29	35, 33, 2[0	šú]	29	49	kur	→

FIGURE 2 *b*

	XIII	XIV	XV	XVI	XVII	
						obverse
1	[kúr 30 9, 26]	[be] 14, 15	bar 1 15, 40	26 17, 30 k[ur]	[⊠⊠ be]	1
	kúr 30 11, 35	[be] 17, 40	gu₄ 1 17, 30	26 17, 50 kur	[2]3, 4[0 be]	
	kúr 29 7, 57	[be] 13, 10	sig 30 13	27 17, 30 kur	18, 3[0 be]	
	[kúr 30 1]0, 42	[be ⊠⊠]	[šu 1 ⊠ 40]	27 16, 40 kur	16 be	
5	[kúr 29] 7, 39	[be 12, 30	[izi 30 ⊠⊠]	27 23, 10 kur	2[2] be	5
	kúr 3[0] 10, 44	be 20, 30	kin 1 [1]7[⊠]	27 19, 30 kur	17 be	
	kúr 29 7, 48	be 15, 50	kin 2-kam 30 12	27 30, 40 kur	22, 20 be	
	kúr 29 4, 58	be 10, 10	du₆ 30 8, 40	28 16, 30 kur	14 be	
	kúr 30 8, 23	be 17, 10	apin 1 16, 50	27 20, 30 kur	19 [⊠] be	
10	kúr 29 5, 28	be 9, 20	gan 30 9, 10	28 14 kur	1[2 ⊠ b]e	10
	kúr 30 8, 37	be 15, 50	ab 1 16, 30	27 13, 30 kur	18 [⊠ be]	
	kúr 29 5, 31	be 9, 30	zíz 30 10 uš	27 12 kur	2[4 ⊠ be]	
	kúr 30 8, 21	be 14	še 1 15, 10	27 10, 30 kur	1[8 ⊠ be]	
	[kúr 3]0 11, 8	be 18, 40	bar 1 19, 20	26 16, 30 kur	2[3 ⊠ be]	
15	[kúr 29 7, 4]2	be 11, 50	gu₄ 30 12, 10	27 17, 10 kur	1[9 ⊠ be]	15
	[kúr 30 9,] 55	be 15, 10	sig 1 15, 20	27 17 kur	1[2 ⊠ be]	
	[kúr 30 12,] 14	be 20, 30	šu 1 19, 20 tab	27 14, 40 kur	[⊠⊠ be]	
	[kúr 29] 8, 41	be 15	izi 30 13	27 25, 40 [kur]	[⊠⊠ be]	
	[kúr 29] 5, 11	be 8, 30	kin 30 8	28 18, 30 [kur]	[⊠⊠ be]	
20	[kúr 30 7, 43]	[be ⊠⊠]	[du₆] 1 12, 20	27 2[8 ⊠ kur]	[⊠⊠ be]	20

	XIII	XIV	XV	XVI	XVII	
						reverse
1	kúr 30 10, 25	be 20, 10	apin 1 17, 20	27 17 [⊠ kur]	[⊠⊠ be]	1
	kúr 29 7, 27	be 14, 50	gan 30 14	27 20, 10 kur	[⊠⊠ be]	
	kúr 29 4, 56	be 10, 10	ab 30 10, 20	28 9, 50 kur	[⊠⊠ be]	
	kúr 30 8, 25	be 15, 50	zíz 1 16, 20	27 11, 10 kur	[⊠⊠ be]	
5	kúr 29 5, 51	be 11, 30	še 30 12	27 12 kur	2[3 ⊠ be]	5
	kúr 30 9, 17	be 16, 30	bar 1 17, 40	27 11, 20 kur	17 [⊠ be]	
	kúr 29 6, 34	be 10, 10	gu₄ 30 10, 50	27 20, 10 kur	2[1 ⊠ be]	
	kúr 30 9, 30	be 15	sig 1 15, 40	27 15, 10 kur	1[6 ⊠ be]	
	kúr 30 12, 3	be 18, 30	šu 1 18, 50	27 13, 20 kur	1[2 ⊠ be]	
10	[kúr] 29 8, 2	be 11, 50	izi 30 11, 10	28 11 kur	[⊠⊠ be]	10
	[kúr] 30 9, 57	be 16, 50	kin 1 14, 20	27 20, 40 kur	16, 30 be	
	[kúr] 29 5, 51	be 10 uš	du₆ 30 7, 30	28 16, 1[0] kur	14 be	
	[kúr] 30 7, 52	be 14, 30	apin 1 11, 50	27 22, 10 kur	21 be	
	[kúr] 30 10,] 10	be 19, 40	gan 1 17, 30	27 14, 50 kur	16, 10 be	
15	[kúr 2]9 [6,] 53	be 13, 50	ab 30 13, 30	27 19, 30 kur	23, 20 be	15
	kúr 30 10, 8	be 21, 50	zíz 1 21, 50	27 9, 40 kur	14 be	
	kúr 29 8, 2	be 17, 10	še [30 18,] 20	27 11, 50 kur	19, 30 be	
	kúr 29 6, 4	be 11, 30	dir-[še 30 12,] 20	27 17, 50 kur	23, 50 be	
	kúr 30 10, 1	[b]e 2[0, 1]0	bar 1 [2]2, 50			

FIGURE 2c

notice that in the first year there is a month VI_2, and at the end of the last year we have a month XII_2, indicating that here it has been necessary to insert an extra month in the normal 12-month lunar year in order to keep it in step with the solar year. Incidentally, these months are at this late date introduced according to a rigid pattern based on what I earlier called the Metonic cycle.

Each of the subsequent columns describes in an entirely numerical fashion some particular facet of the behaviour of the Sun or the Moon or both; thus the first preserved column (labelled I in figure 2*a*) gives the progress in degrees of the Sun, and, since we are dealing with conjunctions, also of the Moon (less 360°) in the ecliptic per month. The next column (II) gives the position of Sun and Moon at conjunction in the ecliptic. The first line tells that at the end of month XII of year 207 of the Seleucid Era the conjunction took place at 2° 2′6″20‴ of the sign Aries (a zodiacal sign means, by this time, simply a 30° section of the ecliptic). The next line says that the position of the subsequent conjunction of Sun and Moon was at 0° 52′45″38‴ of the sign Taurus. You get to the second position from the first by adding the number in line 2 of column I, thus, writing in the now standard sexagesimal notation:

$$\text{column II, l. 1: Aries} \quad 2;\ 2,\ 6,20$$
$$\text{column I, \ l. 2: +} \quad \underline{28;50,39,18}$$
$$\text{column II, l. 2: Aries} \quad 30;52,45,38 = \text{Taurus } 0;52,45,38$$

and so on.

To dispel any notion that these positions are the results of incredibly accurate observations (to sixtieths of seconds, or thirds, of arc) I need only point out that the Moon is invisible for quite an interval before and after conjunction with the Sun, so that there is nothing to observe unless there happens to be a solar eclipse at that conjunction.

But even if the phenomenon had been observable – we have completely analogous texts for full moons which can be seen very well – the structure of column I, the difference column, completely rules out observations. The entries, you will note, decrease regularly by 18 in the second, i.e. the minutes' place, until a minimum is passed between lines 4 and 5. From line 5 the values increase, again by 18, until a maximum is passed between lines 10 and 11, whence they begin to decrease, and so on.

One gets from an ascending to a descending branch of this kind of function by following a very simple rule: if the application of the line by line difference d (here 0;18,0,0, to express it with the number of digits of the column) lead to a value larger than a certain fixed maximal value M, then the excess over M is subtracted from M to yield the next value of the function, and symmetrically about the minimum m. For column I we find

$$M = 30;\ 1,59,\ 0$$
$$m = 28;10,39,40$$

so that the reflexion in the maximum between lines 10 and 11 is executed thus:

$$\text{column I, l. 10:} \quad 29;54,40,\ 2$$
$$+d: \quad \underline{0;18,\ 0,\ 0}$$
$$30;12,40,\ 2$$
$$-M: \quad \underline{30;\ 1,59,\ 0}$$
$$0;10,41,\ 2$$
$$\text{which subtracted from } M \text{ gives} \quad 29;51,17,58$$

obverse reverse

FIGURE 4. A.C.T. no. 20, actual size.

													I		II			III		
1069	obv.	XII	2, 13, 34, 15, 33, 20	6; 11, 15	♈	2, 57, 27, 30	+ 44, 44, 42	15, 3, 36, 33, 45	2, 40	18, 38, 25, 30	lal	−9, 22, 9	2, 12	še				bar		
1070		2, 47* I	2, 10, 48, 20	4; 18, 45	♉	3, 16, 12, 30	−2, 48, 23, 21	14, 21, 36, 33, 45	2, 53, 18, 45	57, 3,45	lal	−9, 22, 30	1, 46, 53	bar				gu_4		
1071		II	2, 8, 2, 24, 26, 40	2; 26, 15	♊	3, 28, 58, 30	−4, 47, 9, 3	13, 39, 36, 33, 45	3, 17, 50, 51, 51, 6, 40	57, 3, 45	lal	−6, 23	2, 14, 24	gu_4				sig		
1072		III	2, 5, 16, 28, 53, 20	0; 33, 45	♋	3, 34, 44, 30	−6, 45, 54, 45	12, 57, 36, 33, 45	3, 43, 39, 30, 22, 13, 20	57, 3, 45	lal	−2, 53	2, 43, 43	sig		3, 34, 5	šú	šu	1	
1073		IV	2, 2, 30, 33, 20	28; 41, 15	♋	3, 33, 30, 30	−5, 39, 19, 33	12, 15, 36, 33, 45	4, 9, 28, 8, 53, 20	57, 3, 45	lal	+ 37	3, 13, 2	šu	28	., 21, 3	šú	izi	1	15, 30
1074		V	1, 59, 44, 37, 46, 40	26; 48, 45	♌	3, 25, 16, 30	−3, 40, 33, 51	11, 33, 36, 33, 45	4, 35, 16, 47, 24, 26, 40	57, 3, 45	lal	+4, 7	3, 42, 20	izi	28	2, 38, 43	šú	kin	1	15, 30
1075		VI	1, 58, 37, 13, 20	25; 44	♍	3, 9, 30, 40	− 53, 14, 18	11, 16, 32, 48, 45	4, 56	32, 50, 33, 30	lal	+7, 52, 55	4, 31, 2	kin	28	4, 7, 41	šú	du_6	30	
1076		VII	2, 1, 23, 8, 53, 20	25; 44	♎	2, 49, 30, 40	+2, 51, 38, 33	11, 58, 32, 48, 45	4, 43			+10	4, 53	du_6	29	5, 14, 41	šú	apin	1	
1077		VIII	2, 4, 9, 4, 26, 40	25; 44	♏	2, 33, 42, 24	+4, 57, 54, 15	12, 40, 32, 48, 45	4, 17, 11, 21, 28, 53, 20			+7, 54, 8	4, 25, 5	apin	28	., 49, 36	šú	gan	30	
1078		IX	2, 6, 55	25; 44	♐	2, 25, 54, 8	+7, 4, 9, 57	13, 22, 32, 48, 45	3, 51, 22, 42, 57, 46, 40			+3, 54, 8	3, 55, 16	gan	29	2, 54, 20	šú	ab	1	
1079		X	2, 9, 40, 55, 33, 20	25; 44	♑	2, 26, 5, 52	+5, 13, 34, 21	14, 4, 32, 48, 45	3, 25, 34, 4, 26, 40			− 5, 52	3, 25, 29	ab	29	5, 28, 51	šú	zíz		
1080		XI	2, 12, 26, 51, 6, 40	25; 44	♒	2, 34, 17, 36	+3, 7, 18, 39	14, 46, 32, 48, 45	2, 59, 45, 25, 55, 33, 20			−4, 5, 52	2, 55, 40	zíz		2, 33, 11	šú	še		
1081		XII	2, 15, 12, 46, 40	25; 44	♓	2, 50, 29, 20	− 21, 54, 6	15, 28, 32, 48, 45	2, 40, 52, 57, 46, 40			−8, 5, 52	2, 32, 47	še				dir-še		
1082		XII_2	2, 16, 10, 55, 33, 20	23; 56, 15	♈	3, 9, 17, 30	−3, 22, 1, 45	15, 43, 15, 56, 15	2, 40	54, 38, 34, 30	lal	−9, 24, 5	1, 35, 58	dir-še				bar		
JN		T	Φ	B		C	E	F	G	J		C′	K		M			P_1		
4193	rev.	2, 47* I	2, 2, 41, 28, 53, 20	20; 15	♎	3, 6, 50	+1, 14, 1	12, 18, 22, 30	4, 30, 48, 53, 20	47, 10, 18	lal	−9, 29	3, 34, 9	bar						
4194		II	2, 5, 27, 24, 26, 40	18; 22, 30	♏	3, 23, 21	+3, 47, 46, 12	13, 0, 22, 30	4, 5, 0, 14, 48, 53, 20	57, 3, 45	lal	−8, 15, 30	2, 49, 41	gu_4						
4195		III	2, 8, 13, 20	16; 30	♐	3, 32, 52	+5, 46, 31, 54	13, 42, 22, 30	3, 39, 11, 36, 17, 46, 40	57, 3, 45	lal	−4, 45, 30	2, 37, 22	sig						
4196		IV	2, 10, 59, 15, 33, 20	14; 37, 30	♑	3, 35, 23	+6, 38, 42, 24	14, 24, 22, 30	3, 13, 22, 57, 46, 40	57, 3, 45	lal	−1, 15, 30	2, 15, 4	šu						
4197		V	2, 13, 45, 11, 6, 40	12; 45	♒	3, 30, 54	+4, 39, 56, 42	15, 6, 22, 30	2, 48, 17, 36, 17, 46, 40	57, 3, 45	lal	+2, 14, 30	1, 53, 28	izi						
4198		VI	2, 16, 31, 6, 40	10; 52, 30	♓	3, 19, 25	+2, 41, 11	15, 48, 22, 30	2, 40	57, 3, 45	lal	+5, 44, 30	1, 48, 41	kin						
4199		VII	2, 14, 52, 35, 33, 20	10; 44	♈	2, 59, 30, 40	−1, 13, 1, 24	15, 23, 26, 15	2, 40	4, 18, 41	lal	+9, 57, 10	2, 45, 39	du_6						
4200		VIII	2, 12, 6, 40	10; 44	♉	2, 39, 42, 24	−3, 54, 46, 24	14, 41, 26, 15	2, 43, 8, 53, 20			+9, 54, 8	2, 53, 2	apin	1					
4201		IX	2, 9, 20, 44, 26, 40	10; 44	♊	2, 27, 54, 8	−6, 1, 2, 6	13, 59, 26, 15	3, 5, 39, 45, 11, 6, 40			+5, 54, 8	3, 11, 33	gan	30	12		destroyed		
4202		X	2, 6, 34, 48, 53, 20	10; 44	♋	2, 24, 5, 52	−6, 16, 42, 12	13, 17, 26, 15	3, 31, 28, 23, 42, 13, 20			+1, 54, 8	3, 33, 22	ab	1	13				
4203		XI	2, 3, 48, 53, 20	10; 44	♌	2, 28, 17, 36	−4, 10, 26, 30	12, 35, 26, 15	3, 57, 17, 2, 13, 20			−2, 5, 52	3, 55, 12	zíz		13				
4204		XII	2, 1, 2, 57, 46, 40	10; 44	♍	2, 40, 29, 20	−1, 44, 21, 36	11, 53, 26, 15	4, 23, 5, 40, 44, 26, 40			−6, 5, 52	4, 17	še						
4205		XII_2	1, 58, 17, 2, 13, 20	9; 52, 30	♎	2, 59, 55	+2, 21, 17, 48	11, 11, 26, 15	4, 48, 40, 14, 48, 53, 20	26, 7, 19	lal	−9, 42, 50	4, 12, 50	dir-še						
4206		2, 48 I	2, 0, 4, 48, 53, 20	8	♏	3, 18, 40	+4, 21, 24, 36	11, 38, 43, 7, 30	4, 53, 20	57, 3, 45	lal	−9, 22, 30	3, 46, 54	bar						

FIGURE 5

parameters of this zigzag function since they will be of importance in the following discussion; they are

$$\text{maximum} \qquad M = 4,29;27, 5°$$
$$\text{minimum} \qquad m = 1,52;34,35°$$
$$\text{monthly difference} \quad d = \qquad 22;30, 0°/m$$

$$\text{mean value } \mu = \tfrac{1}{2}(M+m) = 3,11;0,50° = 0;31,50,8,20 \text{ day}$$

to convert the time degrees into days (1 day = 6,0°). Thus column G implies that in the mean the Babylonian value of one synodic month is equal to

$$29 \text{ days} + \mu = 29;31,50,8,20 \text{ days.}$$

Column VIII (H) contains the differences of column IX (J). The values of J are a further correction to the length of the interval between conjunctions, this time depending on solar anomaly; thus we have that the length of the true synodic month is 29 days + G + J (since J's mean value is 0, the mean synodic month remains 29 days + μ). The values of G + J are listed in column X (K) which is the difference column of column XI (L).

Column L gives us the time of the conjunctions. Line 1 tells that the conjunction in month XII of the year 208 of the Seleucid Era took place on the 29th day at 1,2;43,50 time degrees after midnight (i.e. at about 4h11min a.m.). The hour in the next line is simply this augmented by the value in line 2 of the preceding column, and so on to the end of the text. To say what the date is, one must know whether the previous month had 29 or 30 days, the only two possibilities in a lunar calendar; this information is found in column XV (P_1). To give you a sense of the quality of the text, I shall compare the first and the last lines with the result of modern computations, according to which the conjunction in obverse, line 1 took place on 104 B.C., 23 March at 3h23min (text has about 4h11min) and that in reverse, line 19 on 101 B.C., 18 April at 0h45min (text has about 2h22min). Throughout the entire text the difference between ancient and modern computed values remains in the interval $1\tfrac{1}{2}$h ± 1h, where part of the 1.5h is accounted for by the deviation of the initial value.†

Let me just mention one more column before abandoning the text, namely, column XV (P_1), which concerns the first visibility of the new crescent which marks the beginning of a new month. Thus line 1 tells us that month I (bar) began on the 31 of month XII (hence the 1), and that the time from sunset to moonset was 15;40 time degrees, i.e. about 1h3min. These are precisely the two pieces of information given about the new Moon in the Astronomical Diaries which contain the observational material upon which the arithmetical theories were constructed (see Sachs, this volume, p. 43). Indeed, one of the principal aims of these theories is to generate forecasts of the astronomical events recorded in these diaries, and this column is an example.

By analysing texts such as this, and from another class of texts called procedure texts – they contain rules for computing the various columns of ephemerides – we have gained control over the theories underlying the procedures. To emphasize this, and also that these astronomical texts are entirely computed without the occasional injection of observations except possibly as ultimate initial values, I shall show you another text. Figure 4, plate 2, is a photograph of obverse and reverse of a fragment of a clay tablet, and it is, unfortunately, more typical of our material than the other.

† The modern values for the moments of conjunction are taken from a set of tables of all syzygies from 1000 B.C. to A.D. 1651 computed for the meridian of Babylon by Herman H. Goldstine. The tables will be published shortly by the American Philosophical Society. Dr Goldstine kindly put a copy of the original computer print-out at my disposal.

Figure 5 shows our reconstruction of the text, first published by Neugebauer (1955) as A.C.T. No. 20, from which the fragment came (the preserved surface is outlined with dotted lines). I assure you that this reconstruction is quite secure though it admittedly is a *tour de force* which would not have been attempted without the aid of an electronic computer.

This text is also a lunar ephemeris, but here the obverse concerns new moons, while the reverse treats of full moons, both for the year 167 of the Seleucid Era (145/4 B.C.). Both the previous text and this come from Babylon and both are entirely arithmetical in structure, but they are computed according to two quite distinct systems: the former belongs to system B, this to system A, to use the now standard terminology for the two major Babylonian astronomical systems.

A characteristic difference between the two systems – though far from the only one – lies in the manner of computing longitudes. As I have already mentioned, the monthly progress in the ecliptic of a syzygy is determined by a zigzag function in system B. In system A the approach is quite different. Here the monthly progress of the syzygy in longitude, $\Delta\lambda$, is, to be sure, not explicitly displayed as in column A of system B, but a glance at the longitude column (B) of our text should suffice to reveal that the longitudes fall in groups, within which $\Delta\lambda$ alternately assumes the value 30° (i.e. the syzygy advances precisely one full zodiacal sign per month), and the value 28;7,30°. In my restoration of the text in figure 5, the groups are separated by dotted lines. The transitional values of $\Delta\lambda$ across these dotted lines lie between 30° and 28;7,30°.

The underlying scheme may be described in terms of the following model for the motion of the Sun (it happens that the lunar anomaly has so small an effect on the *position* of syzygies that it may be ignored in first approximation): the Sun's velocity is a piecewise constant function of solar longitude, namely,

from Virgo 13° to Pisces 27°: $V = 30°$/month,
from Pisces 27° to Virgo 13°: $v = 28;7,30°$/month.

The monthly progress $\Delta\lambda$ of the Sun, and hence of the syzygy, is then mostly either V or v, but it assumes an intermediary value, which is readily computed, when a boundary between the two zones is transgressed in the course of a month. Figure 6 shows the character of this step function (heavy horizontal lines); the values $\Delta\lambda$ agree mostly with this generating or velocity function except in intervals of length V and v preceding the discontinuities. Here $\Delta\lambda$ is drawn as a lighter skew line segment, for $\Delta\lambda$ is a piecewise linear, continuous function of λ as can easily be shown. Such step functions, together with a rule for deriving progress in longitude from them, is the alternative Babylonian arithmetical device for describing periodic phenomena. It seems, at first, cruder than the zigzag functions but is actually much more flexible.

My description of the system A scheme for finding longitudes of consecutive syzygies is frankly in the modern kinematical idiom, using a notion like the instantaneous velocity of the Sun; I shall presently give an explanation of a system A model which avoids such anachronisms, but before leaving our text I should like to identify its columns briefly.

Column T gives year (in the Seleucid Era) and month and column Φ is a function in phase with lunar velocity and the basis of finding the later column G. Column B, as just said, lists on the obverse the common longitude of Sun and Moon at conjunction, and on the reverse the longitude of the Sun at midmonth increased by 180°, which is the longitude of the full Moon. Column C gives the length of daylight and column E is actually lunar latitude (in units

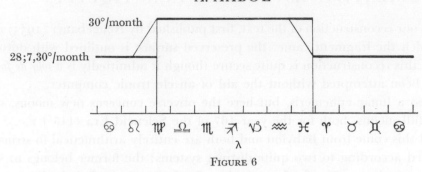

FIGURE 6

'barleycorns', 72 of which equal one degree; the integral barleycorns occupy the first two sexagesimal places). Lunar latitude turns out to be a very simple function – a slightly modified zigzag function – of the Moon's elongation from the ascending node; this nodal elongation is easily found from column B giving lunar longitude combined with the underlying assumption that the node retrogresses by the constant amount $1;33,55,30°$ per month, and some initial position of the node (v. d. Waerden 1966). Again, eclipse warnings would be issued whenever the new or full moon has smallest latitude at a nodal crossing; this is done with an eclipse magnitude function depending simply on column E, but which I have not bothered to reconstruct.

Columns G and J together give, as before, the excess over 29 days of the time from conjunction to conjunction, or opposition to opposition, where G depends on lunar, and J on solar anomaly. Their structures are more complicated than those of their counterparts in system B, and I shall pass them by except for pointing out that J's mean value here is negative (for details see Aaboe 1971). Column C' gives a correction due to the variation in length of daylight, column K the sum of G, J and C', and column M lists the moment of syzygy (on the obverse date and time degrees of conjunction *before* sunset of the day), and the moments proceed from line to line by the amount K. Finally, column P, preserved on obverse only, gives information about the visibility of the new crescent; '1' means that the previous month turned out to be full (30 days long), '30' that it was hollow (29 days long), and the following numbers give the computed time from sunset to moonset.

I realize well that this cursory presentation of two lunar texts is unsatisfactory and, in particular, that it fails to show the beautiful, simple, yet highly sophisticated manner of treating technical details. I hope, however, to have conveyed some sense of the complexity of Babylonian lunar theory, and that the Babylonian theoretical astronomers had succeeded in isolating the essential periods of lunar and solar motion and in putting them to proper use.

My last example from Babylonian astronomy is not a text, but rather the result of an analysis of a planetary model belonging to system A. The scheme is for the longitude of Mars at one of its characteristic synodic phenomena, say, at first stationary point, and I have chosen it for several reasons: to display the role of an unexpected branch of mathematics in such schemes, to emphasize the flexibility of the approach to astronomy of system A, and finally, and most importantly, to point to a possible connexion between the sort of observations we know the Babylonians to have recorded and their theoretical constructs.

The first, and to us unfamiliar, feature of Babylonian planetary theory lies in the very question one asks.

Since Ptolemy's *Almagest* we have wanted our planetary theories to enable us to answer the question: given the time, where is the planet? Thus we consider *time* the independent variable and seek means of deriving all other information from it.

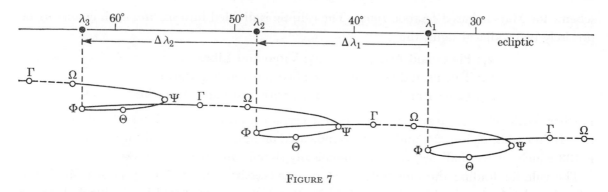

FIGURE 7

The Babylonian approach is entirely different. First, almost all interest, at least primarily, is focused upon the planet only when it is in one of its characteristic synodic situations: for an outer planet they are – see figure 7, which represents a run of Saturn with latitude exaggerated four times – (i) first appearance (Γ), (ii) first stationary point (Φ), (iii) opposition (Θ), (iv) second stationary point (Ψ), (v) last appearance (Ω) (the capital Greek letters are the now standard manner of referring to these synodic phenomena).

The next bold simplification is that we disregard all but one of these, say the first stationary point Φ, and we now ask the question: if we are given the longitude and the time at which a certain planet happens to be at a first stationary point, where and when will it next be at a first stationary point? What we, in the Babylonian mode, consider and wish to reproduce is then, a sequence of discrete points, in time and longitude.

Figure 8 is a graphical representation of the arithmetical model upon which the system A

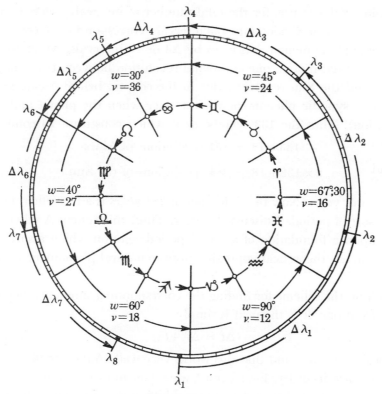

FIGURE 8. Mars, system A: 133 synodic phenomena ~ 18 revolutions ~ 2.133+18 years.
1 synodic arc = 18 intervals.

scheme for Mars is based (Aaboe 1964). The ecliptic is divided into six arcs each consisting of precisely two zodiacal signs, thus:

α_1: Pisces and Aries, α_4: Virgo and Libra,
α_2: Taurus and Gemini, α_5: Scorpio and Sagittarius,
α_3: Cancer and Leo, α_6: Capricornus and Aquarius.

Further, each of these arcs is divided into a certain number (ν) of intervals, thus α_1 into $\nu = 16$ intervals, α_2 into 24, and so on, as indicated on figure 8. The total number of intervals is 133 which, as we shall see, is an astronomically significant number for Mars.

The rule for finding the longitude of Mars at successive first stationary points (Φ) is the following: let the longitude of the initial stationary point be represented by the black dot near the end of Sagittarius at the bottom of figure 8 (it is, actually, 26° 40′ of Sagittarius); we now progress in steps consisting always of 18 intervals each, whatever the length of the intervals. Thus the second Φ for Mars will be near the middle of Pisces at the end of the step, *synodic arc* is the proper term, marked $\Delta\lambda_1$; the third in Taurus at the end of $\Delta\lambda_2$, etc., and the eight Φ, the last indicated on the figure will be at the beginning of Sagittarius (actually at Sagittarius 3° 20′), which has brought the phenomenon almost once around the ecliptic.

The length of a synodic arc depends only on where in the ecliptic it begins, i.e. on the lengths of the 18 small intervals that constitute it, and it varies considerably (the values w, which are 18 times the length of the intervals, i.e. the length of the synodic arc if a new zone were not encountered, are the parameters given in the Babylonian texts, and they play the same role as the two values of solar velocity mentioned above).

It should be clear that after 133 synodic arcs, but not earlier, the phenomenon will return precisely to its point of departure, for the total number of intervals covered will be 133 times 18 which, since the ecliptic is divided into 133 intervals, correspond to 18 complete revolutions. It is only the synodic phenomenon that leaps by $\Delta\lambda$ or 18 intervals; Mars itself, being a swift planet, moves in its actual motion one additional revolution between two synodic phenomena of the same kind, and the Sun revolves twice in the ecliptic before it catches up with Mars after a further $\Delta\lambda$ (a synodic phenomenon takes place when the planet and the Sun have a certain fixed relation). Thus the 133 synodic arcs, which constitute 18 complete revolutions correspond also to

$$133 + 18 = 151 \text{ revolutions of Mars}$$

in the ecliptic and

$$2 \times 133 + 18 = 284 \text{ revolutions of the Sun}$$

or, with another word, years. Now the time span of 284 years is an excellent period of Mars, a refinement of the sort of periods I discussed earlier. Thus, this system A scheme, as all similar schemes, is built upon a foundation of a sound period relation which assures that the errors inherent in this, as in all, theoretical approximations to natural phenomena do not accumulate arbitrarily.

I shall not touch on the scheme for finding the corresponding moments, beyond stating that the time interval from one Φ to the next is simply

$$\Delta t = \Delta\lambda + C,$$

where C is a suitable constant, and that this works very well (Aaboe 1958).

Several comments are in order. First I must warn you most emphatically not to be misled by my graphic representation of the scheme into thinking that there may have been some sort of geometrical model underlying the Babylonian procedures. This is not so; all the Babylonian astronomical models, as I still feel justified in calling them, were entirely arithmetical in character. It is not very difficult to give the essence of my description of the Mars model completely in arithmetical or perhaps number-theoretical language.

Secondly, except for the choice of initial values, the independent variable is here not time, but the longitude of a synodic phenomenon. It is that which entirely determines the length of the step forward, in longitude as in time. Similar, indeed mostly identical, models are applied to the other synodic phenomena. If, finally, one wishes to know where the planet in question is at a given arbitrary moment, this is determined by interpolation between the appropriate synodic phenomena by means of various interpolation schemes which may be of as much as third order (Neugebauer 1955; Huber 1957).

Lastly, I should like to point to a possible way of deriving theoretical schemes of this sort from the kind of observations we know the Babylonians to have made and recorded. Professor Sachs will describe the texts he calls Astronomical Diaries which survive in large numbers, all deriving from unscientific excavations of what must have been an astronomical archive somewhere in the ruins of Babylon. These texts date roughly from the last seven centuries B.C. They are unique as a corpus of ancient historical documents, but what is of importance in this context is that they contain observations of precisely the sort of phenomena reproduced by the Babylonian theoretical schemes. These observations are at first sight disappointing, since, for the phenomena in which we are interested, they are of the form: *In year n of King N, month x, day y, Mars reached its first stationary point; it was in the zodiacal sign Z*. This means that the longitude of the phenomenon which, as we have seen, is at the basis of the theory, is given only as a zodiacal sign (except in unusual circumstances), i.e. the planet is fixed only within an interval of length 30°.

This, then, is the problem, if not dilemma, which has bothered students of Babylonian astronomy for a number of years: how can one from observations of such crudity derive such excellent schemes?

The answer I shall suggest presupposes first, that one had arrived at a good period relation like

133 synodic phenomena $\sim$ 18 revolutions of the phenomenon
in the ecliptic $\sim$ 151 revolutions of Mars $\sim$ 284 years,

to keep to Mars as our example. May I here insert the remark that one need not observe Mars for a full 284 years to reach these relations; it is very possible to construct such periods from much shorter ones by correcting for their deficiencies, but I cannot go into details with this in this brief time.

Thus one knows, to begin with, that there are to be 133 intervals in all, and that there must be 18 of them from phenomenon to phenomenon. The model is determined as soon as these 133 intervals are distributed properly on the ecliptic, so the aim is to fix the values v, whose sum is already known.

To that end one may now take from the observational records a run of observations of consecutive synodic phenomena of a certain kind, preferably for some reasonable subperiod, and sort them out according to zodiacal sign; i.e. one keeps a tally, sign by sign, of the phenomena. It appears very soon that certain signs, particularly Cancer and Leo, are very popular, while others, Capricorn and Aquarius, are not. First, this is an indication that longitude is a sensible choice of independent variable and, second, one gets from such a tally a feeling for the proportionate distribution of the intervals in the several signs.

Not everything is explained by such a tally; thus, the decision to consider six zones of length precisely two signs each is certainly not the only possible one. Further, it is necessary that the values ν be nice numbers – nice in the sense that their reciprocals have finite sexagesimal expansions – in order that the number-theoretical properties be strictly obeyed by the derived ephemerides. But granting this choice and these limitations, it turns out that one really has very little freedom left for distributing the intervals. A comparison between modern planetary tables and ancient ephemerides based on such models shows excellent agreement even in the case of as difficult a planet as Mars.

I shall finish my discussion of Babylonian mathematical astronomy with a few remarks of a more general nature. First, I wish to emphasize the lateness of the relevant texts. The vast majority of them belong to the Seleucid period, i.e. roughly the last three centuries B.C., and though it is impossible to point to an earliest date, I feel reasonably sure that all the presently known and understood texts, some 400 in number, come from the last five centuries B.C., to give a very generous earlier limit. By the beginning of the Seleucid Era the theories were certainly fully developed, and our texts continue to the very end of cuneiform literacy. The creation of mathematical astronomy is thus one of the last, as well as one of the finest, original efforts of Mesopotamian culture, an event without precedent anywhere, and with great consequences.

Secondly, I want to stress that the creators of these astronomical theories, whoever they were, were able to draw from, and combine, a peculiar set of available ingredients which we know of from other textual material. Principal among them are, first, a particular kind of mathematics and, secondly, a body of continuously recorded observations

As to the former the mathematical cuneiform texts fall into two groups in respect of chronology: one Old-Babylonian from the beginning of the second millennium B.C. and the other Seleucid (Neugebauer & Sachs 1945). There is, however, hardly any difference between these two groups of documents in content or character. At the basis of Babylonian mathematics is the sexagesimal number system which you have seen amply in use in the astronomical tables above, a system which reduces the four basic arithmetical operations to trivialities. Though much of what we would call geometrical knowledge is incorporated in Babylonian mathematics – foremost among it a command of the Pythagorean theorem, as we are wont to call it, though it antedates Pythagoras by a millennium – its chief concern is still with numerical problems, algebra, and perhaps number theory, even when the problems at first sight may seem to wear geometrical garb. It is, then, not surprising to find an *arithmetical* treatment of astronomical phenomena into which, further, period relations are built through simple number-theoretical devices.

As to the second, the recorded observations are embodied principally in the Astronomical Diaries as I have already mentioned; here I shall just add that there is little doubt that the Babylonian astronomers had available to them an archive of sustained observations going back to about 700 B.C. The existence of a connexion between the observational and theoretical material is clear, for the goal of the theories is, in the large, to reproduce and predict precisely the kind of phenomena and quantities recorded in the Diaries.

It is idle to ask if the combination of this kind of mathematics and observations was necessary for the creation of mathematical astronomy. The historical fact is that it was sufficient, and that wherever else we encounter scientific mathematical astronomy we can detect, directly or indirectly, the influence of the Babylonian forerunner.

I shall try to justify this, necessarily very briefly, and first I shall point out a passage from the beginning of *Almagest* IV (written about A.D. 150) in which Ptolemy is concerned with establishing relations between the various periods of solar and lunar motion. He says that Hipparchus (about 150 B.C.) had shown that the smallest interval between two eclipses which produces repetition in solar and lunar anomaly is

126 007 days 1 h = 4267 synodic months = 4573 anomalistic months = 4612 rotations of the Moon in long. less $7\frac{1}{2}°$.

Hence it is found by division, says Ptolemy, that

1 synodic month = 29;31,50,8,20 days.

The difficulty with this is that if we do divide the time interval, in days and hours, by the number of synodic months, we do not get 29;31,50,8,**20** but 29;31,50,8,**9**,... days as the length of the mean synodic month. (This discrepacy has been noted repeatedly in the astronomical literature, e.g. by al-Bīrūnī and by Copernicus.)

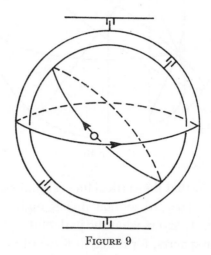

FIGURE 9

Now 29;31,50,8,20 days is precisely the value of the synodic month used in Babylonian lunar system B texts, as we saw above in the comments on column G (already Kugler was aware of this (Kugler 1900, p. 24)). There can, then, be no doubt whatever that Hipparchus got this parameter from Babylonia, and the rest of his relation can be shown to have Babylonian origins as well (Aaboe 1955).

All the works of Hipparchus are lost but one, a not too informative commentary on an astronomical poem by Aratus. Thus all we say about Hipparchus has essentially to be based upon secondary sources, principally Ptolemy who, however, is generous with references to his predecessor. It appears that what Hipparchus was engaged in was to adapt geometrical astronomical models of a certain kind to new purposes.

There was, of course, a tradition for such models in the Greek-speaking world. May I remind you first of the homocentric spheres of Eudoxos (*ca.* 370 B.C.), an aesthetically pleasing theory which was rescued from various secondary references and restored in a now classic paper by Schiaparelli (1925).

The basic device in this scheme for simulating planetary behaviour is two spheres, the inner being able to rotate relative to the outer, and the outer being able to rotate relative to some fixed frame; the two axes of rotation are inclined to each other, and the planet is affixed to the 'equator' of the inner sphere. This arrangement is schematically shown in figure 9. The two spheres are made to revolve in opposite directions, but equally swiftly. If the axes coincided, the planet would not move at all, for the two equal, but opposite rotations would cancel each other. But when the axes are inclined the planet will travel in a path shaped like a figure 8, as is shown in figure 10. Here only one sphere is drawn, but both axes and their equators are represented. The curve is called a hippopede, a horsefetter, and, as can be shown by elementary means, it happens to be the intersection between the sphere and a right cylinder as indicated in figure 10 (Neugebauer 1953, 1957).

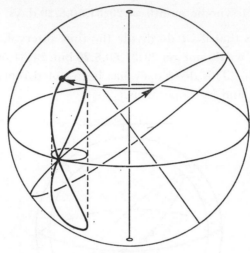

FIGURE 10

The apparatus is now placed so that the vertical line of symmetry of the figure 8 is the ecliptic, and the whole thing is given a forward motion in longitude (this may be achieved by yet another rotating sphere). Figure 11 shows what the path of the planet looks like, when observed from the common centre of the spheres, for different forward motions. At the top we have the hippopede itself, and as the forward motion increases we see a curve which indeed looks somewhat like the apparent path of a planet with its stationary points and retrogradations (cf. figure 7). If, however, the forward motion is too swift then the planet can no longer manage to become retrograde, but is merely slowed down, as shown in the two lowest graphs.

When we toy like this with various velocities, we assume that we are free to assign them at will. But when we deal with a specific planet, that is not so, for the period of the motion in the hippopede, i.e. of the rotations of the first two spheres, must be the planet's synodic period, and the forward motion in longitude must be the mean synodic arc $\Delta\lambda$ per synodic period. Both of these are determined within rather narrow bounds by even crude period relations of the sort I mentioned above. Our only really free choice in this model is, then, the inclination between the two axes in figure 9.

I have already mentioned the simple period relation for Venus:

5 synodic cycles = 8 years = 8 revolutions in the ecliptic of Venus.

If we, then, are to consider an Eudoxian model for Venus, we must let each of the two inner spheres revolve 5 times in 8 years, in opposite directions, to make Venus travel 5 times through the hippopede. Further, the hippopede must be carried 8 times around the ecliptic in the same span of 8 years. It can, however, be shown that this motion is so fast that no matter what inclination between the axes of the first two spheres we choose, we have a situation like that represented in the last graph in figure 11: Venus just cannot become retrograde. Precisely the same holds for Mars.

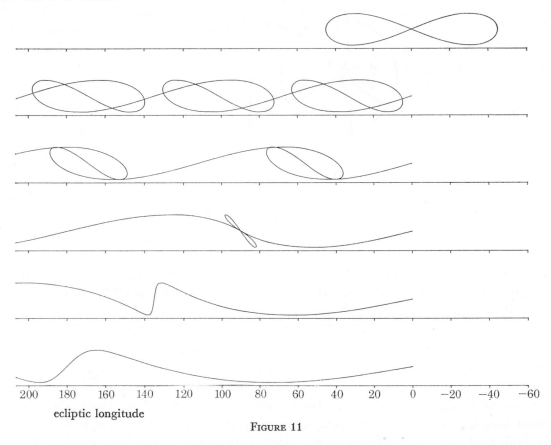

FIGURE 11

I wish to emphasize this point strongly because I believe it reveals that the purpose of these models was to serve as a *qualitative* description of planetary motion. Indeed, it cannot be anything else, for if one wants to get a quantitative result out of such a model one has little latitude in the choice of periods; thus, no matter how one manages to follow the resultant motion of the planets, the models for Venus and Mars will fail most drastically by their inability to produce retrograde motion, one of the most conspicuous phases of planetary behaviour, and the very phenomenon, I suspect, that the models were created to account for. There are other difficulties in putting these models to quantitative use, e.g. that all retrograde arcs for a planet that does become retrograde are of equal length, and that the mathematical techniques (spherical trigonometry) necessary for deriving numerical results from them were not created until some two centuries after Eudoxos.

May I further remind you of another type of geometrical planetary model, one which is much better known, for it became supremely successful in Ptolemy's refined version. In figure 12, the paper is the plane of the ecliptic viewed from the north, O the observer at the centre of a circle called the deferent, C a point travelling on the deferent and centre of a second circle called the epicycle upon which the planet P moves. As before, the periods of the motions are determined: P must move once around the epicycle *relative to the line* OC in one synodic period,

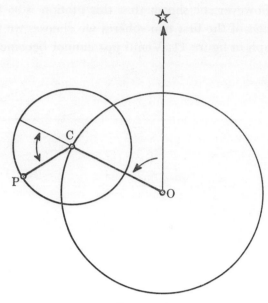

FIGURE 12

and C must move the synodic arc $\Delta\lambda$ per synodic period on the deferent relative to a line from O to, say, a fixed star. If we normalize and set the radius of the deferent OC = 1, there remains then to determine the epicyclic radius CP to suit the specific planet's requirements.

There is, however, a second, often ignored, free choice, and that is of the sense of rotation of the planet on the epicycle. We now take it as a matter of course that the planet has to move counterclockwise on the epicycle, for then the model corresponds closely to a correct first-order representation of geocentric planetary motion; but the contrary choice is in principle possible, for it, too, will make P behave like a planet, except that now it will become retrograde when farthest from O. There is evidence in a Greek papyrus that this choice was, in fact, made by some; this is of interest for a kinematical analysis shows that a model with P going clockwise, and otherwise given the proper periodic motions, is also incapable of making Venus and Mars retrogress, no matter what value is given to the epicyclic radius (Aaboe 1963).

Though our evidence of pre-Hipparchian astronomy is scant and fragmentary, I feel justified by such analyses in my belief that the purpose of early geometrical astronomical models was not to serve as a basis for determining positions of celestial bodies at a given time with some numerical accuracy, but that it rather was like that of the orreries: to display in a qualitative manner how the motions of celestial bodies may be generated.

We can now see more clearly what Hipparchus set out to do: he took one variety of qualitative models, epicyclic models with the right sense of rotation on the epicycle, refined them by employing eccentric deferents, determined reasonable parameters for these models, and made them yield numerical results through the application of trigonometry, a branch of mathematics which he probably created for the purpose. He was successful with models for the Sun, and for the Moon at syzygies, but failed to solve the planetary problem to his satisfaction. His work was completed by Ptolemy some three centuries later.

We saw that Hipparchus had at his disposal for this task important and refined astronomical Babylonian parameters. It is further beyond doubt that he also had access to Babylonian observational material going back to *ca.* 700 B.C. – in what form we do not, and probably shall not know – and so had even Ptolemy. Important though this is, I believe that the Babylonian influence on Hipparchus was even more fundamental. I am convinced, though I cannot offer proof, that whether or not he understood the technical details of the Babylonian ephemerides, it was from them that he got the idea of the possibility and desirability of a quantitative description of astronomical phenomena that could yield fine numerical predictions.

It is on these facts and on this belief that my case for the Babylonian origin of all endeavour in the exact sciences rests. This is a substantial claim. To make it seem more plausible may I try to trace hurriedly the development and the paths of transmission of astronomy from the time of Hipparchus.

We have very little evidence of astronomical activities in the Hellenistic world between Hipparchus and Ptolemy – it is largely due to the excellence of Ptolemy's work, I am sure, that so little primary evidence of pre-Ptolemaic Greek astronomy survives. It seems certain, however, that it was during this interval that astronomical models and parameters were transmitted to India. In the Hindu astronomical writings we find no trace at all of Ptolemy's refined models; the elements we recognize are, to be sure, first geometrical models of the Greek variety, but there is a large admixture of arithmetical schemes of obvious Babylonian origin. (Among them are some based precisely on the Mars model I discussed above in § 4, with its six two-sign zones, admittedly badly corrupted and probably useless, but still unmistakably recognizable (Neugebauer & Pingree 1970).)

Pingree (1963) maintains that, though direct contact between Mesopotamia and India cannot be ruled out, it is still overwhelmingly likely, on linguistic grounds among others, that these schemes reached India via Hellenistic Greece. This, incidentally, gives us indirect evidence of a wider knowledge of Babylonian techniques in the Greek-speaking world than is indicated by the fragile direct evidence.

The culmination of the Greek high tradition in astronomy is Ptolemy's *Almagest*, the importance and influence of which can hardly be overrated. Ptolemy's goals were those of Hipparchus – which, if I am right, were inspired by the Babylonian example – and he succeeded where Hipparchus gave up: in constructing a satisfactory planetary theory and a lunar model that worked at quadrature as well as at syzygy.

Islamic astronomy began with adaptations of Indian schemes, but when Ptolemy's *Almagest* became known and understood – not only do we have translations of it into Arabic, but though, e.g. al-Battānī's astronomical work follows the *Almagest* closely it is intelligently up-dated in essential places – it became the standard against which all advanced astronomical efforts were measured, even by those who disagreed with Ptolemy on various points, such as al-Ṭūsī, Quṭb al-Dīn, and ibn al-Shāṭir.

In the Latin West we find first translations of Arabic works based on Indian mixtures of Greek and Babylonian elements, later, versions in the Ptolemaic tradition, and finally proper translations of Ptolemy himself. Here again Ptolemy's methodology becomes the accepted standard, and so it remains until Kepler, and it is clear that mathematical astronomy was the principal motivation for the continued study of various branches of mathematics, among them trigonometry.

Thus the astronomical tradition in the West is linked to Babylonian astronomy. Mathematical astronomy was, however, not only the principal carrier and generator of certain mathematical techniques, but it became the model for the new exact sciences which learned from it their principal goal: to give a mathematical description of a particular class of natural phenomena capable of yielding numerical predictions that can be tested against observations. It is in this sense that I claim that Babylonian mathematical astronomy was the origin of all subsequent serious endeavour in the exact sciences.

The photographs of the cuneiform texts, figures 1 and 4, are published through the courtesy of the Trustees of the British Museum. Figures 2*a*, *b*, *c* are reproduced from the Astronomical Cuneiform texts with Professor Neugebauer's kind permission.

The reconstruction of A.C.T. no. 20 (figure 5) is published here for the first time. It was made possible by tables produced by the Yale computer according to programs by Miss Janice Henderson and by Mr Christopher Anagnostakis. This was done as part of a study of Babylonian astronomy supported by a grant from the National Science Foundation. Mr Anagnostakis also somehow caused the same computer to draw figures 10 and 11.

I wish to acknowledge my debt of gratitude to all.

REFERENCES (Aaboe)

Aaboe, A. 1955 On the Babylonian origin of some Hipparchian parameters. *Centaurus* **4**, 122.

Aaboe, A. 1958 Babylonian planetary theories. *Centaurus* **5**, 209.

Aaboe, A. 1963 A Greek qualitative planetary model of the epicyclic variety. *Centaurus* **9**, 1.

Aaboe, A. 1964 On period relations in Babylonian astronomy. *Centaurus* **10**, 213.

Aaboe, A. 1971 Lunar and solar velocities and the length of lunation intervals in Babylonian astronomy. *K. danske Vidensk. Selsk. (Mat. -fys. Medd.)* **38**, 6.

Aaboe, A. 1972 Remarks on the theoretical treatment of eclipses in antiquity. *J. Hist. Astron.* **3**, 105.

Huber, P. 1957 Zur täglichen Bewegung des Jupiter nach babylonischen Texten. *Z. Assyriologie* N.F. **18**, 265.

Kugler, F. X. 1900 *Die Babylonische Mondrechnung.* Freiburg im Breisgau.

Neugebauer, O. & Sachs, A. 1945 *Mathematical Cuneiform texts.* New Haven (American Oriental Series vol. 29).

Neugebauer, O. 1953 On the 'Hippopede' of Eudoxos. *Scr. math.* **19**, 225.

Neugebauer, O. 1955 *Astronomical Cuneiform texts.* 3 Vols. London: Lund Humphries.

Neugebauer, O. 1957 *The exact sciences in antiquity,* 2nd ed. Providence: Brown University Press. Reprint by Dover Press 1968.

Neugebauer, O. & Pingree, D. 1970 The Pañcasiddhāntikā of Varāmihira. *K. danske Vidensk. Selsk. (Hist.-filos. Skr.)* **6**, 1 (2 parts).

Parker, R. A. & Dubberstein, W. H. 1956 *Babylonian Chronology 626 B.C.–A.D. 75.* Providence: Brown University Press.

Pingree, D. 1963 Astronomy and astrology in India and Iran. *Isis* **54**, 229.

Sachs, A. 1948 A classification of the Babylonian astronomical tablets of the Seleucid period. *J. Cuneiform Studies* **2**, 271.

Schiaparelli, G. 1925 *Scritti sulla storia della astronomia antica.* P.I -Scritti editi. vol. II. Bologna.

v. d. Waerden, B. L. 1966 *Anfänge der Astronomie.* Groningen.

Note added in proof, October 1973

On P.Mich. 149, the Greek papyrus with evidence of wrong choice of sense of rotation on the epicycle, see now also

Neugebauer, O. 1972 Planetary motion in P.Mich. 149. *Bull. Am. Soc. Papyrologists,* **9**, 19.

and on the relation between Babylonian and early Indian astronomy see further

Pingree, D. 1973 The Mesopotamian origin of early Indian astronomy. *J. Hist. Astron.* **4**, 1.

Phil. Trans. R. Soc. Lond. A. **276**, 43–50 (1974)   [43]

Printed in Great Britain

Babylonian observational astronomy

By A. Sachs

Brown University, Providence, Rhode Island, U.S.A.

[Plates 3–10]

The cuneiform texts from ancient Assyria and Babylonia that are preserved offer direct evidence for systematic astronomical observation in two widely separated periods. From the first half of the second millennium B.C., later tradition has transmitted the dates of successive Venus appearances and disappearances in the reign of a king of the First Dynasty of Babylon. From the middle of the eighth century B.C. to the middle of the first century B.C. are preserved a large number of fragments of astronomical diaries attesting extensive daily observations of naked-eye astronomical phenomena.

Some 125 years separate us from the pioneering stage of the decipherment of the cuneiform script used in ancient Mesopotamia (modern Iraq) for about 3000 years until the first century of our era. Many thousands of clay tablets with this sort of cuneiform writing have been published, and several hundreds of thousands are known to be stored in museums all over the world. It is a fair indication of the expansion of knowledge in this field that a current dictionary of Akkadian, one of the two major languages of ancient Mesopotamia, contains more than 3500 pages in 12 volumes, and is only about two-thirds complete.

And yet, despite this immense accumulation of texts, one is always painfully conscious of the haphazard character of the archeological activities that have unearthed these clay tablets. Relatively few sites have been dug, and none of any size has been completely excavated. For some periods there are no (or virtually no) documents; for others, there are many more than the field of cuneiform studies can digest. A similar unevenness is evident when one considers the regional distribution of the preserved texts or the subject matter of their contents. One must learn constantly to keep in mind the temporal, geographical, and topical lacunae of the primary sources.

When we limit our view to the cuneiform clay tablets containing records of serious astronomical observation (i.e. we exclude schematic calendaric astronomy or lists of names of constellations, and the like), we are left with only two groups of extant documents, separated by a gap of about a thousand years.

The first group is essentially only one document which in the Babylonian calendar (using lunar months, the first day of which begins on the evening of the first visible crescent after conjunction) lists the dates for the consecutive first and last appearances of Venus as an evening star and as a morning star during the 21-year reign of a certain King Ammiṣaduqa of the First Dynasty of Babylon. The most famous member of this dynasty from the first half of the second millennium B.C. is King Hammurapi of the famous law stele. The list of Venus dates, to which omen predictions were secondarily appended, was copied and recopied for many centuries, and, in fact, we have it only in the form of much later copies made in the eighth and later centuries B.C. (and with partly corrupt details) embedded in one of the tablets of a standard collection of astronomical and meteorological omens. How, when, and why omen predictions – for example, 'the harvest will be normal' or 'a king will send messages of peace to another king' – were attached to the Venus dates are questions that we cannot begin to answer in the present state of our knowledge. Indeed, it is quite clear that the scribes who made the much later copies that we happen to have preserved did not have the faintest idea that the reign of King Ammiṣaduqa was involved, and in fact the ascription of the dates to this king was the result of the brilliant reading of a critical line in the text by F. X. Kugler. Scholars are still arguing about the absolute chronology of the reign of King Ammiṣaduqa and, with him, the whole First Dynasty of Babylon as well as preceding dynasties of several centuries' duration; the so-called Middle Chronology places Ammiṣaduqa's reign between −1645 and −1625. It is astonishing to find that somebody or other, for the whole of King Ammiṣaduqa's 21-year reign at so early a period, observed and recorded the Venus dates. Who was this observer? Did he have some reason to observe only Venus, or is it by chance that we do not have preserved his record of the dates of the other planets visible to the naked eye? Why just the reign of King Ammiṣaduqa? We have, alas, no answers to any of these questions either.

The second group of documents that record serious astronomical observations comes from the ancient capital city Babylon in southern Iraq. I have applied the term 'astronomical diaries' to these texts, on an edition of which I have been working for some years. There are more than 1200 fragments of astronomical diaries of various sizes and in diverse states of preservation. With very few exceptions, these texts are now in the British Museum, where, for the most part, they arrived in the 1870s and 1880s. A few dozen were excavated by H. Rassam for the Trustees of the British Museum, but all the rest were bought from antiquities dealers in Baghdad. It is fairly clear that the purchased tablets were quite accidentally excavated by gangs of workmen who, in fact, were primarily intent on removing the excellent baked bricks for re-use in modern construction in a nearby town. It is most fortunate that the British Museum was interested in acquiring these documents at a time when no other institution was purchasing antiquities in Iraq. Needless to say, there is no record of archeological context for any of our tablets.

As we shall soon see, the earliest datable fragment of an astronomical diary comes from −651, the latest six centuries later, from around −50. We have several different reasons to believe that the astronomical diaries began about −750 with the reign of the Babylonian king Nabonassar, a century before the earliest datable piece. From the second century A.D., Ptolemy's *Almagest* (book III, chapter 7) reports that records of observations (at Babylon) beginning with the reign of Nabonassar were still available. Furthermore, we actually have some fragments of cuneiform tablets from Babylon containing records of lunar eclipses going back roughly to this period. It is all but certain that these eclipse records could have been extracted only from the astronomical diaries. Finally, it is highly significant that the so-called Babylonian Chronicle, a record of historical events, begins with the reign of King Nabonassar since, as we shall see, the astronomical diaries contain historical reports.

When we begin to have datable astronomical diaries preserved in significant numbers, from about −400 on, we find that the basic patterns are fairly well fixed. The normal astronomical diary covers the 6 or 7 months comprising the first or second half of a particular Babylonian year.

Certain categories of astronomical events are always recorded within each month. Some of these astronomical phenomena are precisely those that are predicted by the mathematical astronomical cuneiform texts of the Hellenistic period. For the Moon, these characteristic significant phenomena are:

(1) At the beginning of each monthly paragraph, a statement about the length of the

previous month (i.e. 29 or 30 days), followed by an observation made the evening of the first visible lunar crescent measuring the time between sunset and moonset. When clouds or mist prevent the observation, an estimate of the time interval is recorded – how such estimates were made is unclear – followed by such remarks as 'because of clouds I did not observe'.

(2) Around full moon, two pairs of time intervals are recorded (or estimated whenever weather conditions prevented observation): one pair for moonset to sunrise and sunrise to moonset, the other pair for moonrise to sunset and sunset to moonrise. For details, see figure 1.

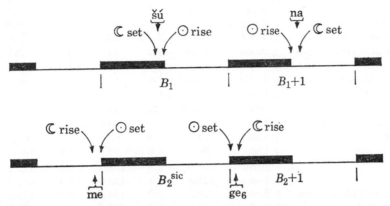

FIGURE 1. The Babylonian day (B) begins at sunset. Night is indicated by a heavy black line. The four significant time intervals around full moon are called šú, na, me, ge₆ in the texts.

(3) Toward the end of the month, the morning of the last visible crescent, with the time interval from moonrise to sunrise (or a prediction when the weather prevented observation).

(4) Lunar and solar eclipses, not only those which took place at Babylon but also those which, as the texts put it, 'passed by'. Various details are reported for lunar eclipses: when the eclipse began, its magnitude, the time from beginning to greatest magnitude, the planets that were visible during the eclipse, sometimes the name of a culminating star, the prevailing wind during the eclipse, etc.

For the outer planets Mars, Jupiter, and Saturn, the diaries always report the dates of first visibility, first stationary point, acronychal rising, second stationary point, and last visibility; and, for first and last visibility, also the zodiacal sign. For Mercury and Venus one always finds recorded the dates and zodiacal signs for first and last visibility as a morning star and as an evening star. All of these characteristic planetary phenomena, again, are also the goals of Babylonian mathematical astronomy.

Furthermore, the dates of the equinoxes and solstices as well as the dates of the significant appearances of the star Sirius are all given as computed by a known scheme.

In addition to all these significant lunar, planetary, and seasonal phenomena, the 'conjunctions' of the Moon and each of the planets with some thirty so-called 'normal stars' (i.e. reference stars) scattered about the zodiacal belt are recorded as they occur, and the distance 'above' or 'below' is given in cubits of 2° and fingerbreadths of 5'. Table 1 lists these reference stars by the transliteration of their Babylonian names and by their modern identifications, together with their longitudes and latitudes for part of the period that concerns us. The reference stars are fairly well distributed in longitude until approximately 230°, after which there is a gap of more than 40°; after about 290° there is an even bigger gap of more than 60°.

All sorts of meteorological events are reported as they take place in the rainy winter season.

TABLE 1. THE STANDARD BABYLONIAN REFERENCE STARS AS THEY APPEAR IN
TEXTS AFTER APPROXIMATELY −300

Babylonian name	star	−600 λ	−600 β	−300 λ	−300 β	0 λ	0 β
múl kur *šá* dur *nu-nu*	η Pisc	350.7°	+5.2°	354.9°	+5.2°	359.0°	+5.3°
múl igi *šá* sag ḫun	β Arie	357.9	+8.4	2.0	+8.4	6.2	+8.4
múl *ár šá* sag ḫun	α Arie	1.5	+9.9	5.7	+9.9	9.8	+9.9
múl-múl	η Taur	23.9	+3.8	28.0	+3.8	32.2	+3.8
is da	α Taur	33.7	−5.7	37.8	−5.6	42.0	−5.6
ŠUR gigir *šá* si	β Taur	46.5	+5.2	50.6	+5.2	54.8	+5.2
ŠUR gigir *šá* u$_x$	ʒ Taur	48.7	−2.5	52.8	−2.5	57.0	−2.5
múl igi *šá še-pít* maš-maš	η Gemi	57.4	−1.2	61.5	−1.2	65.7	−1.1
múl *ár šá še-pít* maš-maš	μ Gemi	59.2	−1.1	63.3	−1.1	67.5	−1.0
maš-maš *šá* sipa	γ Gemi	63.0	−7.1	67.1	−7.0	71.3	−7.0
maš-maš igi	α Gemi	74.2	+9.9	78.4	+9.9	82.5	+9.9
maš-maš *ár*	β Gemi	77.5	+6.5	81.6	+6.5	85.7	+6.5
múl igi *šá* alla$_x$ *šá* u$_x$	θ Canc	89.7	−1.0	93.8	−1.0	98.0	−0.9
múl igi *šá* alla$_x$ *šá* si	γ Canc	91.5	+3.0	95.6	+3.0	99.8	+3.0
múl *ár šá* alla$_x$ *šá* u$_x$	δ Canc	92.6	0.0	96.7	0.0	100.9	0.0
sag A	ε Leon	104.6	+9.5	108.7	+9.5	112.9	+9.6
lugal	α Leon	113.9	+0.4	118.0	+0.4	122.2	+0.4
múl tur *šá* 4 kùš *ár* lugal	ρ Leon	120.3	0.0	124.4	0.0	128.6	0.1
GIŠ-KUN A	θ Leon	127.3	+9.7	131.4	+9.7	135.6	+9.7
gìr *ár šá* A	β Virg	140.5	+0.6	144.7	+0.7	148.9	+0.7
dele *šá* igi absin	γ Virg	154.4	+3.0	158.5	+3.0	162.6	+3.0
sa$_4$ *šá* absin	α Virg	167.8	−1.9	171.9	−1.9	176.1	−1.9
rín *šá* u$_x$	α Libr	189.0	+0.7	193.2	+0.6	197.3	+0.6
rín *šá* si	β Libr	193.3	+8.8	197.4	+8.8	201.6	+8.7
múl múrub *šá* sag gír-tab	δ Scor	206.5	−1.7	210.6	−1.7	214.8	−1.7
múl *e šá* sag gír-tab	β Scor	207.1	+1.3	211.2	+1.3	215.4	+1.3
si$_4$	α Scor	213.7	−4.2	217.8	−4.3	222.0	−4.3
múl kur *šá* kir$_4$ *šil* pa	θ Ophi	225.3	−1.5	229.4	−1.5	233.6	−1.6
si máš	β Capr	267.9	+4.9	272.1	+4.9	276.2	+4.8
múl igi *šá* suḫur-máš	γ Capr	285.6	−2.3	289.7	−2.3	293.9	−2.4
múl *ár šá* suḫur-máš	δ Capr	287.3	−2.1	291.5	−2.2	295.6	−2.2

Good weather is never mentioned. Various kinds of cloud and storm conditions are couched in abbreviated or laconic technical terms, many of which, I must confess, I do not really understand. Occasionally the terminology borders on the quaint: some texts distinguish between rain followed by the 'removal of sandals' and obviously milder rain after which sandals were not removed. Elsewhere in cuneiform literature, this expression about the sandals and rain occurs, to my knowledge, only in an Old-Babylonian proverb. Archeologists who have excavated in southern Iraq tell me that the workmen today hang their shoes around their necks when the ground becomes too muddy after a rain. Rainbows, thunder, lightning, rain, cold, haloes, wind directions and velocities (the last in a terminology that escapes me), etc., are commonly reported in the diaries.

Between the astronomical phenomena – especially the Moon, which passes by a reference star or some planet on many days of the month – and the meteorological events of the rainy season, one often finds one or more events reported for every day of the month.

After the last astronomical or meteorological happening of the month is duly recorded, each monthly paragraph always continues with a statement about the respective amounts of barley, dates, pepper(?), cress(?), sesame, and wool – i.e. the necessities of life in ancient

Mesopotamia – that could be bought for one shekel of silver. These commodities are invariably listed in the same order. If the amounts changed during the month, this is indicated. In one extreme case, the varying amounts for the morning, the middle of the day, and the afternoon in the course of a single day are recorded.

Following the commodity prices, each monthly paragraph proceeds to list the zodiacal signs in which the various planets were to be found during the month.

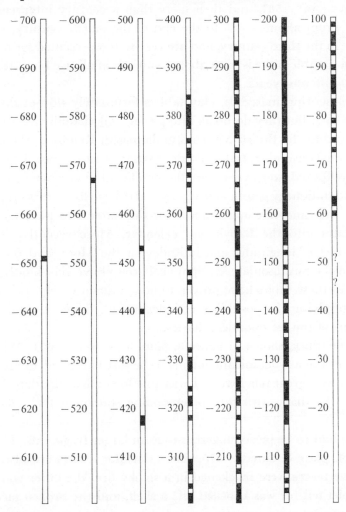

FIGURE 2. The extant datable astronomical diaries, from − 700 to 0.

This information, in turn, is followed by a report on the change of river level at Babylon in the course of the month. The rise and fall are measured in cubits and fingerbreadths. The reading of a river gauge of some sort is also given, with one unit on the gauge scale equal to four fingerbreadths – probably to be interpreted as the thickness of a layer of bricks.

The river-level data are often followed by a report of secular events or rumours of such events. These can be of very local interest, such as an outbreak of fire in some quarter of Babylon, or the discovery of a theft in a temple in Babylon. Or the events can be of great historical importance, such as a report about the joy with which the cities of Babylonia greeted the soldiers of Alexander the Great as liberators after the battle of Arbela. A few years later, the death of Alexander the Great is recorded, allowing us to date it with precision. Battles, military expeditions, the change of reigns, the overthrow of the Seleucid Empire by the Arsacids, raids on Babylon by Arabs in the Arsacid period are all recorded in astronomical diaries.

The distribution of the diaries over the seven centuries before the beginning of our era is uneven. In figure 2, the seven vertical columns, from left to right, represent the time scale from − 700 to 0. Each blackened square signifies the existence of at least one datable fragment of an astronomical diary for that year. After the earliest piece at − 651, almost a century elapses before the next in − 567, and then more than a century intervenes before the next fragment in − 453. From about − 385 to the end of the fourth century, the diaries become fairly dense, more so in the third century, and are very well represented for more than a century after − 200. The last datable piece is from about − 50. All in all, we have at least one fragment from more than 180 different years.

Unlike most other writing materials, clay tablets fortunately do not decay in the ground after thousands of years. But they do break into pieces, and virtually no astronomical diary of any decent size arrived at the British Museum undamaged. Rejoining the broken pieces is an important, if often time-consuming, task. In order to use the astronomical data in a fragment for dating purposes, I computed, some years ago, the significant phenomena (first and last appearances, first and second stationary points, and acronychal risings) for Saturn, Jupiter, Mars, Venus, and Mercury (only first and last appearances for the latter two planets) from − 600 to 0, transforming the dates into the Babylonian calendar. More recently, B. Tuckerman has published I.B.M. tables of longitudes and latitudes for the Moon and the planets at 5- and 10-day intervals, and the Smithsonian Astrophysical group have very kindly computed various lunar data for me. All these tables have proven to be of immense help not only in dating fragments but also in supplying the possibility of independent, detailed control in following the astronomical contents of the astronomical diaries.

The earliest datable diary, shown in figure 3, plate 3, is from − 651. When I first tried to date this text, I found the astronomical contents to be just barely adequate to make this date virtually certain. It was a great relief when I was able to confirm the date by matching up a historical remark in the diary with the corresponding statement for − 651 in a well-dated historical chronicle.

Some diaries have had to be pieced together to form larger fragments. The diary illustrated at the top of figure 4, plate 3, dated − 375, shows that the texts were already broken up in antiquity. Two of the pieces were blackened in a smoky fire, the other was not. Presumably, the original clay tablet fell (or was knocked off) a shelf, and the broken pieces were scattered over the floor of a burning room.

The three pieces of a diary dated − 253 are shown at the top of figure 5, plate 4, as separate fragments which were later joined to make the much bigger fragment shown at the bottom.

Beautiful, and very good, copies made by T. G. Pinches at the British Museum in the 1890s, many hundreds of which I was privileged to publish in 1955, are shown in figure 6, plate 5, for the obverse of an astronomical diary for − 324. The broken lines indicate the relationship between the five fragments which Pinches copied separately. The photograph shows the text as now pieced together.

Very large pieces are rather rare. A beautifully written diary dated − 77 (or year 234 of the Seleucid Era, more than 60 years after the beginning of the Parthian period) is illustrated in figure 7, plate 4. The obverse of a diary for − 346/− 345 (year 12 of Artaxerxes III, Babylonian months IX to XII) is photographed in figure 8, plate 6. The obverse of the large

Sachs

Phil. Trans. R. Soc. Lond. A, volume 276, *plate* 3

FIGURE 4. Top: B.M. 34642 + 35417 + 78829. Bottom: B.M. 132279 joined to tablet in the Wellcome Historical Medical Museum and Library. $\frac{2}{3}$ actual size.

FIGURE 3. B.M. 32312. $\frac{2}{3}$ actual size.

Sachs

Phil. Trans. R. Soc. Lond. A, volume 276, plate 4

FIGURE 7. B.M. 45689. ⅔ actual size.

FIGURE 5. B.M. 34105 + 41901 + 42041. ⅔ actual size.

Sachs

Phil. Trans. R. Soc. Lond. A, volume 276, plate 5

FIGURE 6. B.M. 34794 + 34919 + 34990 + 35071 + 35329. Actual size.

FIGURE 9. B.M. 45708. $\frac{2}{3}$ actual size.

FIGURE 8. B.M. 46229, pieced together by T. G. Pinches from SH. 81–7–6,
691 + 82–7–4, 83 + 98 + 113. $\frac{2}{3}$ actual size.

FIGURE 11. B.M. 31476; B.M. 34000; B.M. 34562. ⅔ actual size.

FIGURE 10. B.M. 32149+...; B.M. 32252; B.M. 32529. ⅔ actual size.

FIGURE 12. B.M. 31474; B.M. 40095+55572; B.M. 45722. ¾ actual size.

FIGURE 14. Top row: B.M. 34694, 34504, 34521. Second row: B.M. 34520, 34700, 34697. Third row: B.M. 46154, 46080, 34670. Bottom row: B.M. 46110, 34699, 34492. $\frac{2}{3}$ actual size.

FIGURE 13. B.M. 41529 + 41546 + 132278 joined to a fragment in the Böhl Collection (Leyden). $\frac{9}{10}$ actual size.

FIGURE 16. CBS 17. $\frac{2}{3}$ actual size.

FIGURE 15. B.M. 36600. $\frac{2}{3}$ actual size.

chunk shown in figure 9, plate 6, dated − 125, is well preserved; the deplorable condition of the reverse is unfortunately all too common.

The arrangement of a diary into two columns is unusual, and is never attested after the beginning of the Hellenistic period. Figure 10, plate 7, illustrates the reconstruction of a two-column diary for − 366, using non-joining pieces that can be placed exactly, thanks to a duplicate tablet.

The normal astronomical diary, as has already been mentioned, covers the 6 or 7 months of the first half or the second half of a Babylonian year. From the Hellenistic period we also have a large number of shorter diaries which refer to a period of anywhere from several days to 2 or 3 months. These short diaries presumably formed the basis for the preparation of the larger astronomical diaries of standard half-year size. Three typical short diaries are to be seen in figure 11, plate 7. Dating from 3 years in the neighborhood of − 170, one covers 6 days, another 5 days, the third 16 days. The three short diaries illustrated in figure 12, plate 8, are for somewhat longer intervals: 1 month and 3 days in − 186, 1 month in − 181, and 2 months and 5 days in − 182. The hand-writing shows that these three short diaries were written by the same scribe.

Although the vast majority of the extant diaries are in the British Museum, a few isolated pieces are stored in various museums all over the world. Occasionally, it can be shown that some of these odd fragments must be part of the same original tablets from which other pieces are now in the British Museum. Thanks to the cooperation of museum officials, it has been possible to bring together, momentarily, fragments of this sort which belong together. The lower part of figure 4 illustrates two pieces rejoined for a few hours, one from the British Museum, the other from the Wellcome Historical Medical Museum and Library, 5 minutes' walk from the British Museum. The date is − 277. Figure 13, plate 9, is a photograph of three pieces in the British Museum joined temporarily to a fragment that crossed the Channel from the Böhl Collection in Leyden; date, − 90.

The mention of so many dated diaries may have led the reader to the false impression that everything is placed chronologically and that nothing remains to be done. Actually, the datable pieces come from 183 years and comprise only about one-third of the available 1200 fragments. Many of the undated pieces are small and fairly insignificant, like the 12 pieces illustrated in figure 14, plate 9, but there still remain many as yet undated pieces that are rather large. Generally, this means that the astronomical data are simply not of the right kinds to warrant a large-scale hunt for the date.

We also have other texts, which extract special kinds of information from the astronomical diaries. For example, the tablet shown in figure 15, plate 10, contains nothing but Mercury data, year after year, from at least − 389 to − 374. Similar texts for Jupiter, Mars, Venus, and lunar eclipses are preserved.

From about − 250 on and continuing for two centuries, we have a fairly large number of so-called 'goal-year texts'. Each contains materials for the making of predictions of lunar and planetary phenomena for some specific year which we call the goal year. The data for each planet and the Moon are extracted from an astronomical diary that precedes the goal year by a period appropriate for the particular planet. There are two Jupiter paragraphs 71 and 83 years before the goal year, 8 years for Venus, 46 years for Mercury, 59 years for Saturn, 79 and 47 years for Mars, and 18 years for the Moon. All these periods are reasonably good, but we do not know how the necessary adjustments were made for the predictions. Figure 16, plate 10,

shows a goal-year text for − 86, in the University Museum of the University of Pe
vania.

Finally, mention should be made of a group of letters written to the royal court of As
in the seventh century B.C. by various astrologers/astronomers who report astronomical e
and their possible astrological importance for the empire. To what extent these people ar
same as the observers of the astronomical diaries is unclear.

Photographs of texts in the British Museum are published by courtesy of the Trustees of
British Museum.

BIBLIOGRAPHICAL NOTE

S. Langdon, J. K. Fotheringham, and Carl Schoch, *The Venus tablets of Ammizaduga* (Oxford
University Press 1928) contains an edition of the texts and a full analysis of the problem as
viewed before 1930. The later history of the controversy is sketched in the two most recent
discussions: B. L. van der Waerden, *Die Anfänge der Astronomie* (Basel 1968), pp. 34–49; John
D. Weir, *The Venus tablets of Ammizaduga* (Istanbul and Leiden, 1972). E. Reiner is preparing an
edition of all the Venus omens in the standard astrological composition.

For the astronomical diaries and related texts, a large number of copies are published and
catalogued in *Late Babylonian astronomical and related texts copied by T. G. Pinches & J. N. Strass-
maier prepared for publication by A. J. Sachs with the co-operation of J. Schaumberger* (Brown University
Press, Providence, R.I. 1955). P. Huber's treatment of this material in B. L. van der Waerden,
Die Anfänge der Astronomie (Basel 1968) is very competent.

Phil. Trans. R. Soc. Lond. A. **276**, 51–65 (1974)   [51]
Printed in Great Britain

Ancient Egyptian astronomy

BY R. A. PARKER
Department of Egyptology, Brown University,
Providence, Rhode Island 02912, U.S.A.

[Plates 11–14]

The early astronomy of ancient Egypt is known to us from its practical application to time measurement, in the large sense of a calendar year and in the smaller of the 24 h day. The earliest calendar year was lunar, kept in place in the natural year by the star Sirius. From this lunistellar year evolved the well-known calendar year of 365 days (three seasons of four 30-day months and 5 days added at the end). The division of the 30 day month into three 10-day 'weeks', combined with the observation of stars called decans rising at nightfall, eventually resulted in our 24 h day of fixed length. Constellations, except for decanal stars, and planets figured only in mythology. The zodiac was introduced into Egypt apparently in the Ptolemaic period and the decans finally became merely names for thirds of a zodiacal sign. In this latest period true astronomical texts also appear but they cannot be counted Egyptian in origin.

1. INTRODUCTION

When toward the end of Ramesside rule in Egypt, about 1100 B.C., a scribe of sacred books in the House of Life, by name Amenope, composed a catalogue of the Universe to be made up of 'heaven with its affairs, earth and what is in it, what the mountains belch forth, what is watered by the flood, all things upon which Re' has shone, all that is grown on the back of earth' he began his list with 'sky', followed by 'sun', 'moon' and 'star'. He then listed five constellations, only two of which can be certainly identified, those of *Sꜣḥ* 'Orion', *Msḫtyw* 'Foreleg', earlier 'Adze', which corresponds to our Big Dipper *'I'n* 'Ape', *Nḫt* 'Giant', and *Rrt* '(female) Hippopotamus'. Surprisingly there is no mention of Sirius which from other texts seems to have been the most important star for the Egyptians. Nor are the planets enumerated. The list goes on to other matters in over 600 entries. Only once again is astronomy mentioned and that is the title *imy-wnwt* 'hour-watcher' or 'astronomer' (Gardiner 1947).

This somewhat casual and rather negative approach to the sky and 'its affairs' suggests that to the ancient Egyptian they were of much less importance than terrestrial matters with which he was intimately involved. This conclusion may be illusory but it is, none the less, a present fact that it is not until the Ptolemaic period, when Egypt was open to and influenced by Hellenistic science, that we have anything approaching a theoretical astronomical treatise. Throughout the three millennia of recorded Egyptian history we have nothing whatever to suggest that the movements of the Moon and planets were systematically observed and recorded as they were in Babylonia. To be sure there are many references in ordinary texts to the Sun, Moon and stars, especially Sirius (the Egyptian Sothis), but except for one cosmological text to which we shall refer later, these convey little or nothing of astronomical import. Moreover, there is a complete absence of technical terms except for the common ones which are found everywhere.

The earliest account we have of what were the concerns of an Egyptian astronomer is as late as the third century B.C. On his statue Harkhebi describes himself as follows (Neugebauer & Parker 1969, pp. 214–15):

'Hereditary prince and count, sole companion, wise in the sacred writings, who observes everything observable in heaven and earth, clear-eyed in observing the stars, among which there is no erring; who announces rising and setting at their times, with the gods who foretell the future, for which he purified himself in their days when Akh (decan) rose heliacally beside Benu (Venus) from earth and he contented the lands with his utterances; who observes the culmination of every star in the sky, who knows the heliacal risings of every...in a good year, and who foretells the heliacal rising of Sothis at the beginning of the year. He observes her (Sothis) on the day of her first festival, knowledgeable in her course at the times of designating therein, observing what she does daily, all she has foretold is in his charge; knowing the northing and southing of the sun, announcing all its wonders (omina?) and appointing for them a time (?), he declares when they have occurred, coming at their times; who divides the hours for the two times (day and night) without going into error at night...; knowledgeable in everything which is seen in the sky, for which he has waited, skilled with respect to their conjunction(s) and their regular movement(s); who does not disclose (anything) at all concerning his report after judgment, discreet with all he has seen.'

A few of these accomplishments can go back to early Egypt, as we shall see, but they are of an elementary level. The astrological concept of the exaltation of Venus (see below) is of known Babylonian origin (Neugebauer & Parker 1969, p. 214) and the astrological flavouring of much of the text is obvious.

What then are we to talk about when the subject is ancient Egyptian astronomy? Little enough it may seem, but that little is not devoid of interest since it still endures as a legacy to us in the measurement of time. We begin with the early Egyptian calendars.

2. THE EARLY EGYPTIAN CALENDARS

Like all ancient peoples, the protodynastic Egyptians used a lunar calendar, but unlike their neighbours they began their lunar month, not with the first appearance of the new crescent in the west at sunset but rather with the morning when the old crescent of the waning moon could no longer be seen just before sunrise in the east. Their lunar year divided naturally, following their seasons, into some 4 months of inundation, when the Nile overflowed and covered the valley, some 4 months of planting and growth, and some 4 months of harvest and low water. At 2- or 3-year intervals, because 12 lunar months are on the average 11 days short of the natural year, a 13th or intercalary month was introduced so as to keep the seasons in place. Eventually the heliacal rising of the star Sirius, its first appearance just before sunrise in the eastern horizon after a period of invisibility, was used to regulate the intercalary month. Sirius, to the Egyptians the goddess Sopdet or Sothis, rose heliacally just at the time when the Nile itself normally began to rise, and the reappearance of the goddess heralded the inundation for the Egyptians. The 12th lunar month, that is the 4th month of the 3rd season, was named from the rising of Sothis and a simple rule was adopted to keep this event within its month. Whenever it fell in the last 11 days of its month an intercalary month was added to the year, lest in the following year Sothis rise out of its month (Parker 1950, chap. 3).

This luni-stellar year was used for centuries in early Egypt and indeed lasted until the end of pagan Egypt as a liturgical year determining seasonal festivals. Early in the third millennium B.C. however, probably for administrative and fiscal purposes, a new calendar year was invented. Either by averaging a succession of lunar years or by counting the days from one heliacal rising of Sothis to the next, it was determined that the year should have 365 days, and these were divided into three seasons of four 30-day months each, with 5 additional 'days upon the year' or 'epagomenal days'. This secular year which is conventionally termed the 'civil' year remained in use without alteration to the time of Augustus when a 6th epagomenal day every 4 years was introduced. That the natural year was longer than the civil year by a quarter of a day was of course known to the Egyptians fairly soon after the civil year was inaugurated but nothing was ever done about it. Still it is a great achievement of theirs to have invented a calendar year divorced from lunar movement and to have been the first to discover the length of the natural year, which eventually led to the Julian and Gregorian calendars (Parker 1950, chap. 4).

3. CALENDAR WEEKS

The first year in Egypt, the lunar one, had divided the month into 4 'weeks' based on 'first quarter', 'full moon', and 'third quarter'. The new month of 30 days was not divisible into 4 even parts but conveniently divided into 3 'weeks' of 10 days each, from later texts called 'first', 'middle' and 'last'. Thus in the entire year there were 36 weeks or decades, plus the 5 days upon the year.

4. STAR CLOCKS

We do not know exactly when the stars were first used to tell time at night but it was certainly by the 24th century B.C. and quite possibly soon after the introduction of the civil calendar. One of the texts (P.T. 515) in the pyramid of Unas, last king of the Fifth Dynasty, has the king clearing the night and dispatching the hours and the plural writing of 'hours' is effected by three stars.

It is not until roughly 2150 B.C. that we know for certain that these night hours totalled 12. This we learn from diagrams of stars on the inside of coffin lids, which were earlier termed 'diagonal calendars' but which are better called 'star clocks'. Figure 1 is a schematic version of such a clock, followed in some degree by all our examples. It is to be read from right to left. A horizontal upper line T is the date line, running from the 1st column, the 1st decade of the 1st month of the 1st season, to the thirty-sixth column, the last decade of the 4th month of the 3rd season. With decade 26 there begins in the 12th hour a triangle of alternate decans to tell

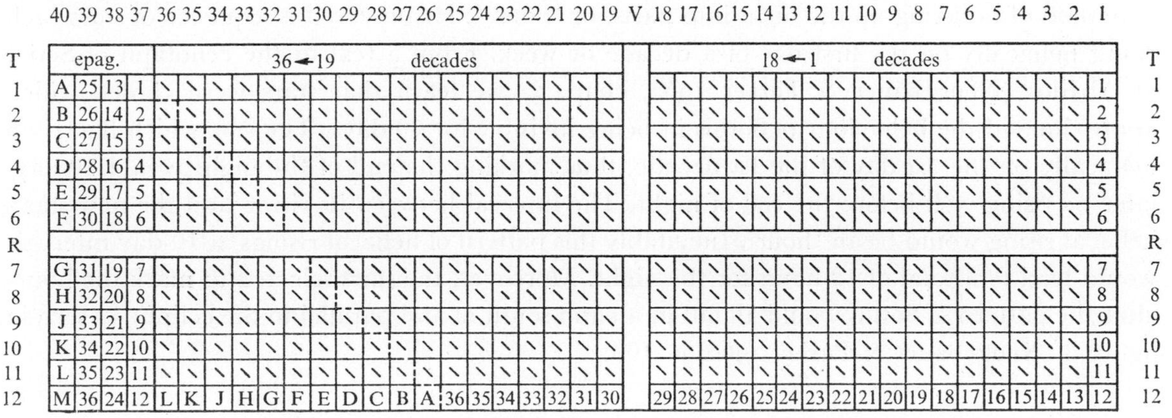

40 39 38 37 36 35 34 33 32 31 30 29 28 27 26 25 24 23 22 21 20 19 V 18 17 16 15 14 13 12 11 10 9 8 7 6 5 4 3 2 1

T	epag.	36←19 decades	18←1 decades	T
1	A 25 13 1	\ \ \ \ \ \ \ \ \ \ \ \ \ \ \ \ \	\ \ \ \ \ \ \ \ \ \ \ \ \ \ \ \ \ 1	1
2	B 26 14 2	\ \ \ \ \ \ \ \ \ \ \ \ \ \ \ \ \	\ \ \ \ \ \ \ \ \ \ \ \ \ \ \ \ \ 2	2
3	C 27 15 3	\ \ \ \ \ \ \ \ \ \ \ \ \ \ \ \ \	\ \ \ \ \ \ \ \ \ \ \ \ \ \ \ \ \ 3	3
4	D 28 16 4	\ \ \ \ \ \ \ \ \ \ \ \ \ \ \ \ \	\ \ \ \ \ \ \ \ \ \ \ \ \ \ \ \ \ 4	4
5	E 29 17 5	\ \ \ \ \ \ \ \ \ \ \ \ \ \ \ \ \	\ \ \ \ \ \ \ \ \ \ \ \ \ \ \ \ \ 5	5
6	F 30 18 6	\ \ \ \ \ \ \ \ \ \ \ \ \ \ \ \ \	\ \ \ \ \ \ \ \ \ \ \ \ \ \ \ \ \ 6	6
R				R
7	G 31 19 7	\ \ \ \ \ \ \ \ \ \ \ \ \ \ \ \ \	\ \ \ \ \ \ \ \ \ \ \ \ \ \ \ \ \ 7	7
8	H 32 20 8	\ \ \ \ \ \ \ \ \ \ \ \ \ \ \ \ \	\ \ \ \ \ \ \ \ \ \ \ \ \ \ \ \ \ 8	8
9	J 33 21 9	\ \ \ \ \ \ \ \ \ \ \ \ \ \ \ \ \	\ \ \ \ \ \ \ \ \ \ \ \ \ \ \ \ \ 9	9
10	K 34 22 10	\ \ \ \ \ \ \ \ \ \ \ \ \ \ \ \ \	\ \ \ \ \ \ \ \ \ \ \ \ \ \ \ \ \ 10	10
11	L 35 23 11	\ \ \ \ \ \ \ \ \ \ \ \ \ \ \ \ \	\ \ \ \ \ \ \ \ \ \ \ \ \ \ \ \ \ 11	11
12	M 36 24 12	L K J H G F E D C B A 36 35 34 33 32 31 30	29 28 27 26 25 24 23 22 21 20 19 18 17 16 15 14 13 12	12

FIGURE 1. Schematic version of a star clock.

the hours through the epagomenal days. What was probably once only 1 column for the epagomenal days has been expanded to 4, the last of which (40) is for the epagomenal days while the intrusive 3 merely list the thirty-six stars found in the decades. The 18th and 19th columns are separated by a space (V) in which are to be found representations of the goddess of the sky, Nut, the Foreleg of an Ox (our Big Dipper), Orion and Sothis. A horizontal inscription (R) divides the 6th and 7th hours. In it funerary offerings are invoked for the deceased from Re', the sun god, the deities of the vertical strip and various stellar deities.

5. THE STAR CLOCK MECHANISM

The mechanism of such a clock was the risings of certain selected stars or groups of stars, that conventionally are termed 'decans', at 12 intervals during the night, and at 10-day intervals through the year. If we now examine the star clock of Idy (figure 2, plate 11)) even though it lists only 18 decades we can see very graphically how the name of a decanal star in any hour is always in the next higher space in a succeeding column, so that a star in the 12th hour rises over 120 days to the 1st hour and then drops out of the clock.

Behind this are the simple astronomical events of the rotation of the Earth on its axis and the travel of the Earth about the Sun.

We have already seen in connexion with the star Sirius–Sothis that it eventually disappears, because it gets too close to the Sun, and then after some days it reappears on the eastern horizon just before sunrise, its heliacal rising. And we have noted that the heliacal rising of Sothis was a very important event. It heralded the inundation and it regulated the original lunar calendar. We have also noted that the lunar month, for the Egyptians, began when the last crescent could no longer be seen just before sunrise, another event associated with the eastern horizon. Moreover, in the Pyramid Texts, the great body of religious literature from the third millennium B.C., there are many references to the Morning Star, with which the King wished to be identified, but none at all to an evening star. The focus is all on the eastern horizon, since a morning star is one which has just, or recently, risen heliacally.

Of course, a star which has just risen heliacally does not remain on the horizon but every day, because of the Earth's travel about the Sun, rises a little earlier and is thus a little higher in the sky by sunrise. Eventually another star, rising heliacally is likely to be called the Morning Star. In the early third millennium B.C., we may conjecture, the combination of 10-day weeks in the civil calendar and the pattern of successive morning stars led some genius in Egypt to devise a method of breaking up the night into parts, or 'hours'. He observed a sequence of stars each rising heliacally on the first day of a decade or week. From a text in the cenotaph of Seti I at Abydos (Neugebauer & Parker 1960, chap. 2) we learn that stars were chosen which approximated the behaviour of Sothis in being invisible for 70 days. The star which had risen heliacally on the 1st day of the 1st decade, thus marking the end of the night would 10 days later be rising well before the end of night. The interval between its rising and the new star's heliacal rising would be an 'hour'. Inevitably this pattern of heliacal risings at 10-day intervals would lead to a total of 12 hours for the night. Moreover, the stars selected to mark the hours after the pattern of Sothis would all fall in a band south of and parallel to the ecliptic as seen in figure 3 (Neugebauer & Parker 1960, p. 100).

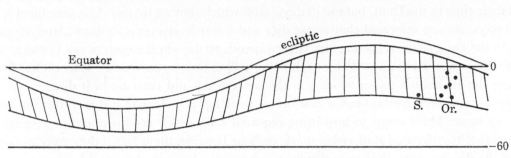

FIGURE 3. Orion and Sirius as hour stars in the decanal band south of the ecliptic.

6. DECANAL HOURS

It can readily be seen that such hours were not all of equal length. As the night grew longer and dawn was postponed the last hours of the clock would also grow longer. Conversely they would shorten as the night grew shorter. It is this lengthening and shortening, combined with the periods of morning and evening twilight and the oscillation of the star clock itself, that explains why the night was divided into 12 hours though there were always 18 decanal stars from horizon to horizon in the night sky. Civil twilight, when very bright stars become visible, averages about half an hour, but astronomical twilight, from sunset until all the stars are to be seen may be well over an hour and the decans measured only the time of total darkness. The 1st hour on the star clock was then of indeterminate length since it began with darkness and ended only when a particular decanal star rose in the eastern horizon. This 1st hour would be longest at the beginning of a decade and would be a little bit shorter each night. At the other end the decanal star rising heliacally and so marking the end of the 12th hour and of night was on the 1st day closely followed by morning twilight. At the end of the decade the 12th hour would be followed by a period of darkness. But we may suppose that this was not of importance to the observer, and the 12th hour may have been considered as running until light.

7. THE COSMOLOGY OF SETI I AND RAMSES IV

The star clocks that we have preserved to us are funerary in purpose and more or less corrupt in makeup. None the less it is clear that after the 36 decans another 12 decans had been chosen to carry the hours through the 5 epagomenal days after which the 1st column of the clock would again become effective. But since the Egyptian civil calendar lacked a leap year, inevitably the clock would require adjustment by shifting decans in place and making new ones. There is some evidence of a revision as late as the Twelfth Dynasty (1991–1786 B.C.) but by this time as well the risings of stars had been abandoned in favour of their transits. This we learn from later texts which are known as the Cosmology of Seti I and Ramses IV (Neugebauer & Parker 1960, chap. 2). Accompanying a vignette of the goddess Nut, bending over the Earth and supported by Shu, god of the air, are numerous texts, many of which are purely mythological but a few of which represent the first astronomical thinking that we have from the ancient Egyptians.

One important insight is the statement that the disappearance and reappearance of Sun and stars are common phenomena. When the Sun disappears at sunset he goes to the Duat and spends the hours of the night travelling from west to east. When the decanal stars disappear they also spend their time in the Duat, but for 70 days, after which they again rise. In a simplified scheme of 360 days, we are informed that a star dies and a star lives every 10 days. After 70 days of death in the Duat a star is born again. It then spends 80 days in the eastern sky before it works, after which it passes 120 days (10 for each hour) telling time by its transit. When it has finished marking the 1st hour it passes 90 days in the western sky and then again it dies.

Since a star spends 80 days in the east before working, it is clear that it is transitting when it marks an hour. This change to transitting required a wholesale readjustment of decanal stars. When rising, two decanal stars could mark an hour between them but if they were at opposite sides of the decanal belt, thus some distance apart on the horizon they could, when transitting, either pass the meridian together or so far after one another as to result in an extremely long hour. Of the 36 decans of the star clocks nearest in time to the transit scheme only 23 remain exactly in place in the transit list. The other 13 have each a different place or drop out and are replaced by new decans.

At the same time the simplification of the decanal round to 360 days instead of 365 reveals the strongly schematic character of such a star clock and poses the question of its real utility. In actual practice one may speculate that all an observer would need to do would be to memorize a list of 36 decanal stars, preferably rising ones as being the easier to observe, watch to see which one was in or near the horizon when darkness fell and then use the risings of the next 12 to mark off the night hours. Whether this was done in fact cannot be said with certainty, but it is true that while we have lists of decans on various astronomical ceilings or other monuments to the end of Egyptian history, we have nothing at all approaching a star clock in form after the time of Merneptah (1223–1211 B.C.), and that that was a purely funerary relic is indicated by the fact that its arrangement of stars dates it 600 years earlier into the Twelfth Dynasty.

The most probable cause for the disappearance of star clocks was the invention of the water clock, the earliest example of which is dated to Amenophis III of the New Kingdom (1397–1366 B.C.), but before we consider all the implications of this discovery we must look into the hours of the day, and see how they were influenced by the night hours.

8. THE TWELVE-HOUR NIGHT

We have discussed above how the diagonal star clock was based on the 36 10-day divisions or decades of the civil calendar and that this resulted in a division of darkness into 12 hours, with 2 to 3 'hours' of twilight before sunrise and after sunset. Now had the 'weeks' been only 5 days instead of 10 we should have had 36 pentades from horizon to horizon at night with perhaps 4 to 5 for twilight on either side, leaving about 26 for the 'hours' of darkness. We can see then that the 12-hour night was the direct result of the 10-day week, and was not an arbitrary choice.

9. THE HOURS OF THE DAY

We have much less early information about the hours of the day. Again it is a text from the cenotaph of Seti I which helps us to suggest a plausible line of development. This funerary text gives directions for making a shadow clock consisting of a base with an elevated cross bar at its head (figure 4). With the head to the east 4 hours are marked off by decreasing shadow

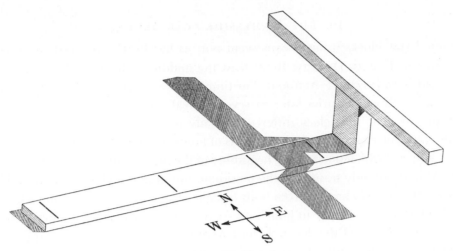

FIGURE 4. Shadow clock.

lengths after which the instrument is reversed with head to the west to mark 4 afternoon hours. In an illuminating statement the text concludes by saying that 2 hours have passed in the morning before the Sun shines on the clock and another 2 hours pass in the evening before the hours of the night (Neugebauer & Parker 1960, pp. 116–18). It is the most plausible assumption that the 2 hours before and after the clock is in use were divided by the observable phenomena of sunrise and sunset, and that the hour before sunrise was the whole period of morning twilight and the hour after sunset the whole period of evening twilight. Again it is plausible that the entire period of light between total darkness of one night and that of the next was divided into 12 parts on analogy to the 12 parts of the night. The day hours would obviously be, on the average, much longer than the night hours in this initial stage of the 24 hour concept.

The next development, and this can be documented from a shadow clock of the time of Thutmose III (1490–1436 B.C.), and so earlier than the funerary text of Seti I, is the division of the sunrise to sunset interval into 12 hours. A complementary division of the sunset to sunrise interval could only be done by means of a water clock. The earliest water clock, however, was still adapted to dividing the period of real darkness.

The concept of hours of constant length, and not merely 'seasonal' hours that are always only one-twelfth of the time from sunset to sunrise or sunrise to sunset, is attested by a Ramesside papyrus of the twelfth century though it describes the calendar situation of a somewhat earlier time. This text gives the hours of daylight and darkness for each month of the year. The total is always 24 and the extreme figures are 18 hours to the day and 6 to the night and conversely 6 hours to the day and 18 to the night. Such figures are impossible for Egypt unless they start from the basis that the 6 night hours correspond to the shortest night of 12 decanal hours which do in fact total about 6 equal hours (Neugebauer & Parker 1960, p. 119).

With the concept of 24 equinoctial hours to the day we have come to the end of the Egyptian contribution to the telling of our time. The division of the hour into 60 minutes and the minute into 60 seconds comes from Babylonia with its sexagesimal system.

10. The Ramesside star clocks

The diagonal star clocks using decans seem not, as has been remarked, to have lasted in use much past the Twelfth Dynasty. By at least the middle of the second millennium B.C. the water clock had been invented. At about that time also was introduced a new star clock which is preserved to us three centuries later in several royal tombs of the Ramesside period as a ceiling adornment. The new clock differed in many ways from the old. Where the decanal clock had 36 tables of 12 stars marking the ends of hours, the new had 24 tables (2 to a month) of 13 stars, the first of which marked the beginning of the night. The decanal clock used transits, the new clock used not only transits of the meridian, but of lines before and after. The decanal clock had stars that never changed from hour to hour. The new clock had stars which moved in very irregular fashion from hour to hour and frequently skipped 2 or 3 hours, being replaced by other stars. The hours of the decanal clock were fixed in length, once established; those of the new clock varied in length. The stars of the two clocks are very rarely the same. Of the 36 decanal stars listed on the ceilings of Seti I and Ramses IV only 3 can be found in the new clocks. There can be little doubt that the stars used in the new clocks, in the main, lay outside the decanal belt and, as we shall see, south of that belt.

We have noted that the new clocks used lines before and after the meridian. These are indicated by notes to each star and also by a chart accompanying each table. The chart has 13 horizontal lines and 9 vertical lines above a seated figure (figure 5, plate 12). Stars are located in the inner 7 vertical lines to agree with a location as given in the table. Thus a star on the central vertical line corresponds to the text *r ʿḳ3 ἰb* 'opposite the heart'; one on the first vertical to left or right is *ḥr ἰrt wnmy* 'on the right eye' or *ḥr ἰrt ἰ3by* 'on the left eye', with 'right' and 'left' used from the standpoint of an observer facing the target figure. A star on the second vertical left or right is *ḥr msḏr wnmy* 'on the right ear' or *ḥr msḏr ἰ3by* 'on the left ear'. Finally, a star on the third vertical left or right is *ḥr ḳʿḥ wnmy* 'on the right shoulder' or *ḥr ḳʿḥ ἰ3by* 'on the left shoulder'.

The procedure is clear. On a suitable viewing platform, probably a temple roof, two men would sit facing one another on a north–south line. The northernmost would hold a sighting instrument like a plumb bob (called by the Egyptians a *mrḫt*) before him and would call out the hour when a star had reached either the meridian or one of the lines before or after as sighted against the target figure. The effort for such precision points to the use of the water clock as an independent means of marking when an hour had ended, and emphasizes as well the reluctance of the Egyptians to abandon telling time at night by the stars. Indeed, a water clock of the Ptolemaic period has an inscription on its rim that its purpose is to tell the hours of the night only when the decanal stars cannot be seen (Borchardt 1920, p. 8).

Unfortunately the texts of the new clock are all more or less corrupt, but the fact that we have four sets of tables from three tombs helps considerably in the effort to establish the prototype. New constellations such as the 'giant' and the 'hippopotamus' which appear in the Amenope onomasticon are found besides others such as *m3ἰ* 'lion', *mnἰt* 'mooring post' and *3pd* 'bird'. The presence of Sirius–Sothis in the new clock ensures that, while most of the new hour stars are out of the decanal belt they must appear in the southern sky and probably in a belt parallel to, or slightly overlapping, and south of the decanal belt.

11. Astronomical monuments

Before proceeding further it will be found useful to review briefly the sources of what we do know about Egyptian astronomy. We have already discussed the star clocks found on the inside of coffin lids. These are 12 in number and they range from the Ninth to the Twelfth Dynasties. Other than these there are some 81 monuments in great or less part concerned with astronomy. In the main these are ceilings, the majority in tombs, though there are many in temples. There are a few water clocks with astronomical depictions on the outside, and in the Graeco–Roman period zodiacs are found on the inside of coffin lids. Many of these monuments include lists of decans, either rising or transitting hour stars. From variant decans it is possible to group these lists into families, so that we have (after the earliest monument in each case) the Senmut family, the Seti I A family and the Seti I C family, all three of rising decans. The Seti I B family is surely of transitting decans, since it accompanies the texts we have referred to above. One other family, Tanis, is not classifiable with certainty. It has the appearance of being a mixed and artificial list and is probably indicative of the decline of the decans as time measurers.

Besides decanal names an astronomical monument may include decanal stars, one or more with each name, and also deities associated with the individual decans as well as decanal figures. Decans of the triangle, usually replaced in part by planets, are also listed. Constellations, notably a group usually referred to as 'northern', may appear and these may be accompanied by deities. Other less frequent elements may be present, such as calendar years, cosmic deities, hours of the day and night. Lastly the zodiacs become frequent in the Graeco–Roman period.

12. The astronomical ceiling of Senmut

An example of an astronomical ceiling, and the earliest one we presently know of, is from the tomb of Senmut, the architect and favourite of Queen Hatshepsut (about 1473 B.C.). At the top of figure 6, plate 13, which is to the south, is found the decanal list beginning on the right. The method of presentation of the decans shows unmistakeably that a star clock was copied. After the first six columns the horizontal line dividing decans from deities is nothing other than the 12th hour line of a star clock and the 11 decans listed in the first six columns are hours 1 to 11 of the first column of a star clock. The last decan is Sirius–Sothis and her position is such that the star clock copied on the ceiling must be dated to the last revision in the Twelfth Dynasty, four centuries earlier. Accompanying the names of the decans are stars, and below the 12th hour line are various deities and more stars. Figured also are the constellations of the ship, the sheep, Osiris in a bark representing Orion and Isis in a bark representing Sothis. On the left of the decans, where a star clock would continue with the triangle decans for the epagomenal days, Senmut has a mixture of decans and planets. Jupiter and Saturn precede the triangle decans as two falcon-headed figures in barks and Mercury and Venus follow the decans, which are only 6 instead of the normal 12, Mercury is shown as a small Seth and Venus as a heron. Mars was omitted, whether intentionally or by error we do not know. One of the decans is figured as two turtles.

Correctly on the northern half of the ceiling is the group of constellations called 'northern' (see below). They are flanked on the base line by two rows of deities, headed on the right by Isis. These are deities of the days of the lunar month in origin (Parker 1950, § 222), and they are present because the 12 circles, divided by the northern constellations into seasonal groups of 4 each, represent the original lunar calendar with the name of each month above its circle. The circles themselves are divided into 24 segments, no doubt for the 24 hours. Later ceilings omitted the lunar calendar, though retaining the lunar day deities and the northern constellations, thus suggesting a relationship which had no basis in fact.

13. The planets

How early all five planets were identified and named is not known to us. The first monument on which they appear is the astronomical ceiling of Senmut but they surely were known well before then. By choice or by error Senmut omitted Mars, but that planet is present with the others on the astronomical ceiling of the Ramesseum, two centuries later. The usual order is Jupiter, Saturn, Mars, Mercury and Venus, that of most distant planet to the one nearest the sun. The first three planets are frequently separated from the last two on the monuments and these were as well considered aspects of the sky god Horus. Thus Jupiter was 'Horus-who-bounds-the-Two-Lands' or 'Horus-who-illuminates-the-Two-Lands' with later variants 'who-illuminates-the-land' or 'who-opens-mystery'. Saturn was always 'Horus-bull-of-the-sky' or 'Horus-the-bull'. Mars was 'Horus-of-the-horizon' or 'Horus-the-red'. As Horuses these planets, when figured, were normally falcon-headed with human bodies.

Mercury had the simplest name *Sbg(w)* but its meaning is unknown. Also unknown are the reasons Mercury was identified with the god Seth, who was an enemy of Horus. Frequently Seth's animal-headed figure was mutilated or replaced on the monuments. Venus in the earlier texts was 'the crosser' or 'the-star-which-crosses', and was pictured as a heron. Later Venus was often termed 'the-morning-star', and had human representation, sometimes with falcon head and occasionally two-headed or two-faced. The indentification of all the planets is secure from the planetary tables of the Graeco–Roman period.

The scraps of texts which usually accompany Jupiter and Saturn tell us little. However, 'Horus-the-red', as the name of Mars, identifies it securely. Another epithet, 'he travels backwards', speaks to be sure of the planet's retrograde movement, but all planets share this peculiarity. A Ramesside text about Mercury is instructive. 'Seth in the evening twilight, a god in the morning twilight'. This shows conclusively that by Ramses VI (1148–1138 B.C.) Mercury was recognized as both evening and morning star. Presumably as evening star and Seth, Mercury was of a malevolent disposition. As morning star and unidentified, it may have been the opposite. We do not know how much before Ramses VI the planet was known to be both evening and morning star.

That Venus also was both evening and morning star was surely known as early as Mercury but there is no textual proof of this. The name 'the crosser' may indicate movement back and forth about the sun and this name is found on the ceiling of Senmut. The late depiction of Venus as two-headed or two-faced also points to this knowledge. It is at least probable that by the middle of the second millennium B.C. the Egyptians had come to the realization that both inner planets could be evening or morning star and that these stars were one.

14. The northern constellations

The term 'northern' has been given to a group which includes one securely identifiable constellation, our 'Big Dipper'. While we are reasonably sure that they are all north of the ecliptic, they may not necessarily be all circumpolar. The number and arrangement vary from monument to monument, but like the decan lists they tend to fall into families. One such family begins with the depiction on the ceiling of the tomb of Seti I, and that will serve us as an illustration differing from Senmut of these constellations (figure 7, plate 12).

On early and again on late monuments shown as the foreleg of a bull, the Big Dipper is here depicted as a bull (Meskhetiu) on a platform. On the right is the constellation 'Hippopotamus' with a crocodile on her back and forefeet resting on a mooring-post. The Foreleg or Bull and the Hippopotamus are essential to the depiction of the northern constellations and they are always present. A falcon-headed god, called An, apparently supports the Bull, while an unidentified man holds the cords which link the mooring-post and the Bull. Above, on the left is the goddess Serket. A bird seems to perch on the head of the constellation Lion, shown with many stars on its head and back. Underneath is the constellation Crocodile with a second unidentified man facing it and in the gesture of spearing it, though here the spear has not been drawn.

The peculiar relationship shown between the Foreleg (or Bull) and the Hippopotamus is mentioned in several mythological texts. One may be quoted from the Book of Day and Night (time of Ramses VI) as follows: 'As to this Foreleg of Seth, it is in the northern sky, tied to two mooring posts of flint by a chain of gold. It is entrusted to Isis as a hippopotamus guarding it' (Neugebauer & Parker 1969, p. 190).

15. ZODIACS

To this point we have been dealing with purely Egyptian astronomical concepts. The zodiac was not native to Egypt but was a Babylonian import. Exactly when it arrived is not known, but the first depiction of it we have is from a temple at Esna, now destroyed, dating from 246 to 180 B.C. Though influenced by Egyptian art the basic design of each sign is Babylonian. In combination with the zodiac, however, are to be found all the traditional elements of Egyptian astronomy, the decans, the Sun, Moon and planets, and the constellations. In particular it is the decans which combine with the importation. They lose their old role of telling the night hours by rising or transits and they become mere 10° subdivisions of the zodiacal belt with three decanal names to each zodiacal sign.

More than one decanal list was thus combined with the zodiac. The Esna ceiling has the decans of the Seti I B family in a strip above and those of the Tanis family in a strip below the one bearing the zodiacal depictions. If the two lists are compared, only 12 decans will be found in the same position in both lists, 16 decans are common to both lists but do not correspond in position, and there are 8 variant decans in both lists.

Now the first complete list in Greek, so far known, of the decans in the zodiac comes from Hephaestion of Thebes in the fourth century A.D. When his list is compared with either the Seti I B list or that of Tanis, no complete agreement is found with either. This suggests that Hephaestion's list is an arbitrary and eclectic one and such a conclusion can be documented in this fashion. The Seti I B list begins with the decan *spdt* (Sothis) and that of Tanis with *knm(t)*. If we place the two lists in columns beginning with these decans and place Hephaestion's list in between we have table 1 (the Greek names in the left column agree with Seti I B, those in the right column with Tanis).

TABLE 1

From Neugebauer & Parker (1969), pp. 170–171.

	Seti IB	Hephaestion		Tanis
Cancer	*spdt*	σωθις		*knm(t)*
	št(w)	σιτ		*ḥry (ḫpd) knm(t)*
	knm(t)	χνουμις		*ḥȝt dȝt*
Leo	*ḥry ḫpd knm(t)*	χαρχνουμις		*dȝt*
	ḥȝt dȝt	ηπη		*pḥwy dȝt*
	pḥwy dȝt	φουπη		*tm(ȝt)*
Virgo	*tm(ȝt)*	τωμ		*wšȝt(i)*
	wšȝt(i)bkȝt(i)	ουεστεβκωτ		*bkȝt(i)*
	ipsd	αφοσο	αφοσο	*ipsd*
Libra	*sbḫs*	σουχωε	σουχωε	*sbḫs*
	tpy-ʿḫnt	πτηχουτ	πτηχουτ	*tpy-ʿḫnt*
	ḫnt ḥr(t)	χονταρε		*ḥry-ib wiȝ*
Scorpio	*ḫnt ḥr(t)*		στωχνηνε	*s(ȝ)pt(i)ḥnwy*
	tms(n)ḫnt		σεσμε	*sšm(w)*
	spt(y)ḥnwy		σισιεμε	*sȝ sšm(w)*
Sagittarius	*ḥry-ib wiȝ*	ρηουω		*knm(w)*
	sšmw	σεσμε		*tpy-ʿsmd*
	knm(w)	κομμε		*pȝ sbȝ wʿty*
Capricorn	*tpy-ʿsmd*		σματ	*smd*
	smd		σρω	*srt*
	srt		ισρω	*sȝ srt*
Aquarius	*sȝ srt*		πτιαυ	*tpy-ʿ ȝḥw(y)*
	ḥry ḫpd srt		αευ	*ȝḥw(y)*
	tpy-ʿȝḥw(y)		πτηβνου	*tpy-ʿ bȝw(y)*
Pisces	*ȝḥw(y)*		βιου	*bȝw(y)*
	tpy-ʿ bȝw(y)		χονταρε	*ḫnt(w) ḥr(w)*
	bȝw(y)	πτιβιου		*ḫnt(w) ḥr(w)*
Aries	*ḫnt(w) ḥr(w)*	χονταρε		*ḳd*
	ḫnt(w) ḥrw	χονταχρε		*sȝ ḳd*
	sȝ ḳd	σικετ		*ḥȝw*
Taurus	*ḥȝw*	χωου		*ʿrt*
	ʿrt	ερω		*rmn ḥry*
	rmn ḥry	ρομβρομαρε		*ts ʿrḳ*
Gemini	*ts ʿrḳ*	θοσολκ		*rmn ḥry*
	wʿrt	ουαρε		*wʿr(t)*
	tpy-ʿspdt		φουορι	*pḥwy ḥry*

Agreement between Seti I B and Hephaestion is found in 24 cases and there are 15 agreements between Hephaestion and Tanis. Only three instances are common to both. This result is astonishing, but it clearly emphasizes the arbitrary character of Hephaestion's list. Though the Tanis list is a doubtful one, Seti I B's list of transit decans was surely based on observation, and any attempt to keep the decans in their proper place in the zodiacal belt must have given much closer agreement between Hephaestion and Seti I B. Instead, we have decans which are off by a whole zodiacal sign. It is clear that the decanal names were adopted merely to name the 10° divisions of the zodiac, and little if any attention was given to their actual location in the sky.

16. THE DENDERA CIRCULAR 'ZODIAC'

This astronomical ceiling (Dendera B), now in the Louvre, but originally part of the ceiling of a chapel on the roof of the temple, is the best known of all such depictions and comes closest to an effort to reproduce the heavens with some degree of exactitude (figure 8, plate 14). In the centre are the two most important northern constellations, the Foreleg and the Hippopotamus. These must mark the location of the pole. The 12 figures of the zodiac are in a circle which does not centre at the pole but is properly askew. Among the figures of the zodiac are to be found the five planets. They are in exaltations, that is in signs in which they are supposed to be particularly influential. Thus Venus is in Pisces, Jupiter in Cancer, Mercury in Virgo, Saturn in Libra and Mars in Capricorn. At the perimeter of the sky are the 36 decans, named and figured. A few selected constellations occupy the area between the zodiac and the pole in addition to the two northern ones. These are presumably all north of the ecliptic but none is depicted in the usual group of northern constellations. Between the zodiac and the decans is a crescent-shaped area also occupied with figures of constellations. Of these we can easily identify Orion and Sothis and the presumption is that the other figures, except for the goddesses Satis and Anukis who accompany Sothis, are of constellations in the decanal band or perhaps slightly south of it. The texts about the figures of the goddesses who support the sky are without astronomical significance.

The circular zodiac (Dendera B) dates to before 30 B.C. In another part of the temple of Dendera, in strips on the ceiling of the Outer Hypostyle Hall, and dated to Tiberius (A.D. 14–37), appears much of what is found in the circular zodiac but this time in linear form (Dendera E), with the constellations placed in particular signs. The combination of the two depictions is thus of considerable importance. The circular zodiac locates constellations north or south of the ecliptic and the linear one locates them in specific signs. A brief catalogue follows.

17. CONSTELLATIONS NORTH OF THE ECLIPTIC

A. Dendera B, above Sagittarius and Capricorn and between them and the pole. Dendera E, A and C together are between Sagittarius and Capricorn.

B. Dendera E, in Scorpio.

C. Dendera B, between Gemini and the pole.

D. Dendera E omits.

E. Dendera B, near Gemini. Dendera E, in Gemini.

F. Dendera B, above Leo. Dendera E omits.

G. Dendera B, near Libra and Scorpio. Dendera E omits.

H. Dendera B, near Scorpio and Aquarius. Dendera E, in Aquarius.

J. Dendera B, near Aquarius. Dendera E, in Aquarius.

K. Dendera B, near Aquarius and Pisces. Dendera E, in Aquarius between H and J.

L. Dendera B, nearest to Aries. Dendera E, in Taurus.

M. Dendera B, between Pisces and Aries. Dendera E omits. This may possibly represent the full moon.

18. Constellations south of the ecliptic

N. Dendera B, near Pisces. Dendera E, in Pisces.

O. Dendera B, near Aries and Taurus. Dendera E, in Taurus.

P. Orion. Dendera B, near Taurus and Gemini. Dendera E, after Gemini and before Cancer.

Q. Dendera B, near Gemini. Dendera E omits.

R. Dendera B, under Gemini. Dendera E, after P.

S. Sirius. Dendera B, under Cancer. Dendera E, after R.

T. The goddess Satis, a companion to Sirius–Sothis.

U. The goddess Anukis, another companion to Sirius–Sothis.

V. Dendera B, near Leo. Dendera E, in Leo.

W. Dendera B, under Virgo. Dendera E, in Virgo.

X. Dendera B, near Virgo and Libra. Dendera E omits. This constellation of a lion could
be that of the hour stars. The lion is in the area of the $ḏ3t$ and $tm3t$ decans and these are
in Virgo.

Y. Dendera B, near Libra and Scorpio. Dendera E, in Scorpio.

In Dendera E are two constellations, one in Libra and one in Leo, which are not present in
Dendera B and so may be either north or south of the ecliptic. In our listing, to be sure, north
or south and location in a sign are very imprecise, and do not give us more than a general
idea of where a constellation is located.

19. Late demotic texts

In the Graeco–Roman period were written a number of demotic documents which concern
astronomical or astrological matters. Except for two, these all have their origin in the Hellenistic
world and, but for the accident of being written in the demotic script and on papyrus, have
little or nothing to do with ancient Egypt. The more important are planetary tables listing
the dates of entry of the planets and the Moon into the signs of the zodiac. These were no doubt
primarily used for the casting of horoscopes of which we have a number of examples from this
period (Neugebauer & Parker 1969, chap. 6).

The two late texts exceptionally Egyptian in content are *P. Carlsberg* 1 and *P. Carlsberg* 9.
The former offers a commentary on the texts of Seti I and Ramses IV which we have discussed
above, when we analysed the behaviour of the decans in transit. The second papyrus outlines
a 25-year lunar cycle, with dates for beginning lunar months in terms of the civil calendar.
Though the papyrus itself was written in A.D. 144 or later, the scheme surely goes back to the
fourth century B.C. At that time, the cycle reflected the beginning of the Egyptian lunar month
with the morning of invisibility of the last crescent. Though the text gives only dates in alternate months it has been found possible to establish the rules for the intervening months and
the whole reconstructed cycle may be seen in table 2 (Neugebauer & Parker 1969, pp. 220–225;
Parker 1950, §§ 49–119).

Figure 2*a, b*. Star clock on coffin lid of Idy.

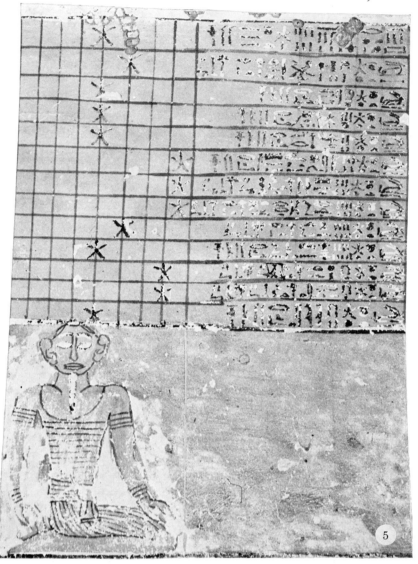

FIGURE 5. Hour table, Tomb of Ramses VII.

FIGURE 7. Northern constellations, Tomb of Seti 1.

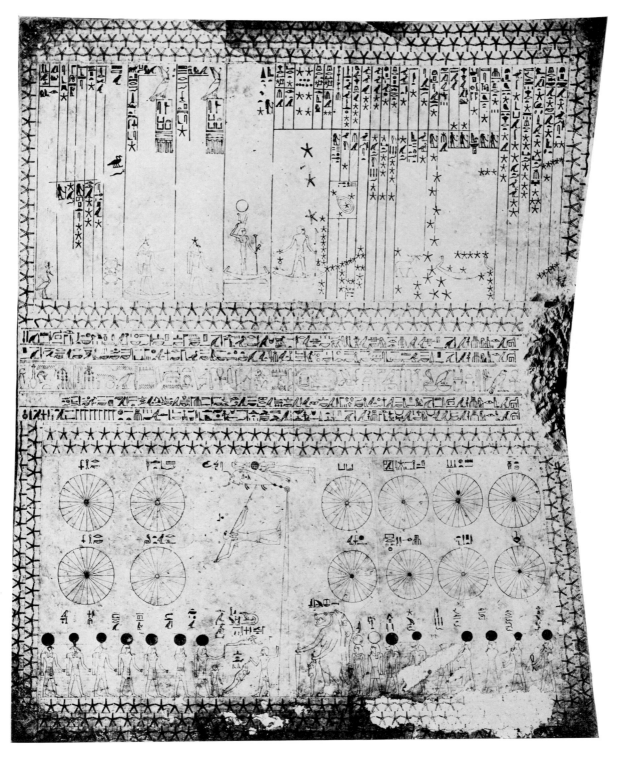

FIGURE 6. Ceiling, Tomb of Senmut.

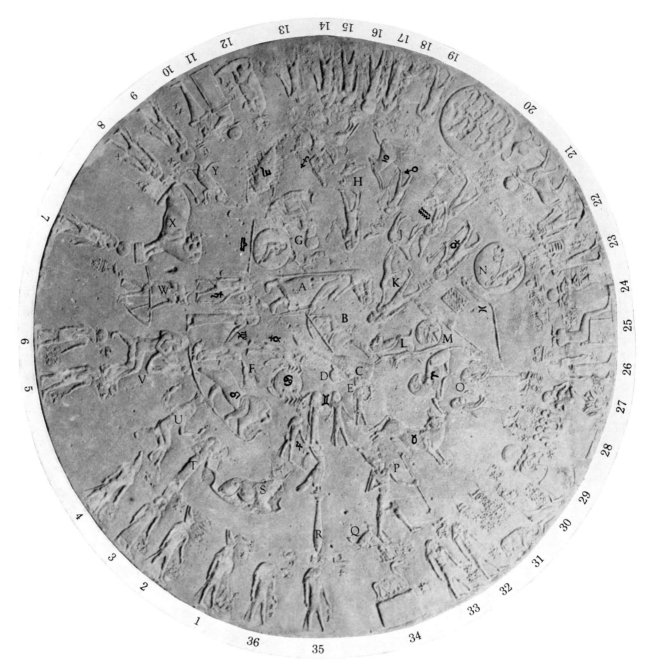

FIGURE 8. Zodiac ceiling, Dendera.

TABLE 2. THE 25-YEAR LUNAR CALENDAR CYCLE

From Parker (1950), p. 25.

months ...	Akhet				Peret				Shomu				Epag
	I	II	III	IIII	I	II	III	IIII	I	II	III	IIII	
year 1	1	1	1–30	30	29	29	29	28	27	27	27	26	–
2	20	20	19	19	18	18	18	17	16	16	16	15	–
3	9	9	8	8	7	7	7	6	5	5	5	4	4
4	28	28	27	27	26	26	26	25	24	24	24	23	–
5	18	18	17	17	16	16	16	15	14	14	14	13	–
6	7	7	6	6	5	5	5	4	3	3	3	2	2
7	26	26	25	25	24	24	24	23	22	22	22	21	–
8	15	15	14	14	13	13	13	12	11	11	11	10	–
9	4	4	3	3	2	2	2	1	1–30	30	30	29	–
10	24	24	23	23	22	22	22	21	20	20	20	19	–
11	13	13	12	12	11	11	11	10	9	9	9	8	–
12	2	2	1	1	1–30	30	30	29	28	28	28	27	–
13	21	21	20	20	19	19	19	18	17	17	17	16	–
14	10	10	9	9	8	8	8	7	6	6	6	5	5
15	30	30	29	29	28	28	28	27	26	26	26	25	–
16	19	19	18	18	17	17	17	16	15	15	15	14	–
17	8	8	7	7	6	6	6	5	4	4	4	3	3
18	27	27	26	26	25	25	25	24	23	23	23	22	–
19	16	16	15	15	14	14	14	13	12	12	12	11	–
20	6	6	5	5	4	4	4	3	2	2	2	1	1
21	25	25	24	24	23	23	23	22	21	21	21	20	–
22	14	14	13	13	12	12	12	11	10	10	10	9	–
23	3	3	2	2	1	1	1–30	30	29	29	29	28	–
24	22	22	21	21	20	20	20	19	18	18	18	17	–
25	12	12	11	11	10	10	10	9	8	8	8	7	–

20. SUMMARY

With *P. Carlsberg* 9 we have returned to time measurement, with which we began our discussion. We have seen that Egyptian astronomy, in a quantitative sense, was almost non-existent. To it we may award the determination of the length of the year, the division of the day into 24 hours and the decan names in the zodiac. Overshadowing all these is the pictorial element of which the Egyptian was a master, and the portrayals of astronomical figures on ceilings and other monuments continue to interest and charm us.

REFERENCES (Parker)

Borchardt, L. 1920 *Altägyptische Zeitmessung*. Berlin.

Gardiner, A. H. 1947 *Ancient Egyptian Onomastica*. Oxford University Press.

Neugebauer, O. & Parker, R. A. 1960 *Egyptian astronomical texts* I. *The early decans*. Brown University Press.

Neugebauer, O. & Parker, R. A. 1964 *Egyptian astronomical texts* II. *The Ramesside star clocks*. Brown University Press.

Neugebauer, O. & Parker, R. A. 1969 *Egyptian astronomical texts* III. *Decans, planets, constellations and zodiacs*. Brown University Press.

Parker, R. A. 1950 *The calendars of ancient Egypt*. The University of Chicago Press.

Phil. Trans. R. Soc. Lond. A. **276**, 67–82 (1974)   [67]
Printed in Great Britain

Astronomy in ancient and medieval China

By J. Needham, F.R.S., F.B.A.
Gonville and Caius College, University of Cambridge

[Plates 15–18]

Chinese astronomy differed from that of the Western world in two important respects: (*a*) it was polar and equatorial rather than planetary and ecliptic, (*b*) it was an activity of the bureaucratic state rather than of priests or independent scholars. Both features had advantages and disadvantages; the first led to the mechanization of celestial models long before the West, but deferred recognition of equinoctial precession till later. The second ensured remarkable sets of celestial observations antedating most of those recorded elsewhere, but discouraged causal speculation, especially in the absence of Euclidean deductive geometry. In cosmology, China developed three doctrines: (*a*) the Kai Thien universe, a domical geocentrism not unlike early Babylonian ideas, (*b*) the Hun Thien universe, essentially the recognition of the primary celestial spherical coordinates, and (*c*) the Hsüan Yeh system, which accepted the Hun Thien as methodologically necessary but viewed the heavenly bodies as lights of unknown nature floating in infinite empty space. Instrumentation developed early, armillary rings being in use by the end of the −2nd century and the complete armillary sphere by the end of the +1st.

Chinese astronomy differed from that of the Western world in two important respects: first it was polar and equatorial rather than planetary and ecliptic, and secondly it was an activity of bureaucratic States rather than that of priests or independent scholars. Both qualities had advantages and disadvantages. The first, the polar–equatorial situation, led to the mechanization of celestial models far earlier than in the West, but it deferred the recognition of equinoctial precession till later. Similarly, the bureaucratic situation ensured the recording of remarkable sets of celestial observations, but on the other hand it probably discouraged causal speculation, especially since the Chinese did not have Euclidean deductive geometry.

I think most people are well aware that from our present point of view nothing much was happening in China until about −1500 when the Shang period started, and after that during the Chou period there remain various records. But we have the fullest details from the Chhin and Han onwards, from about the −3rd century through the many centuries during which the dynastic histories were written, recording an enormous amount of astronomical information. Many people think that it is necessary to rely upon manuscript sources, as for example our friends the Arabists have to do in great part, but fortunately this is not the case in China because broadly speaking one can say that everything there is either printed or lost. Besides this, there probably are archives containing important astronomical data, which have not yet been opened, in China itself and also in countries like Korea, and I think that some considerable discoveries may be expected in the future on the basis of such archives.

Now as regards the polar and equatorial character, it is clear that Chinese astronomy was of this kind from the beginning. The calendrical problem was of course the simultaneous observation of the stars and the Sun, and presumably there are only two possible methods for ascertaining this relation, methods which people have called contiguity and opposability. The method of contiguity was, we know, that of the ancient Egyptians and the Greeks; it involved the observations of heliacal risings and settings, i.e. risings and settings just before sunrise and just after sunset. We all remember the famous heliacal rising of Sirius in ancient Egypt. This kind of observation did not require knowledge of the pole, meridian or celestial equator, nor any system of horary measurement. It naturally led to the recognition of the zodiacal or ecliptic constellations, and of stars appearing and disappearing simultaneously with them nearer or farther away from the ecliptic, the paranatellons as they used to be called. But the opposability method was that which was adopted by the ancient Chinese; they never paid much attention, so far as any of the records show, to heliacal risings and settings, but rather to the pole star (*pei chi*) and the circumpolar stars which never rise and never set. Their astronomical system was associated with the concept of the meridian, which would arise naturally out of the use of the gnomon, and they systematically determined the culminations and lower transits of the circumpolar stars.

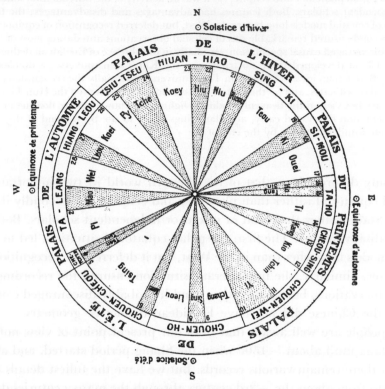

FIGURE 1. Diagram of a projection of the lunar mansions (*hsiu*) on the equator, done for the −24th century (de Saussure). The wide variation in their extensions may be noted.

I think one could really say that the celestial pole was the fundamental basis of Chinese astronomy. It was connected also with the microcosmic–macrocosmic type of thinking, because the pole corresponded to the emperor on earth, around whom the vast system of the bureaucratic agrarian state revolved naturally and spontaneously. I mentioned the gnomon (*piao*) just now, and clearly the meridian concept would arise very easily from this upright stick, because if you looked south you could measure the noon shadow, and if you looked north during the night you could measure the times at which the various circumpolars made their upper and lower transits. We have this in so many words in Chinese texts. Moreover, just as the influence of the Son of Heaven on earth radiated in all directions, so the hour-circles radiated from the pole. During the −1st millennium the Chinese built up a complete system of equatorial divisions defined by the points at which the hour-circles transected the equator (the *chhih tao*); and these divisions contained the 28 'lunar mansions' (*hsiu*), like segments of an orange filling up the celestial sphere, bounded by hour-circles and named from the constellations providing the 'determinative stars' or boundary markers (*chü hsing*). The classical diagram of de Saussure is given in figure 1.

Scholars have sometimes found it almost impossible to believe that a fully equatorial system of astronomy could have grown up without passing through an ecliptic phase, but those of us who study Chinese astronomy are quite sure that this happened. The classical type of Chinese sundial remained equatorial throughout the centuries (figure 2*a*, *b*, plate 15).

Having once established the boundaries of the *hsiu* by means of the characteristic asterisms scattered around the equator, and their determinative stars, the Chinese were in a position to know their exact locations even when invisible below the horizon simply by observing the meridian passages of the circumpolar stars which they could always see. Since they knew where the equator asterisms were at all times, they were able to solve the problem of the sidereal-solar positions, because the sidereal position of the full moon was in opposition to the invisible position of the Sun. In the same way we find growing up quite early in the late Chou, the Warring States period, and from the beginning of the Chhin and Han, from the − 3rd century onwards, a rather full and complete recognition of the great celestial circles. This is known as the Hun Thien cosmological system, but it was not really a cosmology as often so called, it was rather the recognition of these circles. Naturally it accompanied the development of observational and demonstrational armillary spheres.

As regards the origin of the system of the *hsiu* or lunar mansions, this is a very difficult question because we get other systems of the same kind in other civilizations; particularly the *nakshatra* in India and the *manāzil* in the Arabic world. The *manāzil* do not compete of course, but indianists and sinologists have long been disputing about which is the older of the two, the Indian or the Chinese. I cannot today go into the argument on both sides, but one can say that only nine of the twenty-eight *hsiu* determinatives are identical with the corresponding *yogatārās* or 'junction stars' of the Indians, while a further eleven share the same constellation but not the same determinative star. Only eight of the determinative stars and *yogatārās*, however, are in quite different constellations, and of these two are Vega and Altair. On the Chinese side it is possible to say that the *nakshatra* do not show so clearly the coupling arrangements whereby *hsiu* of greater or lesser equatorial breadth stand opposite each other. Indian astronomy moreover, which was far more influenced by Greek astronomy than the Chinese was, does not show that keying of the *hsiu* and the circumpolar stars which is so important in China, in fact the essence of the Chinese system. Besides, the distribution of the *nakshatra* asterisms is much more scattered than that of those of the *hsiu*, following even less closely the position of the equator in the − 3rd millennium.

On the other hand, as regards the documentary evidence, the Indians have very little to yield. There seems to be not much doubt that in the hymns of the *Rig Veda*, which correspond with the Shang oracle-bones and come down from about the − 14th century, two *nakshatra* make their appearance. From that time onwards gradually the system is built up. It becomes complete in the *Atharva Veda*, for instance, and in the black *Yajur Veda* (all three recensions). This may mean that the system was fairly organized in India by about − 800. But in China again there is much the same situation because that ancient calendar called the *Yüeh Ling* may come down from as far back as − 850 and it mentions 23 out of the 28 *hsiu*. We cannot pursue the long argument this morning but nevertheless the problem is a very interesting one and it is not yet solved. I myself have always wanted to believe that the original circle of lunar mansions round the equator was Babylonian, so that in that way I could be happy to agree with Dr Aaboe in what he is saying at this conference on the importance of Babylonian origins in astronomy. The only difficulty is that it may be rather hard to find anything in Babylonian astronomy which could really have given rise to the *hsiu* and *nakshatra* systems.

May I now come to the question of celestial coordinates. In China we have a remarkable text called the *Hsing Ching* or 'Star Manual', the date of which is very unsure, though undoubtedly ancient. Some of its data appeared also in much later works such as the +8th-century *Khai-Yuan Chan Ching*. The *Hsing Ching* is certainly Han, but it may be a little earlier, and it records the positional measurements of many stars made by three astronomers of the −4th century, Shih Shen, Kan Tê and Wu Hsien. Some of them are *hsiu* or lunar mansion stars, and some of them are other stars all over the heavens grouped in asterisms. The epoch of the observations is not at all certain, and calculations have shown varying dates. None of the observations could be earlier than about −350, but some are a good deal later, so the catalogue as a whole may or may not be pre-Hipparchan (−134). Some of the measurements seem to have been made about

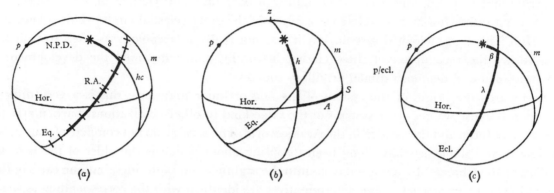

FIGURE 3. The three systems of celestial coordinates: (*a*) the equatorial Chinese, and modern, system; (*b*) the Arabic altazimuth system; (*c*) the Greek ecliptic system.

+130, nearer Ptolemy's time. The text gives the name of the asterism, the number of stars it contains, its position with respect to neighbouring asterisms, and measurements in degrees (on the $365\frac{1}{4}°$ basis of course) for the principal stars in the group; i.e. the hour-angle of the principal star measured along the equator from the first point of the *hsiu* in which it lies, and the north polar distance of the star. The text says, for example, that such and such a star is 2° forward from the beginning of the *hsiu* Hsin, and also that its distance from the north pole is 103°. Obviously the first corresponds to our modern right ascension, and the second gives the complement of our modern declination.

This is really quite interesting because I think it is well known (figure 3) that the Greek coordinates were essentially ecliptic, positions being measured along the ecliptic and towards the pole of the ecliptic. It is equally well known that the Arabic system made use of the horizon, taking the azimuth and altitude. This has the great disadvantage that it applies only to particular individual points on the Earth's surface. In China we never get the horizon system at all, or only perhaps extremely late under Arabic influence. On the other hand, the Greek system does make its appearance in the Thang period, when some of the Indian astronomers were working in China, for example a text may say that a star is so many degrees north or south of the ecliptic; but this is a late thing found only in the +8th, +9th and +10th centuries. All civilizations have used one or other of these three different types of star coordinates, and our modern ones are clearly Chinese. In this matter I would like to doubt, if I might, the impression given by Dr R. R. Newton, if that was what he intended to give, that all ancient systems were primarily horizon ones. I think it would be very hard to say that of Chinese astronomy in ancient times, for the polar-equatorial coordinates come in very early in Chinese culture.

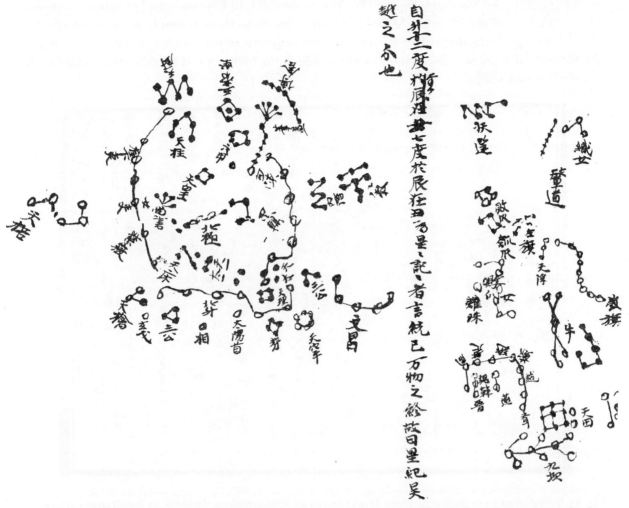

FIGURE 5. The Tunhuang MS. star map of +940 (Brit. Mus. Stein no. 3326). To the left, a polar projection showing the Purple Palace and the Great Bear below it. To the right, on 'Mercator's' projection, an hour-angle segment from 12° in Tou *hsiu* to 7° in Nü *hsiu*, including constellations in Sagittarius and Capricornus. The stars are drawn in three colours, white, black and yellow, to correspond with the three ancient schools of positional astronomers (those of Shih Shen, Kan Tê and Wu Hsien).

Now as for the constellations and the naming of them, it is rather an interesting thing that there is no overlap at all with the nomenclature in other civilizations. There may be a good deal of connexion between Greek and Indian names but practically nothing in China corresponds. From ancient times it was customary to represent constellations in a ball-and-link style, as we see on an inscribed brick from the Han period (± 1st century, figure 4, plate 15). Figure 5 shows an early star map from the Chinese culture-area datable at about +940. This manuscript must be one of the oldest star-charts extant from any civilization. It gives a picture of the north polar region, the 'Purple Forbidden Palace', as it was called, in which you can see down at the bottom the shape of the Great Bear similar to the way everybody else saw it, but nothing else is the same. A European constellation can appear in several different asterisms on the Chinese planisphere; for example, Hydra comprises the three *hsiu* Chang, Hsiang and Liu, together with eight other star groups having no similarity of symbolism. And this is the case all along the line. There is nothing corresponding to Delphinus and Cancer and so on, the only overlaps are the Great Bear, or Northern Dipper (Pei Tou as we call it in Chinese), and of course naturally Orion (Shen) and the Pleiades (Mao), but apart from those there is simply no correspondence. What is rather interesting is that the agrarian bureaucratic nature of Chinese civilization led to a multitude of star and constellation names in which the hierarchy of earthly officials had their heavenly counterparts.

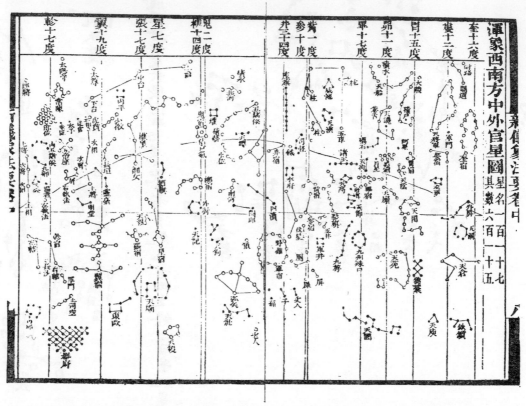

FIGURE 6. Star chart from Su Sung's *Hsin I Hsiang Fa Yao* (+1094), showing 14 of the 28 *hsiu* (lunar mansions), with many of the Chinese constellations contained in them. The equator is marked by the central horizontal line; the ecliptic arches upwards above it. The legend on the right-hand side reads: 'Map of the asterisms north and south of the equator in the southwest part of the heavens, as shown on our celestial globe; 615 stars in 117 constellations.' The *hsiu*, reading from right to left, are: Khuei, Lou, Wei, Mao, Pi, Tshui, Shen (Orion), Ching, Kuei, Liu, Hsing, Chang, I and Chen. The very unequal equatorial extensions are well seen.

Next, figure 6 shows what one can only describe as a star map on a Mercator projection, with the upright bands representing the *hsiu*, the lunar mansions of different widths or stretches, and then the equator running along the middle, the ecliptic also shown over half the heavens, and all the stars placed in their approximately right positions. This was the map of the stars which was made for the celestial globe set up by Su Sung in the Astronomical Clock Tower at Khaifêng, in +1088, some 500 years before Gerard Mercator. I shall want to say a word more about this instrument in a moment. A similar star chart in Japan has been described by Dr Imoto Susumu (figure 7, plate 16, gives the same hemisphere as figure 6). But it has the additional interest of being drawn on squared paper with the aim of greater accuracy. Though the

(a)   (b)

FIGURE 2. Typical Chinese equatorial sundials in the Imperial Palace at Peking. (a) Outside the Wu Mên Gate (photo. A. R. Moore, *ca.* 1925). (b) On the platform of the Thai Ho Tien Hall (photo. Sirén, *ca.* 1910).

FIGURE 4. The Moon passing through a constellation; from a Szechuanese moulded brick of Han date (from Wên Yü). The moon disk shows a toad under a tree (a legendary consequence of the lunar crater markings, etc.), and is borne along by a feathered and winged genius. The constellation is drawn in the usual ball-and-link convention.

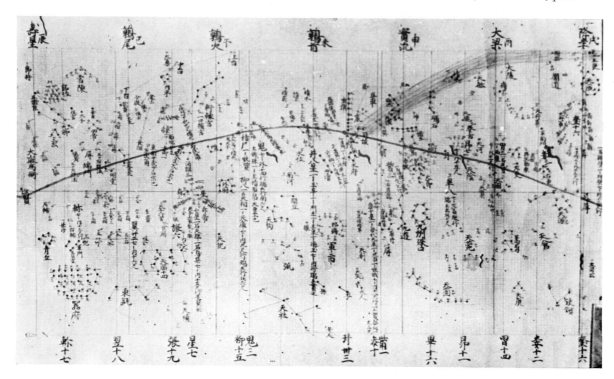

FIGURE 7. The same hemisphere as in figure 6, with the equator again horizontal and central, and the *hsiu* divisions and constellations shown, as before, on a 'Mercator's' projection; from a MS. star map *Kōshi-gettshin-zu* formerly preserved in Japan (Imoto Susumu). Many similarities with the preceding figure will be apparent, but the drawing has been done on squared paper for greater accuracy. Perhaps the original was more likely Sung than Thang.

FIGURE 9. The Tower of Chou Kung for the measurement of the Sun's solstitial shadow lengths at Kao-chhêng (formerly Yang-chhêng), some 80 km southeast of Loyang, and for many centuries the site of China's central astronomical observatory. The present structure is a Ming renovation of the instrument built by Kuo Shou-Ching about + 1276. The 12 m gnomon stood up in the slot, and its shadow was measured along the stone scale projecting on the left, with special arrangements to secure a sharp edge reading. One of the rooms on the platform housed a clepsydra (perhaps a hydro-mechanical clock), and the other probably an observational armillary sphere.

FIGURE 11. The equatorial armillary sphere of Kuo Shou-Ching (*ca.* +1276), now preserved in the grounds of the Purple Mountain Observatory, Nanking.

FIGURE 12. Kuo Shou-Ching's 'equatorial torquetum' (*chien i*, simplified instrument), precursor of all telescope equatorial mountings, in the grounds of the Purple Mountain Observatory, Nanking. Like the armillary sphere in figure 11, this instrument may be one of the identical replicas cast by Huang-Fu Chung-Ho in +1437 (orig. photo., 1958).

FIGURE 13. Detail view of part of the instrument shown in figure 12 (orig. photo., 1958). The bronze mobile declination split-ring or meridian double circle carrying the sighting-tube is well seen. Below, the fixed diurnal circle, and the mobile equatorial circle, with movable radial pointers probably used to demarcate the boundaries of *hsiu*.

FIGURE 14. The famous polyvascular inflow clepsydra at Canton made in the Yuan dynasty (+1316) by Tu Tzu-Shêng (orig. photo., 1958).

FIGURE 18. Model of the Khaifêng clock tower in the Science Museum at South Kensington (J. H. Combridge). Here it is seen from the back and right.

FIGURE 20. Working model of the hydro-mechanical escapement of the Khaifêng clock tower in the Science Museum at South Kensington (J. H. Combridge).

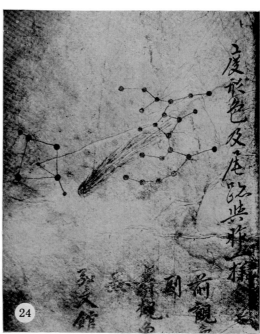

FIGURE 24. A comet of +1664, from the Korean astronomical archives (Rufus).

manuscript is of a later date, Imoto Susumu traces its origin back to the time of I-Hsing in the +8th century, since the R.A. values mostly, though not entirely, agree with those given by him. Su Sung may have drawn from the same source. Finally, we show the famous planisphere of Suchow (+1193) in figure 8.

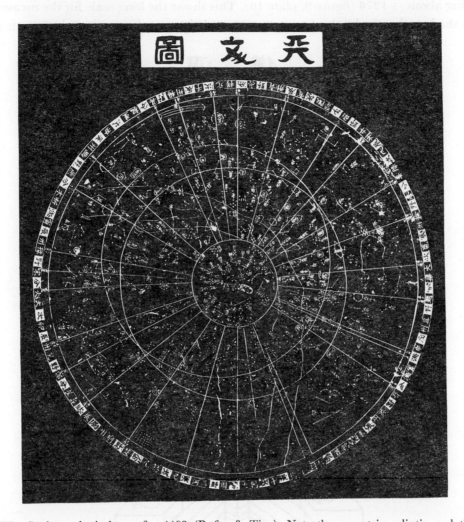

FIGURE 8. The Suchow planisphere of +1193 (Rufus & Tien). Note the excentric ecliptic and the curving course of the Milky Way (*thien ho*, the river of heaven). The map with its explanatory text was prepared by the geographer and imperial tutor Huang Shang, and committed to stone by Wang Chih-Yuan in +1247.

Coming back to the question of instrumentation, the earliest measuring device was undoubtedly the gnomon. We have many references to that in ancient books, such as the *Tso Chuan* and the *Chou Pei Suan Ching*. It must have been used from the Shang period, *ca.* −1500 onwards. The *Chou Pei Suan Ching* (Arithmetical Classic of the Gnomon and the Circular Paths of Heaven) and the *Chou Li* (Institutes of the Chou Dynasty) are both probably early Han texts but based on the usages of the centuries preceding. In many ways, the gnomon was a much used instrument in Chinese culture. Professor Willy Hartner has recently written an interesting new contribution on this, relating especially to the meridian line of observation posts set up about +725 under the Buddhist monk I-Hsing, the greatest astronomer and mathematician of his age. Now this was a chain of stations reaching from Indo-China right up to Siberia, and the seasonal observations of noon sun shadows with the gnomon were done at a dozen places along that line. Measuring just over 2500 km in length, this meridian survey must be one of the greatest efforts of organized scientific research in any medieval civilization.

The gnomon reached its climax in Chinese culture with the giant instrument set up by Kuo Shou-Ching about +1276 (figure 9, plate 16). This shows the long scale for the measurement of the Sun's shadow thrown by the 12 m gnomon, and the star observation platform is to be seen

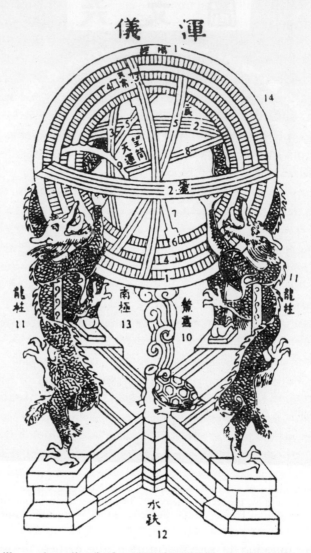

FIGURE 10. Su Sung's armillary sphere (*hun i*) of +1090, described in the *Hsin I Hsiang Fa Yao*, redrawn from the text and labelled by Maspero. For explanation of details see SCC, III, Fig. 159, p. 351.

at the top of the tower. One of the chambers there housed a clepsydra, perhaps a hydro-mechanical clock; the other probably an armillary sphere. This wonderful piece of giant astronomical equipment still exists at a place called Yang-chhêng not far from Lo-yang (in Honan) near the geographical centre of China, and it has been repaired in very recent times, in fact even during the Cultural Revolution. During the war it was used for target practice by the Japanese, and the top of one of the side chambers was knocked off, but no other damage was done, and it has now been completely repaired and restored. It must be regarded as a notable monument even though no doubt it was rebuilt under the Ming. The +13th-century data obtained with this gnomon still exist also, and are highly creditable for that time, especially as a sophisticated device, the 'shadow definer', was used to focus the image of the cross-bar and avoid the difficulty of the penumbra.

The development of armillary rings and spheres is an even more intriguing question. It is fairly certain that the most primitive form of armillary was a simple single ring with some kind of fiducial line or sights which could be set up in the meridian or equatorial plane. Measurement one way gave the north polar distance or declination, measurement the other way gave the position in the *hsiu*, i.e. the right ascension. No doubt that was all that Shih Shen and Kan Tê had at their disposal, and there is some evidence that it came down to about −100 because it may be that Lo-Hsia Hung and Hsien-Yü Wang-Jen had nothing else. But then things happened rather fast. Kêng Shou-Chhang introduced the first permanently fixed equatorial ring in −52, and an ecliptic ring was added by Fu An and Chia Khuei in +84; while with Chang Hêng's apparatus in +125 the sphere was complete with horizon and meridian rings. It is rather remarkable that this rapid evolution should have come about historically parallel with Greek times, and just before the life of Ptolemy himself.

Spheres of all kinds continued to be made without much change for many centuries. The one in figure 10 is the very famous instrument constructed by Su Sung in +1088 or so for the astronomical clock tower at Khaifêng which I mentioned before. In the Thang period, about +630, Li Shun-Fêng made the radical innovation of building not two nests of concentric rings but three, so it got rather over-complicated, but the design that Su Sung used is one which has many similarities with the equatorial spheres of Tycho Brahe in +16th-century Europe. Figure 11, plate 17, shows another sphere, that of Kuo Shou-Ching (+1276) which is to be found at the Purple Mountain Observatory near Nanking today. This was the same astronomer who made the giant gnomon device. But his major achievement was his equatorial torquetum or 'simplified instrument' (*chien i*), which did away with the unnecessary parts of the armillary sphere and achieved what was essentially an equatorial mounting of the sighting tube. One can still see it (figures 12 and 13, plate 17) on the top of the Purple Mountain at Nanking, and I am glad to report that it is in perfect order and very carefully preserved.

This brings me to the question of powered celestial models which I mentioned earlier, and that certainly was an extremely interesting development unexampled in any other culture. Twice already I have referred to the astronomical clock-tower of the late +11th century described in the *Hsin I Hsiang Fa Yao* (New Design for a mechanized Armillary Sphere and Celestial Globe), a book presented to the throne in +1092. These cosmic models, in fact demonstrational armillary spheres and celestial globes, were rotated by water power using a constant-level tank and a driving wheel with buckets. One might call it a mill wheel but it was retained and guided all the time by an escapement, so we use the term hydro-mechanical clockwork, and speak of the hydro-mechanical linkwork escapement. The background of the constant-level tank lay in clepsydra technology, for in the +6th century the older polyvascular trains of compensating tanks (figure 14, plate 18) had been superseded by arrangements of overflow tanks to produce a perfectly steady flow (figure 15).

The escapement was indeed a great invention; it was certainly in use by the time of I-Hsing and Liang Ling-Tsan at the beginning of the +8th century, and the only problem is whether it may go back much further. It is still uncertain whether Chang Hêng in the +2nd century had already got this or not. Thus using the constant-level tank, and the driving wheel retained

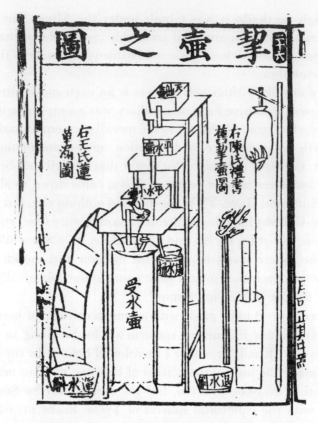

FIGURE 15. Overflow tank clepsydra for constant-level operation seen in a Sung edition of
Yang Chia's *Liu Ching Thu* of +1155.

by the linkwork escapement, one had a real time measuring machine, and of course it may also
be considered the first of all clock drives.

Figure 16 shows the appearance of the clock tower, with the tanks and the driving wheel
laid open on the right; and you can see the celestial globe on the first floor, and on the roof the
armillary sphere. I called it demonstrational just now, but that is not quite fair because it was
certainly an observational one too, and we have speculated that one of the reasons for introducing the clock drive was so as to be able to turn round the 20-odd tonnes of bronze in time to
make an observation just before dawn. But the globe rotating in the intermediate storey would
have been for demonstration in times of clouds and storm when the heavens could not be
observed directly. Figures 17 and 18, plate 18, show reconstructions of what the clock-tower
looked like, and the mechanism follows in figure 19. Here one can see how the drive on the
right-hand side at the bottom rotated the column bearing much jack work, and a celestial
globe on a bevel drive at the top, and how after some time the long shaft was replaced by a
chain drive, in fact mark 2 and mark 3 chain drives, getting progressively shorter and so better
designed. Finally the escapement mechanism is seen in figures 20, plate 18, and 21; this is the
device that is certainly of the +8th century and possibly of the +2nd.

Now these powered models illustrate, I think, the advantage which the Chinese found in
having a polar-equatorial system. After all, the celestial latitude and longitude grid is a purely
conceptional network thrown over the heavens, and along its lines nothing ever actually moves;
but things of course do move on the circles parallel with the equator. Finally, figure 22 shows a

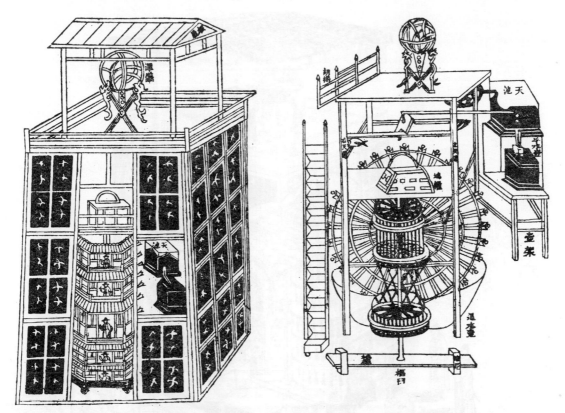

FIGURE 16. General views of the astronomical clock tower built at Khaifêng by Su Sung and his collaborators in +1090. On the top platform, some 11 m above the ground, there was a mechanized armillary sphere for observations; in a chamber on the first floor a mechanized celestial globe was installed, and below, in front of the hydro-mechanical clockwork, numerous jacks appeared at the openings of a pagoda façade, constituting a time-annunciator. The tower and its machinery are fully described in Su Sung's *Hsin I Hsiang Fa Yao*. (Left) External appearance, with a panel removed to show the constant-level tank. (Right) Internal structure. In front, the jack wheels and vertical shaft, behind this the driving wheel. On the right the constant-level tank delivering the water to the driving-wheel scoops or buckets. Above the driving wheel one can see a few traces of the escapement mechanism (cf. figures 20, 21). On the left, the staircase, on the top platform, the armillary sphere.

reconstruction by Dr Liu Hsien-Chou of I-Hsing's orrery arrangement for Sun and Moon models as deduced from the textual descriptions of his early +8th-century astronomical clock.

Let me return lastly to the theme of bureaucratism that I opened with, because it is really a very interesting feature, a characteristic of Chinese civilization which made it quite different from all others in those ancient times. One might say that the majority of the observers who thought and calculated and wrote about astronomical problems through 2000 years were in State service. They were organized in a special department of government, the Astronomical Bureau or Directorate, which went by various different names. The most ancient title of the Director was Thai Shih Ling, and although we have always been very conscious of his astrological function we feel that the true astronomical and calendrical element in the work of his department was amply sufficient to warrant the translation 'Astronomer-Royal'. He did of course have to keep his star-clerks on the watch, every moment of every night, for any unusual developments in the heavens. Celestial phenomena like novae, supernovae, eclipses, comets, meteor streams, sun-spots and all such things were regularly reported to the Imperial Court, because although individual genethliacal astrology came only very late to China, perhaps in

FIGURE 17. Pictorial reconstruction of the Khaifêng clock tower (J. Christiansen). Besides the components already mentioned, the norias which wound up the water back into the tanks are here glimpsed behind the driving wheel.

the Ming, the general belief that 'comets do foretell the death of princes' was a very old Chinese idea. The origin of the Bureau of Astronomy was thus perhaps twofold: the importance of keeping the calendar in order was a very important task, but the watch on the heavens for celestial events was also a strong motive. It is interesting that inauspicious happenings were generally regarded as *chhien kao* or reprimands from heaven, and the emperor or some high official, very often the emperor himself, took the guilt upon him, prayed, fasted and promised to amend. Omens were regarded really as signs of bad government, and there would be trouble if things were not put in order; such was the astrological function of the Bureau over the ages.

One remarkable fact not generally known is that in some periods it was customary to have two observatories at the capital both furnished with armillary spheres, clepsydras and all manner of necessary apparatus. For example, Phêng Chhêng tells us about this in the Northern

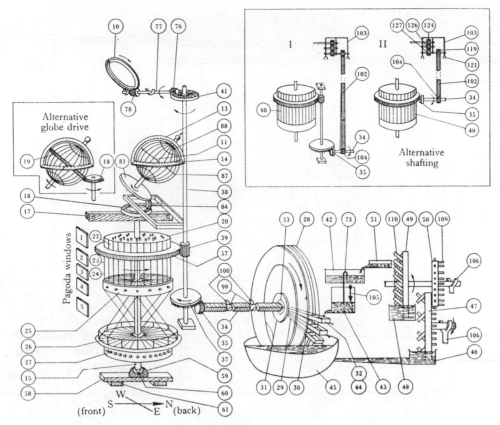

FIGURE 19. Diagram of the power and transmission machinery of the Khaifêng clock tower (J. Needham, L. Wang and D. de S. Price). Norias and tanks on the right, driving-wheel central, then the main vertical shaft (afterwards replaced by successively shorter chain-drives) operating jack-wheels, globe and armillary. For explanation of details see SCC, IV, pt. 2, fig. 652a, p. 452.

Sung. The Astronomical Department called the Thien Wên Yuan was located within the walls of the Imperial Palace itself, while the other one, the Directorate of Astronomy and Calendar, the Ssu Thien Chien, presided over by the Thai Shih Ling himself, was outside the walls. The data from the two observatories, especially concerning unusual phenomena, were supposed to be compared each morning and then presented jointly so as to avoid false or mistaken reports.

I have mentioned the grave political significance of celestial events. It is therefore rather interesting to find exhortations to security-mindedness addressed century after century to the astronomical officials. For example, the *Chiu Thang Shu* (Old history of the Thang Dynasty) tells us that in +840: 'in the twelfth month of the 5th year of the Khai-Chhêng reign-period an imperial edict was issued ordering that the observers in the Imperial Observatory should keep their business absolutely secret. "If we hear", it said, "of any intercourse between the astronomical officials or their subordinates, and officials of any other government departments, or miscellaneous common people, it will be regarded as a violation of security regulations which should be strictly adhered to. From now onwards, therefore, the astronomical officials are on no account to mix with civil servants and common people in general. Let the Censorate see to it".' All one can say about that is that there is nothing new about Los Alamos or Harwell, but whether or not the greatest scientific achievements happen under such conditions is another question. And I suppose that even a Galileo and a Priestley had their difficulties with the powers that were.

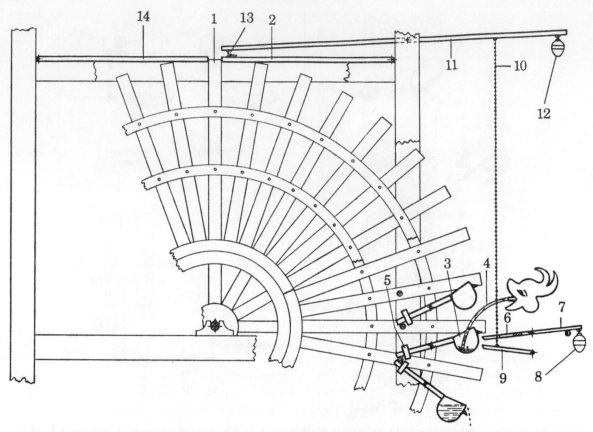

FIGURE 21. Diagram to illustrate the functioning of the hydro-mechanical linkwork escapement (J. H. Combridge). Each scoop bucket on the perimeter of the driving wheel is individually balanced; when it is full it descends, depressing a counter-weighted trip-lever, and operates a chain-and-link connexion which opens a gate at the top of the wheel and allows the next scoop bucket to come into place. One 'tick' took place about every 24 s. For further explanation of details see SCC, IV, pt. 2, fig. 658, p. 460.

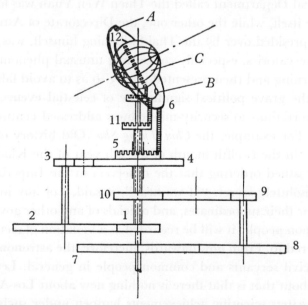

FIGURE 22. Reconstruction of the orrery movement of Chinese hydro-mechanical astronomical clocks (Liu Hsien-Chou). Only the Sun and Moon model movements are shown as well as that for the celestial globe, in a design which would have needed both concentric shafting and gear wheels with odd numbers of teeth. Remaining textual descriptions authorize it, however, for the instruments of I-Hsing (+725), Chang Ssu-Hsün (+979) and Wang Fu (+1124).

Something of what the observatories did has come down to us in extant texts. Records of eclipses start with the Shang oracle-bone material from −1361 onwards, and records of novae from −1300 (figure 23).

FIGURE 23. The oldest record of a nova; inscription on an oracle-bone dating from about −1300.

The oracle-bones are of course difficult to interpret, but nevertheless they provide evidence of great importance. As for supernovae, there was the famous one of +1054, the Crab supernova. There has recently been an interesting paper by our collaborator Professor Ho Ping-Yü and others on the difficulties of locating and identifying that exactly. Then there is the abundance of documentation on comets. Figure 24, plate 18, shows a manuscript drawing of a comet passing between two *hsiu* constellations on the night of 28 October +1664; it comes from the Korean archives. For comets in general we have Chinese records from −613 onwards. Recently a very interesting paper appeared by a friend of ours, Dr Chiang Thao, from Dunsink Observatory, on the recurrences of Halley's comet calculated from Chinese records. Finally sun-spots were being sedulously recorded from −28 onwards, a fact which I think would have been a great surprise to Galileo and Christopher Scheiner if they had ever been aware of it. Here there is cause for warm agreement with Dr R. R. Newton, who spoke so interestingly on the usefulness for modern astronomy of studies in ancient records. Almost every month some new paper comes out dealing with something based on Chinese data. There was one the other day about a nova of +363 which may be a radio source and could be identified from these records.

My last remarks must concern cosmologies. The Kai Thien cosmology, with a domical arrangement of the Earth and the heavens, was rather Babylonian, obviously very primitive (figure 25), and did not play much part after the Chhin and Han in China. At this time there was perfected the Hun Thien cosmology, which (as I said) was really the recognition of the great celestial circles. Finally there was the Hsüan Yeh cosmology, which is of greater interest because it maintained that the stars were lights of uncertain origin, floating in infinite empty space, and that the planets and moving stars were carried round in this dark space by some kind

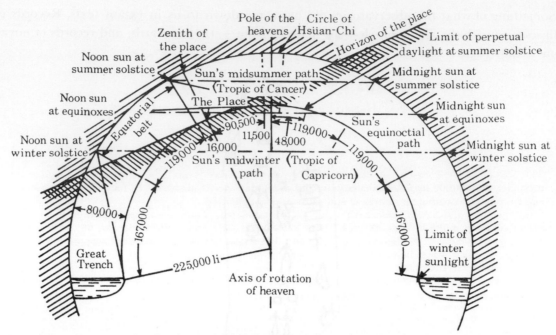

FIGURE 25. The Kai Thien cosmology (Chatley).

of wind. Since Euclidean deductive geometry was not available, no Ptolemaic geometrical model was devised, and all through the ages calendrical calculations were done by algebraic methods, thought about the actual mechanics and geometry of the solar system being laid aside. Recently our collaborator Professor Nathan Sivin has studied how this preference came about, in his interesting monograph 'Cosmos and Computation in Chinese astronomy'. Perhaps that preference was again a Babylonian characteristic. In any case the Hsüan Yeh theory was rather modern and long before its time. One could say that if the Chinese had no Euclidean geometry they had no crystalline celestial spheres either, so they did not have to break out of them at the time of the Renaissance; and it may even be that a knowledge of the Chinese conceptions helped Giordano Bruno, William Gilbert and Francis Godwin to take this great step themselves.

To sum it all up, Chinese astronomy cannot be neglected by anyone who seeks for an oecumenical account of the development of human knowledge of the starry firmament and our own place within it. It is all the more important on account of its extreme originality, influenced indeed a little by Babylonia and India, but unlike the latter culture highly independent of those Greek and Hellenistic discoveries which had so wide-ranging an influence everywhere west of Turkestan.

NOTE: The romanization system adopted in this paper is that of Wade-Giles, with the substitution of an *h* for the aspirate apostrophe. Full documentation, with Chinese characters, will be found in *Science and Civilization in China* (7 vols., Cambridge, 1954–), herein abbreviated as SCC.

Phil. Trans. R. Soc. Lond. A. **276**, 83–98 (1974)
Printed in Great Britain

[83]

Maya astronomy

By J. E. S. Thompson, F.B.A.
Centre of Maya Studies, National University of Mexico

[Plates 19 and 20]

The Maya had three concurrent counts: 365-day years; 360-day 'years' (*tuns*), with named vigesimal multiples to 3 200 000 *tuns*, for calculations; and a 260-day sacred almanac (13 numbers and 20 concurrent names) covering all mundane and astronomical activities.

Solar eclipse and Venus synodical revolutions are tabulated in one hieroglyphic book to reach the lowest common multiple with 260: for Venus 37 960 days ($584 \times 65 = 260 \times 146$ also 365×104); for the Moon 11 960 days (405 lunations $= 260 \times 46$). The Maya successfully predicted eclipses, but were unaware of which would be visible to them. Means were astronomical; ends, astrological.

Ingenious corrections, also retaining the 260-day connexion, occur. The corrected error in Venus revolutions is one day in 6000 years. Lunar corrections similarly had to conform to the sacred almanac. Other planetary tables are very dubiously identified.

Solar data are challengeable. Dates are recorded 5 million, possibly 90 and 400 million years ago.

The Maya occupied the whole peninsula of Yucatan and the area south of it including eastern Chiapas, the highlands of Guatemala and the western fringes of Honduras and El Salvador. The area extends from about 13 to 21° north of the equator. The most advanced centres were in the heart of this territory, the rain-forest lowlands of the Peten and Usumacinta drainage. The highlands in the south, although endowed with products, such as jade, obsidian, quetzal feathers of great value to the Maya economy, lagged behind the lowlands in art, architecture, and science. Maya culture, an outgrowth of earlier cultures, notably the Olmec, with greatest development in the Isthmus of Tehuantepec, southern Veracruz and the Pacific coast of Chiapas and Guatemala, was late, with its greatest period approximately A.D. 200 to 900. I see no evidence that Maya successes in astronomy and calendrics owe anything to Old World influences except perhaps for a basic stratum introduced, one may suppose, by hunter-gatherers crossing from the Old World via the Bering Strait from 10 000 B.C. onward.

By A.D. 950 the great ceremonial centres had been abandoned to the forest, but Maya culture continued in peripheral regions, particularly on the western, northern and northeastern parts of the peninsula. This continuing culture, affected by outside influences, was marked by a rising importance of warfare and a concomitant loss of influence by religion. Deterioration of the arts is very obvious and Maya interest in time was no longer obsessive. The period of decline ended with the Spanish conquest (A.D. 1540).

No people in history has shown such interest in time as the Maya. Records of its passage were inscribed on practically every stela, on lintels of wood and stone, on stairways, cornices, friezes and panels.

Time was pictured as an endless relay march to eternity with the number of each time period from the day up carrying the period as a load on its back. Each sunset the marchers halted. Were it, for example, the fifteenth day, the bearer, the god of number 15, handed over his load to the god of number 16 to carry next day. On the last day of the year, the divine numerical carriers for year, month and day would similarly hand over to a new series (Thompson 1950, pp. 59–61).

6-2

The Maya calendar

The extremely complex Maya calendar comprised three separate but concurrent counts. There was a year of 365 days formed of 18 'months' of 20 days each and an additional and extremely unlucky period of 5 days at its end. An approximate year of 360 days, composed of eighteen 20-day months, called *tun* was used in long-distance calculations. Multiples of this were in the vigesimal system used by the Maya for all forms of counting. There are hieroglyphs and names for the tun itself as well as for 20, 400, 8000, 160000, and 3200000 tuns.

The third count, and to the Maya the most important, was their sacred almanac of 260 days. This was formed of the numbers 1 to 13 and 20 named days which ran concurrently, more or less as though we had 1 Sunday, 2 Monday, etc., and then 8 Sunday, 9 Monday, etc., until the cycle closed with the thirteenth Saturday and went back to 1 Sunday at the end of 91 days. The Maya cycle was 260 days, the lowest common multiple of 13 and 20.

Each of the 20 names and 13 numbers was a god or goddess, not merely influenced by some god, and each day was looked on as a living being. Every activity on earth and in the skies was related to this sacred almanac, with certain days and numbers propitious or otherwise for such activities as planting crops, hunting, marriage, collecting honey, curing disease, making war, or the outgoings and incomings of planets.

Maya dates were recorded in terms of both the 260 and 365 day counts. As the highest common factor of those two numbers is five, a combination of the two counts could not recur until 52 years had passed. In addition, time was fixed by a count of tuns and their multiples from an epoch far in the past, which was not historical, but possibly marked a recreation of the world (the Maya believed the world had been created and destroyed four times and that we are now in the fifth creation). This epoch corresponded, according to the most widely accepted correlation of the two calendars, to 3113 B.C. – 10 August to be precise. It lay 3000 years in the past, when Maya culture, Maya glyphs and distant reckonings began. There are Maya dates far earlier than that, and such was the complexity of the Maya machine, that a date could be fixed in such a way that it would not repeat for 3000000000 million years, give or take a million years. There is some uncertainty about the numbering of the very high periods, but according to one arrangement, dates 90000000 and 400000000 years in the past are recorded. Even should the suggested arrangement be incorrect, there is no doubt that dates 300000000 years apart are inscribed on stelae (Thompson 1950, pp. 314–16).

Correlation of Maya and European calendars

The Maya calendar can be correlated with our own with considerable certainty. The 260-day almanac survives to the present day among some remote Maya communities. All those almanacs agree to the day. Projected back to the sixteenth century, they fail by only one day to synchronize with a Maya year and a European year (for 1553) correlated by a Franciscan missionary. The one-day break arose probably because the friar may have collected his information on the start of the Maya year (16 July o.s.) in 1551 and failed to allow for the intervening leap day of 1552.

The modern Maya almanacs, furthermore, agree to the day with Aztec-European double dates. Clearly, all Middle America had a synchronized 260-day almanac, and as this has not lost a day in the face of Spanish attempts to suppress it during the past four and a half centuries, one must conclude that there was no loss when the priest astronomers were dominant at an earlier period. Moon-age records from then bear out that conclusion.

Other data, notably records of Moon ages and heliacal risings of Venus in Maya sources, events of the Spanish conquest expressed also in terms of Maya new year days or of the katun (20-tun) count which formed yet another cycle of 260 tuns, the findings of Maya archaeology and their relationships to the archaeological sequences in the Valley of Mexico and other parts of highland Mexico, and, finally, carbon-14 readings of wooden lintels and beams dated in terms of the Maya calendar come very close to leaving the exact correlation of the two calendars beyond doubt. It does, however, depend on the assumption that there was never any break in the Maya calendar. Reasons for making that assumption are, as we have seen, very strong. To those one may add that the sacredness of the 260-day almanac was such that any tampering with it would have been unthinkable to the Maya priesthood; in addition, it would have set awry all the interlocking cycles of the moon and Venus and other counts, the importance of which is noted below. Discussions of all factors are in Morley (1920, appendix II), L. Roys (1935), Spinden (1924), Thompson (1935, 1950), and Satterthwaite & Ralph (1960).

The correlation, known as Goodman–Martínez–Thompson, which meets those conditions, calls for the addition to any Maya date, first reduced to days, of the number 584283, to give the Julian equivalent. Thus, as we have seen, the Maya epoch, written 13.0.0.0.0 4 Ahau 8 Cumku, becomes the equivalent of 10 August 3113 B.C., and 9.15.0.0.0 4 Ahau 13 Yax, at the height of the Classic period, corresponds to 20 August A.D. 731.

This correlation has won wide acceptance, but it has been rejected by a few who, with one exception, depend solely on astronomical data as interpreted, often very dubiously, by themselves. They pay little or no attention to the non-astronomical lines of evidence listed above and give no reasons for ignoring them, an attitude which hardly elicits one's respect. Each proclaims his own correlation – one student within a few years has announced six different correlations – based solely on differing and speculative interpretations of Maya astronomical data. None has persuaded any rival that the true light has been vouchsafed him.

Needless to say, an unchallengeable correlation of the two calendars would be immensely helpful in identifying astronomical data in the texts, although I myself am far from convinced that planetary observations were recorded on the stone monuments, unless favourable phenomena perhaps governed a ruler's accession date. Stelae recorded past events whereas Maya astronomers aimed at prediction. Planetary matters would have been noted on work sheets, not on stelae, so sacred in Maya eyes.

Scores of dates with records of contemporary Moon age and the position of the current moon in a group of six (sometimes five?) lunations, as well as a few Moon ages calculated for dates then far in the past, are carved on stone monuments, but the most striking astronomical data are in the Dresden codex, one of only three surviving hieroglyphic books of the Maya (Thompson 1972). This is an edition of around A.D. 1250 of a lost earlier version of perhaps about A.D. 750. Dresden codex, so-named for the city in which it now reposes, contains, in addition to a great number of 260-day almanacs of divination in connexion with all forms of mundane activities, tables for synodical revolutions of Venus and for solar eclipses, each preceded by multiplication tables of the periods involved and entries giving corrections which must be applied.

Synodical revolutions of Venus

It is necessary to emphasize what has already been said, namely that every astronomical mechanism, just like everything else in Maya life, had to be related to the 260-day sacred almanac. In the case of Venus, the position 1 Ahau was *the* day. On that day the cycle of revolutions of Venus started and ended. As it was more important for the Maya to keep that contact with 1 Ahau than to record exactly the calculated day of heliacal rising of the planet, corrections had to be made with that end in view.

The Venus table, occupying six pages of Dresden codex, covers 65 synodical revolutions of the planet (more precisely from one heliacal rising after inferior conjunction to another) averaged at 584 days. The first page, an introduction, gives religious and divinatory data and multiplications of 584, as well as needed corrections. Each revolution has four divisions or stations 230, 90, 250 and 8 days apart, marking roughly disappearance and reappearance before and after superior conjunction, then Venus as evening star, and finally 8 days of invisibility from disappearance to reappearance at heliacal rising after inferior conjunction. Note that no attention was paid to invisible phenomena – inferior and superior conjunctions, an important point in assessing claims regarding other planets.

The total of 65 synodical revolutions was chosen because 584×65 (37 960 days) equal 146 of the 260-day sacred almanacs, that being the lowest common multiple of the two. It is incidentally 104 years of 365 days, which gave the number extra importance, but was not a deciding factor in the choice of the over-all period. The Maya knew very well that the length of the average revolution – it can vary from about 580 to 587 days – was not 584 days, but their equivalent of about 583.92 days (they did not use decimals), and that therefore their table was too long. However, should they make, say, a 1-day deduction when needed, the start of the table would no longer fall on the all-important day 1 Ahau; such a correction would have broken contact with the 260-day almanac, a disastrous happening in Maya eyes. A correction had to be made in such a way that it involved a multiple of 260 days.

At the bottom of the page prefacing the table are given multiples of 5, 10, 15, 20 and similar intervals of 5 up to 60 synodical revolutions of 584 days. The table continues at the top of the page with 65, 130, 195 and 260 multiples of 584, these last representing the length of the table and twice, thrice and four times its length. The last figure equals 365×416. It was not put there for ornament but was utilized in computations (figure 1, plate 19, left page).

The Maya were well aware that their table of 584×65 was slightly over 5 days too long. They corrected the error by formulae given in the centre of the prefatory page. The Maya priest-astronomer knew from his tables that the 61st revolution of Venus ended on the day called 5 Kan, which was precisely 4 days after 1 Ahau. The accumulated error then was between 4 and 5 days. Accordingly, he subtracted 4 days, and, recovering the all-important 1 Ahau, started the count again. The equation is

$$584 \times 61 \ (35\,624 \text{ days}) - 4 \text{ days} = 35\,620 \quad \text{days}$$
$$260 \times 137 \qquad\qquad\qquad\qquad = 35\,620 \quad \text{days}$$
$$583.92 \times 61 \ (\text{true syn. revs.}) \quad = 35\,619.12 \text{ days}$$

The correction was good, but still 0.88 day short in just under a century, and that was not good enough for the Maya. They learned to make that 4-day correction, and then, when the error in their first correction had amounted to 4 days, they made an 8-day correction at the end

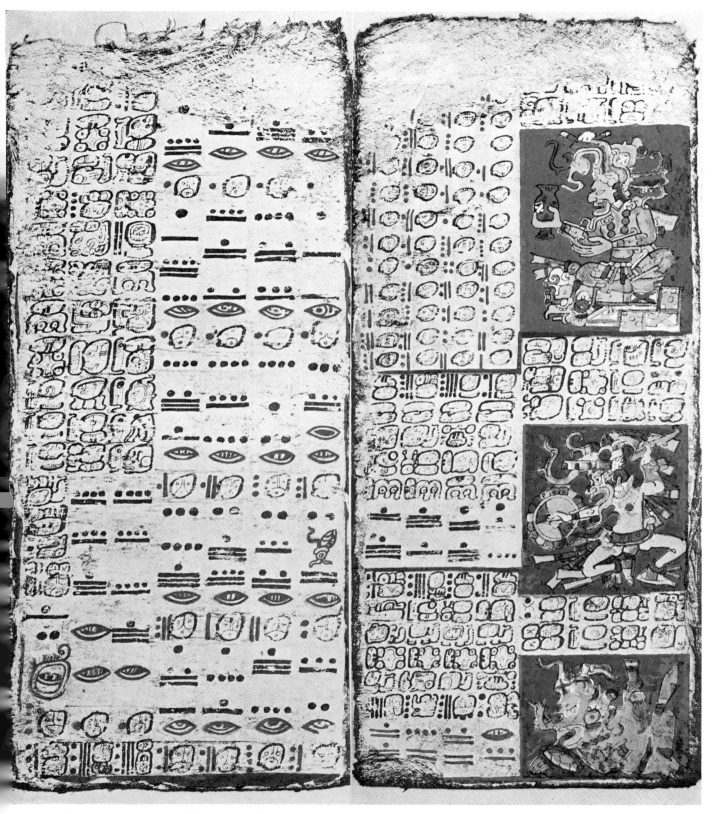

FIGURE 1. Two pages of Venus data, Dresden codex. Left page: top left, gods and prognostications for heliacal risings; right, multiples of synodical revolutions with corrections to them in space below first line of glyphs. A bar represents 5, a dot, 1, a flat oval, 0. Place numeration top to bottom: 400 tuns (each of 360 days), 20 tuns, single tuns, 20-day months, and days. Bars and dots serve as multipliers of suppressed periods. Accordingly, 8.2.0 in bottom right corner represents $8 \times 360 + 2 \times 20 + 0$ days $= 2920$ days, 5 Venus revolutions of 584 days. Columns to left of this are multiples thereof, 10 Venus revolutions, etc. Right page: phases of synodical revolution. Bottom of page records intervals between stations. Left to right: 11.16 (236 days), to disappearance, 4.10 (90 days), to reappearance after superior conjunction, 12.10 (250 days) to disappearance before inferior conjunction, and 8 days to heliacal rising. Total 584 days. On right, patron deity, Venus deity who has hurled spear earthward, and, at bottom, victim with spear driven through him. Also augural glyphs for heliacal rising.

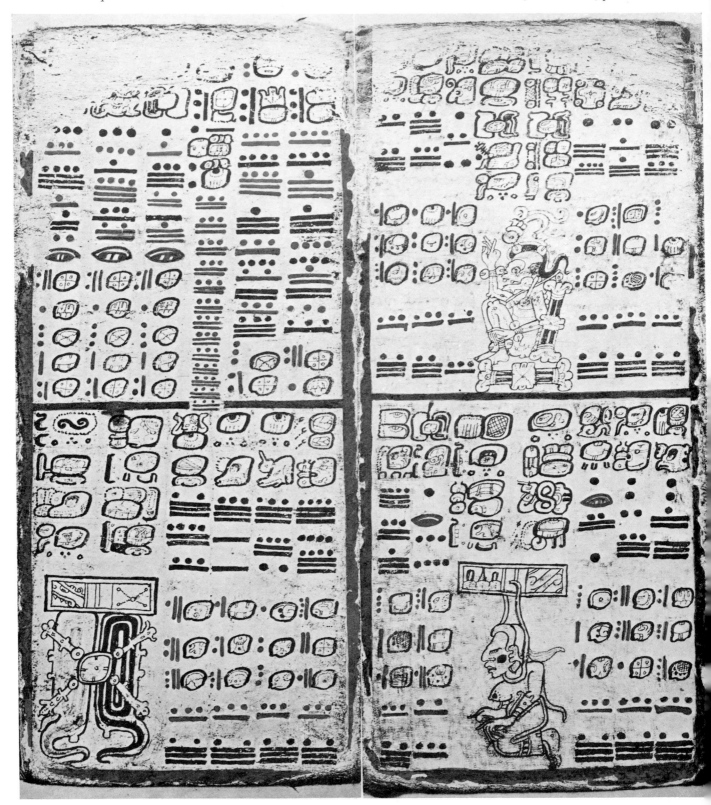

FIGURE 2. Two pages of eclipse tables, Dresden codex. Left page: top left, multiples of 1.13.4.0 (405 lunations). Right page: top, upper sets of bars and dots are accumulated totals of groups of 6 and 5 lunations: 177, 354, 502, picture, 679 (mistakenly written 674), 856, 1033 days (35 lunations). Lower set of bars and dots: 177, 177, 148, picture, 177, 177, 177 (6 and 5 Moon groups). Picture with prognostications only after 5-lunation group. Bottom halves of pages: continuation of lunation count as above six lunations (117 days) and five lunations (148 days). The latter only immediately before pictures.

of the 57th revolution of the planet, which terminated on the day 9 Muluc, 8 days after 1 Ahau, thus again recovering the vital position 1 Ahau by producing a total divisible by 260. In the line of corrections in the centre of the introductory page to the table appears the number 33 280, which is exactly $584 \times 57 - 8$ days, and beside it the number 68 900, which is $584 \times 61 - 4$ days and $584 \times 57 - 8$ days.

From the correctional entries it can be deduced that the second correction was made after the first had been applied four times in the following arrangement:

$$4(584 \times 61 - 4 \text{ days}) + 584 \times 57 - 8 \text{ days} = 175\,760 \quad \text{days}$$
$$260 \times 676 = 175\,760 \quad \text{days}$$
$$583.92 \times 301 = 175\,759.92 \text{ days}$$

After 301 synodical revolutions of Venus have passed and the Maya corrections totalling 24 days have been made, the Maya calculation was still in error, but that error was only 0.08 day in a span of over 481 years. Bearing in mind the variability of the planet's synodical revolution and the hindrances to accurate observation caused by cloudy weather in the rainy season and morning mists in the dry season, the accuracy attained is almost unbelievable. It was based on boundless patience and undoubted cooperation of astronomers of different places and different generations. Students may differ as to the exact times when those corrections were to be made, but there can be no doubt as to how the system operated.

These pages had been identified as a table of Venus revolutions in the early days of Maya research; the credit for recognizing the all-important system of corrections goes to Teeple (1930), a chemical engineer, who took up the study of Maya astronomy to wile away long train journeys across the U.S.A.

From Mexican sources it is known that Venus was much feared at heliacal rising after inferior conjunction; its rays then slew various categories of persons or personified manifestations of nature. Illustrations in the Maya table (figure 1, plate 19) show Venus gods hurling spears earthward, and, below the slain victims. The accompanying glyphs, with very few exceptions are direful: 'Woe to the maize, woe to the corn fields, drought, misery, affliction of war' and so on (Thompson 1972, p. 70). By predicting the day of heliacal rising, the priests were able to warn the threatened group, so that it could take protective measures. For instance, we know from Mexican sources that 'when it [Venus] emerged much fear came over them; all were frightened. Everywhere the outlets and openings [of houses] were closed up. It was said that perchance [the light] might bring a cause of sickness, something evil when it came to emerge' (Sahagún 1950–70, book 7, ch. 3). The spear represents the 'death ray'.

This last brings out a most important point: we must try not to look at Maya astronomy through European eyes. Maya emphasis was almost wholly on heliacal rising after inferior conjunction because of the astrological importance of that position as listed above. The Maya astronomer – astrologer, if you will – completely ignored phenomena which seem important to us – greatest elongation, retrograde motion, greatest brilliance, and conjunction with the Sun – presumably because those phenomena had no affect on mundane affairs. Modern astronomers who have delved into the mathematics of Maya astronomy assume almost without exception that the Maya would have been extremely interested in planetary conjunction with the Sun. Yet, the evidence in this Venus table – and the Maya were certainly more interested in Venus than in any other planet – is that that phenomenon was completely ignored. Indeed, since solar conjunction of any planet is unobservable, there seems no valid reason why we should credit the Maya with such an interest still less in the case of other planets of far less importance to them. We know neither names nor glyphs of any planet other than Venus.

The famous astrologer Dr John Dee used an Aztec obsidian mirror to see into the future. We may look down our noses at his ideas, but one may be sure that in outlook he was far closer to a Maya priest astronomer than is an astronomer of our century.

ECLIPSES AND LUNAR CALCULATIONS

The eclipse table, occupying eight pages, follows immediately the Venus table in Dresden codex, and, indeed, is of the same pattern. It, too, has an introduction with dates leading back to the epoch, and multiples of 11 960 days, the length of the table, a number which is also a multiple of the 260-day sacred almanac. It also has corrections, again like those of the Venus introduction in that they are applied before the whole cycle of 11960 days has run its course, and, again, are made so that 12 Lamat, the day of the lunar table is recovered. These parallels are, in my opinion highly important, for they establish a pattern, to which other planetary tables, if they exist, should be expected to conform. The lunar equation is:

$$405 \text{ moons} = 11\,960 \text{ days}$$
$$260 \times 46 = 11\,960 \text{ days}$$
$$405 \text{ astronomical moons} = 11\,959.89 \text{ days}$$

The table consists principally of totals of 177 days, very occasionally of 178 days, representing six-moon groups. There are also nine five-moon groups, amounting to 148 days, each being followed by a picture and additional glyphs (figure 2, plate 20).

As early as 1910 the astronomer Robert Willson (1924), observing the number 6585 among the figures, concluded that the pages dealt with the saros or at least a series of eclipses. He also noted that the intervals between the pictures were 1742, 1033 and 1211, thrice repeated. He was, naturally, aware from Schram's tables that if there is a central solar eclipse somewhere on Earth on a given date, there will be the same phenomenon after 1033, 1211, 1388 and 1565 days and perhaps after 1742 days. Clearly, then, one function of the table was solar eclipse prediction. Another function, as we know, was long-distance lunar calculation. However, the Maya, not knowing anything of the nature of the world or the Copernican system, could not predict whether any particular eclipse would be visible in the Maya area. It is hard to see how, with their lack of knowledge of the mechanics of eclipses, they could have gone beyond that; they could not know that when an eclipse failed to materialize for them on one of the dates in their table, it had, in fact been visible in Tibet, Timbuctoo or Australia.

Astronomers, who have not been able to steep themselves in Maya attitudes and Maya colonial literature, have supposed that this is a table of observed eclipses and have tried to match the numbers accompanied by pictures with solar eclipses visible to the Maya. However, the multiples of the period (one would carry the original date of the table over 1200 years into the future) and the various introductory dates given at the start of these pages as well as the Maya obsession with divination, particularly in terms of the 260-day almanac, make it clear that this was a table for predictions to be used time after time. Solar eclipses were occasions of dire peril. It was believed the world might end during a solar eclipse – a belief still held by the present-day Maya, and, at the very least, dreaded creatures then descended to earth to cause great havoc. The Maya sought to predict eclipses so that the threatened disaster could be averted by prior ceremonies. Hence the association of the lunar table, like other handling of phenomena, with their sort of 'Old Moore's almanac' of 260-days. The use of the table as a prediction apparatus is supported by accompanying glyphs of death, crop failure and of sky-supporting deities apparently thought to descend to Earth during eclipses; all are prophetic.

John Teeple, whose contributions to understanding of the Venus tables have been mentioned, not only discovered how lunar data were recorded on the monuments, but with equal genius reconstructed the method the Maya almost certainly used in constructing this table and its predecessors, finding how they solved the problem of where to insert each 5-lunation group (Teeple 1930). Because of the retrogression of the node day, the table could not be used indefinitely, so it can be inferred that the present table had various predecessors, in which the positions of the 5-lunation groups would have been different. He found that choice of locations for those 5-lunation groups involved the use of a double 260-day almanac. He discovered that if observed eclipses were plotted on a circle or strip with 520 teeth or gradations corresponding to the day names and numbers of a double sacred almanac, they would be found to cluster in three segments of the circle or sections of the strip, each segment comprising up to 34 days. The clustering and the size of the segment are so because a solar eclipse must fall within approximately 18 days either side of the node. As the paths of Sun and Moon cross every 173.31 days, three of those eclipses half years equal 519.93 days, which by a remarkable coincidence are less than a tenth of a day short of 520 days, the doubled 260-day Maya almanac. Observed eclipses, when plotted on that wheel, would be seen to occur within 18 or 19 days either side of the main spokes or radii 173, 173 and 174 days apart.

As the eclipse interval is 177.18 days, each eclipse advances nearly 4 days, or nearly 12 days at each return to one of the three segments. Observation would show that if, by the addition of the normal moons, 177 days, a position was reached beyond the limits of the segment, it would be necessary to use instead a five-moon grouping of 148 days to keep within the segment which alone would insure that an eclipse might be observable. Thus, with no knowledge of node crossing, the Maya constructed this very accurate table of solar eclipse predictions, but, as remarked, without being able to forecast whether any given eclipse would be visible to them.

This same formula of 405 moons equal 46 sacred 260-day almanacs was used for long-distance lunar calculations at least as early as the seventh century A.D. and probably earlier still. At the important site of Palenque a date then 3812 years in the past and two others, a few days apart, 3051 years in the past, were recorded on stone wall panels, each accompanied by a statement of the age of the Moon and the number of the lunation in a group of six. The above 405-moon formula connects then current lunar observations to those calculated Moon ages in the far past to within a day. However, the formula not being quite accurate, an error of 13 days had accumulated. The earliest date informs us that the Moon was then 5 days old and the Moon was the second in the group of six. The real Moon age would have been 18 days.

There is good evidence that the Maya realized that their formula was too long, and at about the time those Palenque dates were carved, they began to apply a correction of the same kind as they made in the Venus count. That is, they stopped at a point in the 405-moon formula where they could make a correction (of 1 day, not 4 or 8 days as in the Venus formulae) yet still retain the essential link with the 260-day almanac by recovering the day 12 Lamat.

The point for correction was at the end of the 361st moon:

$$361 \text{ Maya moons} - 1 \text{ day} = 10\,660 \quad \text{days}$$
$$260 \times 41 \qquad\qquad = 10\,660 \quad \text{days}$$
$$361 \text{ astronomical moons} = 10\,660.54 \text{ days}$$

The correction was slightly over a half day short of the mark, whereas the 405 moons = 11 960 days was 0.11 day too long.

One backward reckoning at Copan of just over 1000 years comprises 19 405-moon groups and 13 361-moon groups. The long distance from the 12 Lamat nearest the epoch of the calendar to the 12 Lamat which marked the start – A.D. 755 – of (the long-disused) lunar table comprises 111 405-moon groups plus eight 361-moon groups. Among the multiples of 405 moons on the introductory page of the lunar table appears the number 371 020 which resolves itself into twenty-three 405-moon groups and nine 361-moon groups. This number added to 101 405-moon groups and reduced to days (1 578 980) is the distance of over 4300 years from the 12 Lamat nearest the start of the calendar to the base in Maya notation 10.19.6.1.8 12 Lamat 6 Cumku (A.D. 1210), which is written in the introduction to the eclipse tables, and which was the current base without much doubt when the present edition of this hieroglyphic book was made. At that late date 12 Lamat was still *the* Moon day; the essential link with the 260-day almanac was unbroken.

There are other computations which involve a mixture of 405-moon groups and 361-moon groups (Thompson 1972, p. 74). Most probably the Maya never learnt which ratio of one group to the other was most accurate. As the 405-moon cycle accumulates an error of one day only after three centuries, the Maya failure to achieve a perfect solution is understandable. The ideal ratio would have been 5:1:

$$405 \text{ Maya moons} \times 5 \qquad = 260 \times 46 \times 5 = 59\,800 \quad \text{days}$$
$$361 \text{ Maya moons} - 1 \text{ day} = 260 \times 41 \qquad = 10\,660 \quad \text{days}$$
$$\text{total} \quad 2386 \text{ Maya moons} - 1 \text{ day} = 260 \times 271 \qquad = 70\,460 \quad \text{days}$$
$$2386 \text{ astronomical moons} \qquad\qquad = 70\,459.98 \text{ days}$$

One such interval of 2386 moons is recorded, but as there is no specific indication that this is a lunar count, it could be a coincidence, for which one must always be prepared in Maya arithmetic. If it was intended to mark that 5:1 ratio, it shows an accuracy in measuring the length of a lunation of a brilliance never approached by any other people on the same level of civilization.

OBSERVATIONS OF OTHER PLANETS

The Maya surely were interested in the synodical revolutions of other planets, and numbers occur which are multiples or near multiples of such cycles, but there are hundreds of intervals recorded on the stone monuments and in the books. In cases of long intervals, perhaps running into hundreds of thousands of days, anyone can get plenty of planetary data if one allows oneself sufficient latitude in deciding what length the Maya accepted for the synodical revolution of a planet. That of Jupiter is 398.867 days. If one postulates the Maya calculated it as 399 days or 398.7 or 398.85, one may get 'striking results', but even the very close approximation of 398.85 days accumulates a huge error in the 3500 years between the calendrical epoch and the start of the Classic period. The astronomer Ludendorff (1937, p. 19) used 398.883, 398.8842 and 398.8844 for Jupiter's synodical revolution in discussing three related intervals of around 34 000 years. Not unexpectedly, with such leeway he could show that the intervals were divisible by revolutions of the planet without remainder. One of the intervals, incidentally, was wrongly written by the Maya without any doubt, yet produced equally notable phenomena. Yet another warning against 'playing with numbers'.

As noted, planetary data are unlikely to have been recorded on stone monuments except possibly for astrological reasons in connexion with choosing accession dates for rulers; the Venus and lunar tables in Dresden codex have led to the search for other planetary tables in that book.

Claims have been made (Escalona Ramos 1940; Makemson 1943; Smiley 1961, 1967; Spinden 1942; Willson 1924) that pages of that book treat of synodical revolutions of Jupiter, Mars, Saturn and Mercury, of equinoxes and possibly periodicity of hurricanes, and even of sidereal revolutions. No two investigators agree as to which pages cover which planet.

In the absence of introductory multiples of the synodical revolutions and necessary corrections, as on the Venus and lunar pages, of any convincing planetary phases, and of any evidence of relating synodical revolutions to the 260-day almanac, the cases for such tables of Jupiter, Saturn and Mercury are extremely weak.

The associated almanacs show no relationship to the planets either in length or subject-matter. For instance, it has been claimed that dates on pp. 61 and 62 deal with revolutions of Jupiter. They introduce a 7×260-day almanac (1820 days), far from a multiple of the Jupiter revolution. The subject-matter is equally at variance. It comprises thirteen pictures of the rain gods and accompanying glyphs of the deities' activities and food offerings made to them. There are a large number of dates which would have to be planetary stations if the almanac is astronomical, varying from 1 to 13 days apart, but such intervals as stations are quite meaningless. Clearly, there is no connexion with Jupiter or any other planet.

A possible exception to what has been written above is a table of 780 days (3×260), one of several in Dresden codex which is a multiple of 260 days. The number is a very close approximation to the synodical revolution of Mars (779.936 days), and this is preceded by a table of multiples of 78 and 780 days, but with no correctional numbers. Willson (1924) believed this covered revolutions of Mars. The arrangement of the triple almanac is like the general run of divinatory almanacs; it does not resemble the Venus record, and a number of the associated glyphs cover agriculture. It consists of ten sections of 78 days, each in turn made up of intervals of 19, 19, 19 and 21 days. If this is a Mars revolution, it has no less than forty stations compared with Venus's four stations at obvious points – appearance and reappearance of the planet. One cannot conceive of a planet having forty stations at those repeating, short intervals. Other students have rejected the Mars interpretation and suggested the table deals with other planets. It is surely far more likely to be mere chance that this triple almanac so nearly equals in length a revolution of Mars. The 78-day subdivisions, prominent in both the almanac and the table of multiples, have no obvious connexion with Mars or with any other planet for that matter.

I am confident that the only ephemeris in Dresden codex is that of Venus. The failure of students to agree on which pages are to be assigned to other planets strongly supports that view. The question of sidereal revolutions of planets is discussed below.

The Maya zodiac

The Maya had a sort of zodiac, the best example being in the hieroglyphic book called Codex Paris (Spinden 1916, 1924, pp. 55–56, 1941). It comprises thirteen animal signs. Rattlesnake, turtle, scorpion, bat, two birds and frog (?) are identifiable; others are fantastic creatures or are obliterated. Each, suspended upright from the sky, has a sun glyph in its jaws or beak. Intervals of 28 days between the 13 associated day signs form a 364-day year, which is repeated five times to achieve the vital relationship with the 260-day sacred almanac: $364 \times 5 = 260 \times 7$. Unlike most glyphic texts, the sequence reads right to left. It has been suggested that that was because the star groups 'feed into the path of the sun'. How this 364-day count could have been related to a sidereal year is not clear. The concept may have been brought to the New World by hunter-gatherers at a very early date (p. 97).

Stars and constellations

Information on Maya ideas concerning stars and constellations is scarce and few Maya names for the latter have survived. That is surely because no European showed any interest in Maya astronomy before Maya culture collapsed; our knowledge of the wonderful achievements in the matters of eclipses and Venus movements come entirely from pre-Spanish sources. Glyphs of stars or constellations may be drawn in the Venus table, and conjunctions of the Moon with stars may be recorded in the same Dresden codex. That stars affected human life is made clear by two passages in a book of the colonial period, written in the Maya language but using the European alphabet, which recounts much Maya history and mythology: 'In due measure they sought the lucky days until they saw the good stars enter into their reign; then they kept watch while the reign of the good stars began. Then everything was good' and, in contrast, 'ill-omened is the star adorning the night. Frightful is its house' (R. L. Roys 1933, pp. 83, 91).

The Aztec are said to have held their great ceremony of rekindling fire at the end of the 52-year cycle when the Pleiades reached the zenith at midnight, but as the ceremony receded 13 days each time it recurred because of the absence of bissextile years, the ceremony could not have been related to the movement of the Pleaiades indefinitely. In fact, the last time the ceremony was held, in 1507, it would have occurred in or near February, so it is probable that on that occasion it happened to coincide with an overhead position of the Pleiades at midnight. The point is that this is evidence that the peoples of Middle America did pay attention to positions of constellations.

The not too reliable Tezozomoc (1878, ch. 82), writing at the age of 78 – he was born just before Cortés conquered Mexico – tells of how his grandfather Moctezuma was admonished at his induction specially to rise at midnight to observe the firesticks constellation, as they call the Keys of St Peter [perhaps in Gemini], the Ball Court constellation, the Pleiades and the Scorpion [Great Bear] which mark the four celestial cardinal points. Toward dawn he must also observe carefully Xonecuilli, 'the Cross of St. James, which appears in the southern sky in the direction of China and India'. According to Sahagun's Spanish text (1938, book 7, ch. 8), a far more reliable source, this last was the Little Bear. This confused account makes clear the importance of stellar observation among the Aztec and, we may be sure, the same was true of the Maya. It also reveals confusion about identifications, a matter to which I shall return.

The only reference I know, early or modern, to constellations as time markers among the

Maya is that the present-day Lacandon say that the corn fields should be burned preparatory to sowing when the Pleiades have reached tree-top level at dawn. One may conjecture that such practices were once widespread, but modified by choice of a nearby favourable day; the Lacandon have lost the 260-day almanac.

I had intended to list in an appendix names in various Maya linguistic groups for individual stars and constellations. Unfortunately, utter confusion – uncertainties of identification and the same name given by different informants to distinct constellations – makes that impossible.

One reason for the confusion is that since lines joining stars to form constellations are imaginary, there is no reason to suppose that the peoples of Middle America saw the same figures in the night sky (unless they were introduced from Asia and remained unchanged) as we do. Secondly, communication between informant and interrogator is often bad. The latter, conditioned to our names, might point to Taurus, for instance, and ask the Maya name for the constellation, but for the informant, parts of Taurus and Orion may form a single constellation, with resulting misunderstandings. Thirdly, the interrogator may point to a constellation, but the informant mistakes the direction of his interrogator's finger or misunderstands his description in a language which is not shared by both. A drawing, probably seldom employed, may not be of much help, for the Maya peasant is not used to that form of representation. Finally, the old priest-astronomer group, who could have supplied the information, ceased to exist as such a generation after the Spanish conquest – sons of the aristocracy were removed to centres in which they were given a European education.

I would suppose that imagining star groups as pictures is a very ancient introduction from the Old World to the New, probably on the hunter–gatherer horizon, but the old names have been lost. Of well-known constellation names only Scorpio is found also in Middle America, but it is highly doubtful that the same constellation is involved.

Native terms (*xok* and *tzec*), meaning scorpion, are given by the Kekchi Maya and the Chaneabal Maya to the Great Bear, although the latter is queried by the interrogator, perhaps because he was surprised by the reply. Moreover, the great sixteenth-century ethnologist, Bernardino de Sahagún (1938, book 7, ch. 4), wrote that the Aztec called the Great Bear (*el carro*) scorpion. In his Nahuatl writings he illustrates the scorpion (*colotl*) constellation with a perfect drawing of the Great Bear.

The earliest (sixteenth century) Yucatec–Maya dictionary has the entry: '*zinaan*, scorpion (alacrán o escorpión) and also it is escorpio, celestial sign'. *Signo*, sign, as in English, is used in reference to the zodiac. It is possible that the author of the dictionary misunderstood the location of the Maya constellation Scorpion, and assumed it coincided with the Old World Scorpion.

On the other hand, in the Maya 'zodiac' the scorpion is in the fourth position. The first sign is obliterated; the second is *tzab*, rattle of the rattlesnake in Yucatec, Lacandon and Manche Chol Maya (but sandal or heap in other Maya languages). The creature's body, in the fashion of Maya snakes, probably loops down through Aldebaran and the left arm and lion skin pendent from it of Orion, and then up to Bellatrix, then a right-angle turn with Betelgeuze the eye or tip of snout. The third sign is turtle, which the Yucatec Maya placed in Gemini according to more than one source, but, again, one cannot be sure that the two constellations coincided.

With roughly equal spacing – the thirteen signs should be 28° apart (p. 92), the fourth sign should lie between Cancer and Leo; our Scorpio is much too far west. One could hazard a guess that it is in fact Leo. The Maya depicted the scorpion head down with outspread and thrust-forward claws. Regulus would be the tip of the left claw; northern stars the raised tail. Naturally, this is mere speculation.

I have discussed the elusive Middle American constellation of the scorpion in detail to warn of pitfalls awaiting those without knowledge of the native cultures, and to urge caution in identifying Maya star groups and, far worse, theorizing from the positions those shakily identified stars held far in the past. That way one quickly finishes up in a nebula.

ALINEMENTS AND MEASURES

Orientation of buildings presents difficulties; the Maya seldom alined walls correctly and seem to have been incapable of making a true right angle. For instance, the walls of the Castillo, most prominent temple at Chichen Itza, read 18°, 20°, 22°, and 24° (Rivard 1970). The nearby Warriors complex has 17° 35′, 16° 20′ and 15° 15′ for the main structure, the northwest and north colonnades respectively; the underlying Chac Mool temple is 13° 30′ (Morris, Charlot & Morris 1931). One wonders whether the 4° difference between the under and upper and later temple was deliberate.

At Tikal, greatest Maya site, orientations of front walls of the five great pyramid temples (rear-wall readings differ by up to 1° 37′) read 7° 1′, 9° 3′, 9° 51′, 10° 46′ and 18° 16′ (Tozzer 1911, p. 115). One suspects that in the first four cases the Maya sought uniform orientation, and so variation arises from sloppiness, for such variability occurs all over the Maya area and is a warning against crediting the Maya with intention and precision for every significant orientation noted. Divergence of one wall from another presumably resulted from the inability of the Maya to lay out true right angles. Satterthwaite (1935, p. 1, 1944, p. 21), who excavated the site for several seasons, noted that his plans and sections 'are based on the assumption that intended right angles really are such and that intended straight lines are straight lines. Nowhere at Piedras Negras does such an assumption agree with the facts. If there are true right angles in the buildings of the city (we have found one or two) they are probably the result of chance'. He has noted, for example, a difference of 6° between the long and short axes of one ball court at the site.

The reluctance of archaeologists to give the orientation of Maya buildings probably stems from the affect of those factors on complete accuracy.

It has been thought that a round tower at Chichen Itza served as an observatory. Three longish window shafts survive. It would seem most logical to have sighted down the middle of each shaft with the aid of sticks or cords, but such lines produced nothing of obvious import. Diagonal sights from one inner to the opposite outer jamb were more promising. The readings are: N 60° 15′ W, and due west (window 1); S 61° 15′ W and S 48° 0′ W (window 2); S 18° 0′ W and S 2° 45′ W (window 3). Only the due west line seems significant. That shaft is considerably wider (68 cm) compared with 21 cm for the other two (Ruppert 1935, pp. 189, 233–37). Unless the other lines of sight have values not at present obvious, one is reluctant to accept the observatory explanation.

Carnegie Institution of Washington carefully checked an apparent line of sight joining two stelae set on ridges, 6.5 km apart, at the ruins of Copan. It has been claimed that sunset behind one stela on 12 April (81° 09′) informed the Maya when to burn the felled timber on their cornfields (Morley 1926, pp. 277–282). In fact, the Maya know when to burn: just before the rains, heralded by increasingly humid heat. Were any special date chosen for that activity – a day of changing winds is required – it would have been a lucky day in the 260-day almanac (the modern Lacandon who, as noted, observe the Pleiades no longer have a calendar), and present-day 260-day almanacs have days regarded as favourable for maize.

The present-day Ixil Maya of the Guatemalan highlands have a double line of sight comprising stone markers from the town cemetery of Nebaj west [sic] to an indentation at the top of a high hill. The account by Lincoln quoted by Long (1948) is very brief and confused. Sunrise was observed 'on March 19, 1940, two days before the equinox. Sun rose this day at 6° 31.5'. Direction observed with simple adjustable compass. Observations of the Sun are made at the stone today by zahorins [shamans] for planting and harvesting'. Long was not sure that he had correctly deciphered the figure given for the line of sight or the word adjustable. Lincoln was an ethnologist and probably had a simple hand compass, quite probably with its own error, but surely not readable to the third decimal point of the minutes. Presumably one must add that the reading was south of east. The reading was presumably magnetic, a matter of around 7°. *Today* in the context seems to mean *nowadays*. Certainly, the spring equinox marks neither planting nor sowing, and is 20 or more days before the time for burning off the corn fields. One may suppose that this is a line of sight on the sun at the spring equinox, but how it bore on agricultural activities remains in doubt.

Groups comprising a temple on the west side of a court, facing a line of three temples on a single platform on the opposite side, are fairly common in Maya ruins (Ruppert 1940). A bearing from the centre of the west structure to the centre of the middle east structure varies from under 1° S to 10° 15'. The group at the site of Uaxactun shows most promise. A bearing from a stela at the base of the stairway leading to the west temple to the middle of the doorway of the centre temple of the three on the east platform is S 89° 03' E (true east bearing touches the face of the north jamb of the doorway). To the centres of the doorways of the north and south temples of the east platform, bearings are respectively N 68° 00' E and S 65° 18' E. The latter is a good approximation to the winter solsticial line; the former 2° 40' S of the summer solsticial line (Ricketson & Ricketson 1937, pp. 105–108). It must be noted that the lines of sight are quite short – just over 60 m for the solsticial lines, so a small change in the observer's position – and there is no evidence that it was from in front of the stela – would produce widely different bearings, although the true east bearing would remain unchanged. For the Maya, Sun overhead may have been more important than solstices, of far greater interest to dwellers in northern climes.

There is, however, documentary evidence that the lay-out of buildings in some parts, at least, of Middle America, was related to the equinoxes. In a very early source (Motilinía 1971, pt. 1, ch. 16, para. 89) we read: 'This feast [the 20-day month Tlacaxipeualistli] used to fall when the sun was in the middle of the [temple of] Huitzilopochtli, which was the equinox, and because it was a little twisted, Montezuma wished to tear it down and straighten it.' Names have been modernized. In similar passages *the* Huitzilopochtli makes sense only if read as temple or pyramid. Huitzilopochtli was the tribal god of the Aztec. His temple shared a pyramid with that (on the north) dedicated to the rain god, Tlaloc. They faced west. Huitzilopochtli was intimately connected with the Sun and in many respects may be regarded as a solar deity. Whether the Sun passed over the centre of the roof of the temple of Huitzilopochtli, or whether it was visible in the narrow passage between the temples of the two gods, which is more probable, is not of major importance. The point we must bear in mind is that Montezuma was prepared to tear down the temple to get a correct equinoctial alinement. As there were no tall buildings on the west side of the great court, no view of the Sun at rising would have been possible. Nor would such a view have been possible at Uaxactun.

No standard measurement in Maya architecture has yet been recognized.

The Maya did not use a bissextile system, but they were obviously aware of the inadequacy of their 365-day year. There are a number of dates on stelae which can be interpreted as correcting the loss accumulated since the epoch, over 3500 years in the past, at the height of the Classic period. If these are corrections, they rival our Gregorian calendar in accuracy. Some have now proved to record civil events, such as accessions of rulers, but since astrology plays a world-wide part in choice of dates for such occasions, their function of bissextile corrections is not necessarily negatived. I once accepted them as such (Thompson 1950, pp. 317–20), but now, like most students in the field, I am sceptical. No glyph indicative of correction has been isolated and, with so many dates recorded, one must beware of coincidence.

Coincidence is, indeed, a very serious problem in Maya astronomy because of the huge quantities of numbers to 'play with'. Many years ago the German astronomer Hans Ludendorff published astronomical phenomena associated with Maya dates and using what was in all probability a wrong correlation of calendars. Some of the dates he used had, unfortunately, been wrongly read. I was able to demonstrate that the incorrectly read dates produced a higher percentage of astronomical phenomena than those correctly read (Thompson 1935, pp. 83–87).

More recently, another astronomer (Smiley 1968), misled by a poor grasp of the mechanics of the Maya calendar, read two non-existent dates (Satterthwaite 1964, pp. 51–53) which, in his correlation, fall respectively 2–3 days before conjunction of Jupiter and Saturn with the Sun and 1 day after a conjunction of Saturn with the Sun. He cites these as evidence of the outstanding mastery of astronomy and mathematics which enabled the Maya to predict invisible conjunctions, and as support for his correlation. These incorrectly read dates would have fallen at the start of the Maya Classic period, only eighty years after the earliest known Maya text and some four centuries before the peak of Maya astronomy. Why the Maya should have wished to record invisible conjunctions, granting they had the ability to calculate them, is hard to say. Certainly, the Venus tables, to the Maya a far more important planet, pay no attention to either inferior or superior conjunction; they display interest only in heliacal risings and disappearances before conjunction.

Smiley (1961, p. 241), using the same method of reading Maya dates unacceptable for the past 50 years to all students of the Maya calendar, produces five consequently wrongly read dates which in his correlation fall near – they range from 2 days before to 22 days after – conjunctions of Jupiter with the Sun. He concludes that the chance of coincidence is less than one in ten million. Since the dates were never recorded, we are made doubly aware of the dangers of coincidence. Incidentally, four of the dates are very early; one is very late, but that is the one which is farthest off conjunction, so, had the dates existed, one would have to conclude that, after some seven centuries, the Maya ability to calculate this invisible configuration had become considerably less.

SIDEREAL REVOLUTIONS OF PLANETS

Several persons interested in the subject have assumed that the Maya were able to measure sidereal revolutions of the planets, citing lengthy Maya intervals which they claim to be multiples of these. Such intervals, amounting often to some 4000 years can produce all sorts of exciting 'results' when one allows oneself a variation of the second or third decimal point in the length of the sidereal revolution of a planet (p. 90).

Lawrence Roys (1935, p. 92) in a planet-by-planet discussion of the problem, wrote: 'It seems very improbable that they [the Maya] knew those sidereal periods, but a general denial is hardly in order where there is no direct Maya evidence on the subject. However, the difficulties of obtaining the astronomical records necessary for these determinations are so great, and the mathematical logic so advanced, that they appear too difficult for a people in the Maya stage of civilization, and the burden of showing a reasonably simple way of finding sidereal periods lies on anyone who suggests that they were known to the Maya.'

Roys goes on to cite as an example of difficulties subsequent observations of conjunctions of Jupiter with Aldebaran with variations from the true sidereal period of 35, 198 and 24 days. Even averaged out, they are 46 days short. He adds that conjunctions with other stars produce markedly different results. Retrograde motion and proximity of the Sun denying observation increase the difficulties. He concludes 'For the Maya to have discovered the true sidereal period seems very far from likely'.

One must not attribute to the Maya knowledge which is unattainable with the sole aid of the naked eye. In that connexion, it is worth bearing in mind that the Maya had no knowledge of algebra or even fractions (Maya 'hours' were based merely on the approximate position of the Sun), and, as we have seen, they were incapable of measuring a right angle.

One must beware of coincidence, an alarming feature when one 'plays with numbers'.

Surely, to comprehend the aims, attitudes and achievements of Maya astronomers who were first and last priests, an understanding of Maya mentality and outlook is essential. One must appreciate the impact of the divinatory aspects of the 260-day sacred almanac, and never lose sight of the fact that the ends of Maya astronomy were not scientific, but astrological. The Maya were interested in heliacal rising of Venus because then the world was in danger, and carefully recorded that happening; invisible conjunctions of planets with the Sun had no bearing on man's future, and therefore there was no call to try and calculate them. One must try and get in the skin of the Maya priest-astronomer. Also some knowledge of Maya culture and history is essential. The findings of archaeology cannot be ignored.

Had astronomers interested in Maya astronomy such a background, we would be spared assertions that the Maya calendar was in full swing before 3500 B.C., when in fact, agriculture in the New World had hardly got under way, and the Maya were still over 3000 years short of developing an identity. We would also be saved from consequent wild deductions that because the Pleiades were on the celestial equator at that date, the Maya were then probably living in or near Peru (Smiley 1960). Needless to say, there is not the slightest evidence that the Maya or anyone else in Middle America had then any sort of reliable time reckoning, or that the ancestors of the Maya were then living south of the equator. In the light of such deductions, which one can only designate as highly intemperate, and of other material discussed above, one is inclined to say that Maya astronomy is too important to be left to the astronomers.

In conclusion, I believe that Maya calendrical and astronomical achievements were made independently of the Old World, except that giving animal names to constellations in the Maya 'zodiac' and in other parts of the heavens, as well as of some days in Middle American calendars, may have been a custom surviving from very simple systems of counting of hunter-gatherers brought by immigrants to the New World by way of the Bering Strait, perhaps as early as 10000 B.C. They savour of a pre-agricultural horizon. Later immigrants, one may suppose, should be credited with the introduction to the New World of such concepts as dragon-like beings, each with its associated colour and world direction. This is surely too complex an aggregation to have developed independently in two areas.

REFERENCES (Thompson)

Escalona Ramos, A. 1940 *Cronología y astronomía maya-mexica*. Mexico.

Long, R. C. E. 1948 Observations of the sun among the Ixil of Guatemala. *Carnegie Instn Wash. Notes Mid. Am. Archaeol. Ethnol.* no. 87.

Ludendorff, H. 1937 Zur Deutung des Dresdener Maya-Kodex. *Sber. der preuss. Akad. Wiss. (Phys.-Math. Klasse)* **8**.

Makemson, M. W. 1943 The astronomical tables of the Maya. *Publs Carnegie Instn Wash.* **546**, Contrib. 42.

Makemson, M. W. 1957 The miscellaneous dates of the Dresden codex. *Publs Vassar College Observatory* no. **6**. Poughkeepsie.

Morley, S. G. 1920 The inscriptions at Copan. *Publs Carnegie Instn Wash.* no. 219.

Morley, S. G. 1926 Archaeology. *Yr. Bk. Carnegie Instn Wash.* **25**, 259–286.

Morris, E. H., Charlot, J. & Morris, A. A. 1931 The Temple of the Warriors at Chichen Itza, Yucatan. *Publs Carnegie Instn Wash.* no. 406.

Motilinía (Fray T. de Benavente) 1971 *Memoriales o Libro de las cosas de Neuva España y de los naturales de ella.* Edited and annoted by E. O'Gorman. Mexico: Universidad Nacional Autónoma de México.

Ricketson, O. G. & Ricketson, E. B. 1937 Uaxactun, Guatemala. Group E, 1926–1931. *Publs Carnegie Instn Wash.* no. 477.

Rivard, J. J. 1970 A hierophany at Chichen Itza. *Katunob* **7**, no. 3.

Roys, L. 1933 The Maya correlation problem today. *Am. Anthrop.* **35**, 403–417.

Roys, L. 1935 Maya planetary observations. In Thompson 1935 (Appendix II).

Roys, R. L. 1933 The book of Chilam Balam of Chumayel. *Publs Carnegie Instn Wash.* no. 438.

Ruppert, K. 1935 The caracol at Chichen Itza, Yucatan, Mexico. *Publs Carnegie Instn Wash.* no. 454.

Ruppert, K. 1940 A special assemblage of Maya structures. In *The Maya and their neighbors*, pp. 222–231. New York: Appleton-Century Co.

Sahagún, B. de 1938 *Historia general de las cosas de Nueva España*, 5 vols. Mexico: Pedro Robredo.

Sahagún, B. de 1950–70 *Florentine codex. General history of the things of New Spain.* Trans. and ed. A. J. O. Anderson & C. E. Dibble. 11 vols. School Am. Res. Monograph 14. Santa Fe and Salt Lake City.

Satterthwaite, L. S. 1935 *Palace Structures J-2 and J-6 with notes on Structure J-6-2nd and other buried structures in Court 1.* Piedras Negras Prelim. Papers 3. Philadelphia: University Museum.

Satterthwaite, L. S. 1964 Long count positions of Maya dates in the Dresden codex with notes on lunar positions and the correlation problem. *Proc. 35th Int. Congr. Am., Mexico, 1962*, **2**, 47–67. Mexico.

Satterthwaite, L. S. 1965 Calendrics of the Maya lowlands. *Handbk Mid. Am. Indians* **3**, 603–631. Austin: University of Texas.

Satterthwaite, L. S. & Ralph, E. K. 1960 New radiocarbon dates and the Maya correlation problem. *Am. Antiq.* **26**, 165–184.

Smiley, C. H. 1960 The antiquity and precision of Mayan astronomy. *R. Astron. Soc. J.* **54**, 222–226.

Smiley, C. H. 1961 Bases astronómicos para una nueva correlación entre los calendarios maya y cristiano. *Estudios de Cultura Maya* **1**, 237–242.

Smiley, C. H. 1967 A possible periodicity of hurricanes. *Cycles*, pp. 283–285.

Smiley, C. H. 1968 Stationary conjunctions and near-conjunctions of Jupiter and Saturn. *R. Astron. Soc. J.* **62**, 181–184.

Spinden, H. J. 1916 The question of the zodiac in America. *Am. Anthrop.* **18**, 53–80.

Spinden, H. J. 1924 The reduction of Mayan dates. *Papers Peabody Mus., Harvard Univ.* **6**, no. 4. Cambridge, Mass.

Spinden, H. J. 1941 *The zodiacal calendar of the Maya.* Presented 40th Annual meeting Am. Anthrop. Ass., Andover, Mass. Mimeographed, 7 pp.

Spinden, H. J. 1942 Time scale for the New World. *Proc. 8th Am. Scient. Congr., Washington, 1940*, **2**, 39–44.

Teeple, J. E. 1930 Maya astronomy. *Publs Carnegie Instn Wash.* no. 403, Contrib. 2.

Tezozomoc, H. A. 1878 *Crónica mexicana.* Anotada por D. M. Orozco y Berra y precedida del códice Ramírez. Mexico: Biblioteca Mexicana.

Thompson, J. E. S. 1935 Maya chronology: the correlation question. *Publs Carnegie Instn Wash.* no. 456, Contrib. 14.

Thompson, J. E. S. 1950 Maya hieroglyphic writing: an introduction. *Publs Carnegie Instn Wash.* no. 589. (Also Norman: Univ. of Oklahoma Press, 1960.)

Thompson, J. E. S. 1972 *A commentary on the Dresden codex, a Maya hieroglyphic book.* Am. Philosoph. Soc. Mem. **93**. Philadelphia.

Tozzer, A. M. 1911 Prehistoric ruins of Tikal, Guatemala. *Mem. Peabody Mus. Harvard Univ.* **5**, no. 2. Cambridge, Mass.

Willson, R. W. 1924 Astronomical notes on the Maya codices. *Pap. Peabody Mus. Harvard Univ.* **6**, no. 3. Cambridge, Mass.

Phil. Trans. R. Soc. Lond. A. **276**, 99–116 (1974)   [99]

Printed in Great Britain

Two uses of ancient astronomy

By R. R. Newton

Applied Physics Laboratory, Johns Hopkins University,
Silver Spring, Maryland 20910, *U.S.A.*

Observations from ancient astronomy are useful in studying the non-gravitational accelerations of the Earth and Moon, and recent developments in this study are reviewed. Such a study necessarily involves astronomical chronology and simultaneously shows some limitations in its use. Limitations include lack of veracity in many records, errors in dating events, and uncertainty in calculating the circumstances of ancient eclipses of the Sun. These limitations are studied both quantitatively and by example.

1. Introduction and background

It has been recognized for some time that friction in the tides is gradually slowing down the Earth's rotation about its axis and that it is also gradually decreasing the angular velocity of the Moon in its orbit about the Earth. These effects are often described by saying that the lengths of the day and the month are both gradually increasing. The general way in which tidal friction brings about these effects is described in many places (see Munk & MacDonald 1960 or Jeffreys 1970, for examples).

So far as we know, tidal friction is the only phenomenon that tends to change the length of the month at a rate sufficient to concern us here. However, it is almost certain that other phenomena tend to change the length of the day at a significant rate. Possible phenomena include effects of the Earth's magnetic field and slow changes in the average radius of the Earth because of changes in the average temperature of the Earth's interior.

The terms 'acceleration of the Earth' and 'acceleration of the Moon' are commonly used to denote the rates at which the angular velocity of the Earth's rotation and the orbital velocity of the Moon are changing. Both accelerations have negative values, meaning that both angular velocities are decreasing.

Within the past century, the gravitational theory of the solar system has been brought to a state of high accuracy. This fact allows us to use the observations of ancient astronomy in estimating the accelerations, in the following way: Whenever we find an ancient observation, we start by calculating what the observation should have been on the basis of modern theory. We can then confidently ascribe any discrepancy between the calculation and the observation to the accelerations.

The symbol $\dot{n}_M$ will be used to denote the acceleration of the Moon (with respect to ephemeris time) and the symbol $\dot{\omega}_E$ will be used for the acceleration of the Earth. Values of both accelerations will be given in units of seconds of arc per century per century, written as "/century². This is a conventional usage for $\dot{n}_M$. It is not conventional for $\dot{\omega}_E$, but it will be used here in the interest of simplicity.

The main purpose of the work that I have done with ancient astronomy, as well as with medieval astronomy, has been to study the accelerations of the Earth and Moon. However, the work has necessarily brought me into contact with astronomical chronology, and some of the results of that work have significance for astronomical chronology. Thus, in this paper, I shall discuss two and only two uses of ancient astronomy. These uses are in the study of the accelerations and in the study of chronology.

2. The basic idea of astronomical chronology

In many cases, we know the structure of an ancient calendar and thus we know the relative chronology of events that were dated by means of that calendar. Our need is then to be able to relate events in that calendar to those dated in another calendar, and in particular to our calendar. In principle, we can do this if we can establish the date of a single event in both calendars. For the sake of reliability, of course, we should like to have more than one such event.

In using astronomical chronology, we look for an astronomical observation that can be identified uniquely and that is dated in terms of the ancient calendar. We then calculate the date of the observation, using astronomical theory, and thus establish the correspondence between the ancient calendar and ours.

The accuracy with which we can establish the correspondence depends upon the nature of the observation. The position of the Moon among the stars changes by about 13° per day. Thus, if the position of the Moon is given with an accuracy of about 1°, which is easy for an observation made with the naked eye, we can establish the correspondence down to the very day, provided that the necessary data have been preserved. At the other extreme, the same accuracy in the position of the vernal equinox allows us to establish only the century, since the equinox precesses at the rate of about 1.4° per century.

If we are to use an astronomical observation for chronological purposes, then, four conditions must be satisfied:

(1) The record of the observation must be basically truthful.

(2) The observation must have the accuracy that we need for the desired chronological accuracy. The accuracy of calculation from modern theory is usually not a limiting factor.

(3) The observation must be correctly dated in the ancient calendar.

(4) We must be able to identify the astronomical event uniquely. Identification is usually a problem only with eclipses, although one can imagine circumstances in which it is a problem for other kinds of observation.

Discussions of these four conditions will form the subject-matter of the next four sections.

3. The veracity of ancient astronomical records

Many passages in the literature that describe ancient astronomical observations are simply not true. There are several classes of untrue records that will be discussed separately.

(1) *Records that involve a modern, or at least a relatively recent, error.* There are numerous examples of this class, of which only two will be mentioned.

The so-called 'eclipse of Babylon' is an example of this class that has arisen within the twentieth century. This 'eclipse' refers to a cuneiform text that Fotheringham (1920) and others have studied. According to many later writers, Fotheringham concluded that the text describes an eclipse that was total at Babylon. What Fotheringham actually concluded (p. 124) was that 'the phenomenon recorded in the Babylonian chronicle was something other than an eclipse, or, if an eclipse, was total in southern Babylonia and not at Babylon itself . . .'. That is, the one thing that Fotheringham specifically excluded is the thing that he is credited with establishing.

The 'eclipse' that involves the minor Chinese deities Hsi and Ho involves another erroneous reading, in my opinion. Many writers of the past few centuries (see the summary on pp. 62–65, Newton 1970) have tried to identify this with eclipses ranging from −2154 October 12 to −1904 May 12. Aside from the fact that the passage is almost surely a forgery written around +300 or later, it seems to me that the passage does not deal with an eclipse at all. It seems to be concerned with the problem of keeping a lunar calendar in average adjustment with the solar year.

(2) *The 'literary' eclipse.* Mark Twain (Clemens 1889) and Haggard (1886) insert total eclipses of the sun into their action for literary purposes. These 'eclipses' may be taken as archetypes of the 'literary eclipse'. The 'eclipse' described by Anna Comnena (*ca.* 1120, chap. VII. 2) is the earliest example I have found in which a person uses the ability to predict an eclipse in order to confound an uneducated opponent. Plutarch (*ca.* 90) uses an eclipse description for literary purposes, but without using it to confound the ignorant, and he mentions many other 'literary eclipses', such as the 'eclipse of Archilochus', that were already classic in his time.

Probably no one would be more surprised than these writers (with the possible exception of Anna) to find that their literary eclipses were being used as valid astronomical observations by modern scholars, but this has often happened. Sometimes the 'eclipse' is first identified as to date and the result used to date the composition of the literary work. Sometimes the work is dated independently and the result used to 'identify' the eclipse. Analysis of this sort is based upon the tacit assumption that the writer wrote the passage immediately after seeing a total eclipse. This is equivalent to the assumption that the writer had neither imagination nor memory.

Here of course we come to a matter of opinion. I cannot prove that the writer's inspiration was not based upon recently seeing a total eclipse of the Sun. Other students of the subject are entitled to opposite opinions. However, I do not see how these opinions can be given the rank of precise astronomical data. In particular, I do not see how a single opinion of this sort can be given more weight than all the surviving observations made by professional astronomers in all of ancient and medieval times, yet this is what several modern scholars have done. For further discussion of this point, see § 8 below.

A special type of record that we may call 'literary' is a calculated result that forms part of the astronomical or astrological literature. There are many results that were calculated in preparing ephemerides (Neugebauer (1955), for example) or in preparing astrological predictions, for which there are an enormous number of examples. Occasionally we can use calculated results as representative of smoothed observations, if we can be sure when the underlying observations were made. Usually, though, we must avoid anciently calculated results for present purposes, although they may be highly valuable for other purposes.

(3) *The 'assimilated' eclipse.* The human memory is fallible and a large eclipse of the Sun is a dramatic event. It is likely that a person has a tendency to bring two important or dramatic events together in his memory after the passage of time. Thus a writer may unintentionally displace an eclipse to make it coincide with some other event, or vice versa. I use the term 'assimilated eclipse' to designate a record in which an eclipse is displaced in time in order to assimilate it to some other event. The account of the eclipse of 1133 August 2 in the *Anglo-Saxon Chronicle* (*ca.* 1154) is probably an assimilated eclipse. The eclipse is put in 1135, where it is used as a precursor of the death of Henry I, but all details except the year are kept correct. An assimilated eclipse may also be literary.

(4) *The hoax*. It is well known that eclipses, comets, and other astronomical happenings were widely regarded as portents. When we see an eclipse, for example, being used as a portent, we should be suspicious that it is a hoax that has been invented for the occasion. It is possible that the 'eclipse of Stiklestad' (Snorri, *ca.* 1230) is such a hoax, invented as a bit of religious propaganda, although it may be a genuine misunderstanding by Snorri. It seems certain that the eclipse put on the day that Xerxes left Sardis (Herodotus, *ca.* −446) to begin the invasion of Greece is a hoax, at least in the way that Herodotus uses it. There was no eclipse at that time, and events could not have happened as he describes them. It is not clear whether Herodotus was the hoaxer or the hoaxed. Dubs (1938) suggests that the questionable Chinese record that he lists under −183 May 6 was fabricated as an omen for political reasons.

An eclipse that is used magically may none the less be genuine. For example, an eclipse that happened within the year before a person's death is often called an omen of the event. There is a reasonable probability that a solar eclipse will be visible at a given spot within a given year, and hence there are many 'correct' omens. We can only proceed with caution with eclipses that are used magically. If little or no conforming detail is given, and if superlatives are used in the description, we are probably safe in ignoring the record. However, if detail is correctly given, and if there is no exaggeration, we can probably accept the record. For example, the Byzantine historian Gregoras (*ca.* 1359) says that a solar eclipse was an omen of the death of the emperor Andronicus II Paleologus. It occurred as many days before his death as there had been years of his life. Indeed we find an eclipse that was readily visible in Istanbul within a day of the date specified in this way, and I believe that Gregoras has recorded a genuine observation of the eclipse of 1331 November 30.

Unfortunately, we must cope with hoaxes in the writings of professional astronomers. Al-Biruni (1025) quotes the astronomer Abu Sahl al-Kuhi, who worked in Baghdad around the year 990, and whose work appears to be lost. Al-Kuhi claimed that he had made careful measurements to find the obliquity of the ecliptic, and that he found it to be 23° 51′ 20″. Al-Biruni is profoundly suspicious of al-Kuhi's claims, pointing out that his value agrees with Ptolemy's to the second of arc but that it disagrees by about 15′ with other contemporary determinations. Al-Biruni doubts that al-Kuhi made the claimed measurements at all, and he is certainly justified in his suspicions.

It is ironic that al-Kuhi's probable hoax was committed for the purpose of 'confirming' an older probable hoax. Ptolemy (*ca.* 152, chap. I.12) describes two instruments for making the measurements needed to find the obliquity. He then says that he observed the solstices for several years, presumably with one of the instruments described. He found the obliquity to have the value that Eratosthenes had found four centuries before, namely 23° 51′ 20″. The obliquity in Ptolemy's time should have been about 23° 41′. Further, as al-Biruni (1025) points out, the data which Ptolemy quotes would actually lead to 23° 51′ 15″. I think that there is little doubt that Ptolemy's 'measurement' of the obliquity is also a hoax.

I shall not treat the famous question of how Ptolemy obtained his star table. Instead, I shall mention briefly his solar data, which, it seems to me, are unquestionably a hoax.† Ptolemy (*ca.* 152, chap. III.2) gives, to the hour, the times of two autumnal equinoxes, one vernal equinox,

† I published this conclusion in 1969 (Newton 1969). I remarked that J. P. Britton, in an unpublished work, had reached the same conclusion in independent and slightly earlier work. I have since found that Delambre (1819) had already published it. I overlooked this part of Delambre's work earlier because, oddly, it is not in his *Astronomie Ancienne* but rather in his *Astronomie du Moyen Age*. The conclusion may be one that is often discovered and as often forgotten.

and one summer solstice. He says that these times were measured with great care. However, the errors in them are more than a day, whereas Hipparchus three centuries before him had made such measurements with errors of only 2 or 3 h. On the other hand, the data agree exactly, to every numerical digit written down, with what we would calculate from Hipparchus's data and his value for the mean motion of the Sun. It is almost impossible that such errors and such agreement could happen by chance.

In sum, most ancient astronomical data are probably genuine, but a disturbingly large fraction are probably hoaxes. We must beware of accepting data without independent confirmation.

4. THE ACCURACY OF ANCIENT MEASUREMENTS OF POSITION

There are a few sets of data that allow us to estimate the errors made by ancient astronomers in the measurement of a celestial position.

In an earlier work (Newton 1970, chap. VIII), I analysed about twenty observations of conjunctions of Venus with other bodies, made by Islamic astronomers between 858 and 1003. In five instances, the astronomers said that the latitude difference between Venus and the other body was zero, and I found that the standard error in this statement was 0·061°. I did not calculate the standard deviation of the longitude difference at the time of the stated conjunctions. However, it is easy to determine from the results given that the standard error in the longitude determination was about 0.15°. It was interesting that all errors greater than 0.1°, with one exception, were made in the evening. The accuracy of the morning observations is about the same as that of the latitude differences.

Thus it is reasonable to say that the standard error in an observation of stellar position made with the naked eye is about 0.1° or 6'. It must be recognized that the accuracy depends somewhat upon the circumstances. For example, astronomers frequently measured the time by measuring the altitude of a star and calculating the time from the stellar coordinates. Such a measurement must be made quickly, and it is plausible that the error in altitude would be considerably more than 6'. I have not found any data that allow separating this error from other errors.

Another example concerns measuring the time of an equinox passage of the sun by measuring its declination. I have studied two main bodies of equinox data (Newton 1970, chap. II). Hipparchus measured the times of 20 equinox passages between −161 and −127. The analysis of his data indicates that he had a bias of about 4.5' in the establishment of his equatorial circle. After the bias is removed, the standard error in his times is 2 h or less. A sample of eight Islamic equinoxes between 830 and 882 shows errors about half this size. Since the declination of the Sun changes by about 1' per hour near the equinox, the precision of measuring the solar declination was about 2' or less. Thus solar observations, perhaps because of the large amount of light available, seem to be somewhat more accurate than stellar or planetary observations.

The remarks about solar observations probably apply only to observations made with circles or other devices having no obvious bias other than refraction. There is evidence (Beer *et al.* 1961) that solar measurements made with a gnomon may be subject to considerable bias, because the observer does not judge correctly (or may not know that he should judge) where the centre of the penumbra is.

Thus the standard error in an ancient observation of position should allow establishing chronology to a day if the quantity observed changes as rapidly as the longitude of the Sun, and, of course, if the other requisite conditions are met. However, the standard accuracy is not always present. To illustrate this, let us pretend that we do not know the date of the Council of Nicaea, and let us try to estimate it by astronomical chronology.

It is often said that the Council of Nicaea fixed the rules by which Christians determine the date of Easter. There is no evidence that the Council in fact did so (Jones 1943; Newton 1972 a chap. II), but it is certain that the observance of Easter was one of the main subjects on the agenda of the Council, and hence it is plausible that the rules were established within a short time after the Council.

One of the factors that enters into determining the date of Easter is the convention that the date of the vernal equinox is to be taken as 21 March in every year and for every meridian. From this, we may conclude that the equinox occurred on 21 March at the time of the Council. However, because of the difference between the true tropical year and the average length of a Julian year, the date of the equinox moves steadily earlier in the Julian calendar. Thus we should be able to date the Council by finding when the equinox came on 21 March in that calendar.

We must choose a place whose local time is to be used in making the necessary calculations. The most plausible choices are Alexandria and Jerusalem. Since local times at these places differ by only about 20 min, it does not matter much which one is used, and I shall choose Alexandria, since it was the leading centre of astronomy in the Roman Empire. I shall take 3 h as the difference between ephemeris time and universal time at the period of the Council.

In a group of four successive Julian years, the equinox comes earliest in the leap year. It comes later by about 6 h in each succeeding year until the intercalary day in the next leap year causes it to move earlier again. If we are to be justified in saying that the equinox comes on a particular day of the year, the equinox must come on that day in at least two successive years out of a set of four.

In the set of four years beginning in 140, the calculated equinox comes on 21 March in the first two and on 22 March in the last two. This is probably the first occasion on which the equinox came on 21 March in two successive years, and hence about 140 is the earliest possible time for the Council. In the set of years beginning in 292, the equinox comes on 20 March in the first two and on 21 March in the last two. This is probably the last time when two equinoxes came on 21 March. Hence we have determined that the Council of Nicaea occurred between about 140 and 292.

This conclusion is wrong. The Council of Nicaea was in 325. The explanation is probably that the astronomers who fixed the rules for Easter used poor data for the equinox. Elsewhere (Newton 1972a, § II.3), I have speculated that the error, which still survives in the ecclesiastical calendar, was a consequence of using the fudged solar data given by Ptolemy (ca. 152).

5. Dating errors

In a recent study (Newton 1972a), I analysed 629 records of solar eclipses found in medieval European and Byzantine sources. Table 1 is a summary of the errors in the years given for these eclipses, after we make due allowance for differences in conventions about when the year began. There are also many errors in the month and in the day of the month, but I have not analysed them. There are several notable features about the table.

First, the probability that the year is wrong is almost exactly one in four. The probability shows some tendency to improve with the later records, but I have not attempted to study this matter carefully.

TABLE 1. A SUMMARY OF DATING ERRORS IN MEDIEVAL
RECORDS OF SOLAR ECLIPSES

From Newton 1972a. Reproduced by permission of the Johns Hopkins Press.

error, years	no. of records with this error	error, years	no. of records with this error
0	472	9	2
1	88	10	1
2	21	12	1
3	16	16	1
4	6	33	1
5	5	99	1
6	3	533	1
8	1	550	1
unidentifiable	8		

Secondly, in slightly more than one record in a hundred, the data are so badly garbled that we cannot identify the eclipse that is meant. These records pose a serious hazard. It often happens that we know that the eclipse cannot be identified because we know the chronological system used. If we did not know the system, we might think that we could 'identify' the eclipse, but we would be wrong.

Thirdly, there are the errors of 533 and 550 years. These errors have a specifically Christian origin and might not arise in records from other cultures. They result in part from a cycle of 532 years that occurs in the medieval ecclesiastical calendar. However, other cultures may have analogous cycles. For example, there are circumstances in which an Egyptian record might be displaced by one 'Sothic' cycle of 1460 years.

TABLE 2. DATING ERRORS IN RECORDS OF SOLAR ECLIPSES IN THE
ANNALS OF THE FORMER HAN DYNASTY

errors, years	number of records with this error
0	41
0.5	5
1	6
2	1
3	4
unidentifiable	2

Chinese records have a pattern similar to that of the European records. Dubs (1938) has analysed 59 records of solar eclipses found in the annals of the former Han dynasty (approximately − 200 to + 20). Table 2 summarizes the dating errors in these records as Dubs gives them, rounded to the nearest year or half-year, as the case may be. In 13 cases out of 59, not quite 1 in 4, the rounded error is 1 year or more. In two cases out of 59, the data are so badly garbled that Dubs did not venture an identification.

There seems to be a difference between the Chinese and European records in that table 2 shows no errors greater than 3 years, except for the unidentifiable cases. However, this is probably a consequence of Dubs's conventions rather than a real difference. In all but the two unidentifiable cases, Dubs assigned a date for purposes of discussion, and the date he assigned was that of the eclipse nearest to the date of the record that was visible at all in any part of

China. The differences used in preparing table 2 are those between the recorded date and the nearest eclipse, which was not necessarily the eclipse that gave rise to the record.

We should like to assign a standard deviation to the dating error in table 1, but we are hampered in doing so by the unidentified eclipses. If we ignore them, the standard deviation is 31.5 years. This value is dominated by the two errors of more than 500 years. If we ignore these two errors, but assign a nominal value of 20 years to the unidentifiable cases, the standard deviation is 3.0 years. I suspect that the latter value comes closer to representing the typical situation.

There is no apparent reason why the errors in dating eclipses should be different from the errors in dating other events.

6. THE IDENTIFICATION OF SOLAR ECLIPSES

When there is a considerable amount of detail in a record of a solar eclipse, and particularly when the date is given in a chronological system that we know, there is usually no difficulty in identifying the eclipse. Difficulties begin to arise when we can date the record only to a general historical period. When this happens, we begin by calculating the local circumstances, at the point of observation, of all eclipses within the allowed historical period.

If we are lucky, there is only one possible eclipse within the allowed period. We then have a tentative identification, but we should avoid accepting it as certain if there is no corroborative detail. Our main trouble arises when there is more than one possibility within the allowed period.

When this happens, many people have had recourse to considerations about the magnitude. There is a strong tendency to assume that a recorded eclipse was total. For example, Fotheringham (1920) says with regard to the 'eponym canon' eclipse from Assyria: 'As the eclipse is the only eclipse mentioned in this Chronicle, which covers an interval of 155 years, there can be no reasonable doubt that it had been reported as a total eclipse'. Unfortunately, an inspection of annals and chronicles does not support Fotheringham's contention. There are too many counter-examples. To give but one which is familiar to a British audience, the *Anglo-Saxon Chronicle* (*ca.* 1154) records only one solar eclipse between the years 733 and 878. That is the eclipse of 809 July 16, which probably did not attain a magnitude of more than about 3/4 anywhere in England. (Magnitude is used here to mean the fraction of the solar diameter that is covered by the Moon at the centre of the eclipse.)

The usual procedure when there is more than one possible eclipse is to play what I have called the 'identification game' (Newton 1969, 1970). In this game, the player starts by assuming that he knows the accelerations of the Earth and Moon within small uncertainties. Using his assumed accelerations, he calculates the magnitude of each possible eclipse and 'identifies' the eclipse as the one that yields the largest calculated magnitude. Since any eclipse can be made total at a given point by some values of the accelerations, the procedure of the identification game is equivalent to choosing the accelerations that are closest to the ones assumed. The player then completes the game by using the choices just made in making new estimates of the accelerations.

This is, of course, reasoning in a circle. One way to see that the method has no validity is to note that it works just as well if the 'historical periods' and 'places of observation' are chosen with the aid of a table of random numbers.

Quite apart from this error in logic, the identification game is not valid because it is based upon choosing the maximum calculated magnitude. This choice seems to be based upon the assumption that the magnitude must be large if an eclipse is to be recorded. It is instructive to examine the magnitudes in the large sample of European and Byzantine records that was mentioned above. The sample used in studying dating errors contained 629 records. In studying the magnitudes, we can use only those records which, with high probability, are independent and which can be assigned to a particular point of observation rather than to a general region. There are 202 records in this sample.

TABLE 3. CALCULATED MAGNITUDES OF 202 SOLAR ECLIPSES RECORDED IN
MEDIEVAL EUROPEAN AND BYZANTINE SOURCES

	no. of records lying in this range	
magnitude range	if $\dot{\omega}_E = -1068''/$ century^2	if $\dot{\omega}_E = -1424''/$ century^2
$\geqslant 1.00$	23	26
0.98–1.00	16	22
0.96–0.98	24	13
0.94–0.96	17	11
0.92–0.94	12	16
0.90–0.92	21	20
0.88–0.90	13	11
0.86–0.88	18	15
0.84–0.86	4	13
0.82–0.84	8	13
0.80–0.82	12	3
0.78–0.80	6	3
0.76–0.78	2	7
0.74–0.76	4	2
0.72–0.74	3	5
0.70–0.72	4	4
< 0.70	15	18
smallest value	0.30	0.31

Note: The acceleration of the Moon was taken to be $-42.44''/$century^2.

The centre time for this sample is about 1000. For this epoch I have estimated (Newton 1970, p. 272):

$$\dot{n}_M = 42.3 \pm 6.1''/\text{century}^2, \quad \dot{\omega}_E = -1068 \pm 170''/\text{century}^2.$$

The centre column in table 3 gives the number of eclipses whose calculated magnitudes, when calculated with these accelerations, lie within the indicated ranges. (I changed $\dot{n}_M$ slightly for convenience in preparing table 3.) The range of magnitude for those which were total is roughly equivalent to a range of 0.02, the range used in counting the eclipses that were less than total. Thus we see that the number of eclipses in each range is roughly constant from totality down to a magnitude of about 0.85 or perhaps less. Only about one eclipse out of nine recorded was total, and the smallest magnitude in the sample was about 0.30.

The last column in table 3 is calculated with $\dot{\omega}_E$ equal to 4/3 of the best estimate. This change in $\dot{\omega}_E$ changes the values for individual eclipses considerably but it makes little change in the general nature of the distribution. Hence the distribution calculated with $\dot{\omega}_E = -1068$ should be close to correct.

Thus, if the record of an eclipse gives no specific statement about the magnitude of an eclipse, we have no warrant to assume that the eclipse was total or nearly so. We are entitled to a reasonable presumption, say about 11 chances out of 12, that the magnitude is greater than 0.7. In playing the 'identification game', when there is no explicit statement about the magnitude, we must keep all calculated magnitudes greater than, say, 0.7 if we wish to have reasonable confidence in the 'identification'. Yet Ginzel (1899) and others have rejected calculated magnitudes as great as 0.96, even in cases where we do not know the place of observation.[†]

Even if an ancient record should state explicitly that an eclipse was total, we cannot take this statement at face value. The reason for this is connected with the existence of annular eclipses. If an eclipse occurs when the apparent diameter of the Moon is less than the apparent diameter of the Sun, the lunar disk is unable to cover the Sun completely even for an observer on the shadow axis, and the eclipse is called annular. Out of all eclipses for which the shadow axis strikes the Earth, more than half (about 56 %) are annular.

The Cairo chronicler Ibn Iyas (ca. 1522) furnishes an example of the erroneous reporting of an eclipse as total. He clearly states (at least in the cited translation) that the eclipse of 1473 April 27 was total in Cairo and that complete darkness lasted for an appreciable time. However, calculation shows that the eclipse was certainly annular and that it was not total anywhere.

Ibn Iyas was probably not an expert astronomer. We might asign this as the reason for his failure to discriminate between annular and total eclipses except for the fact that many experts also failed to make the distinction. For example, in the eclipse tables of Ptolemy (ca. 152), the minimum apparent diameter used for the Moon is so large that annular eclipses cannot occur. Thus it seems almost certain that annular eclipses were unknown to Ptolemy and to earlier astronomers. That is, ancient astronomers probably did not distinguish between total and annular eclipses, and they may have thought that the corona and the annulus that remains visible in an annular eclipse were the same thing, merely varying in brightness from one eclipse to another. In fact, I have not yet found an ancient or medieval record that unmistakably refers to the corona (Newton 1972a, § XVII.6).

In summary, I have not found a single instance in which the identification game yields a valid identification when there are two or more visible eclipses within the historical period allowed by a record. Unfortunately, Fotheringham, Ginzel, and others have based their conclusions heavily upon the use of the identification game.

We do not have the evidence needed in order to be sure why people have used the identification game. I suspect that acceptance of it rests upon two main factors: (1) The person who did the 'identifying' was usually not the person who used the 'identification' in estimating the accelerations, and thus the circular nature of the game was not as patent as I have made it here. (2) The people who did the identifying probably over-estimated the accuracy with which we can calculate the magnitude of an ancient eclipse. This latter topic will now be discussed.

7. THE ACCURACY OF CALCULATING ANCIENT ECLIPSES

For reasons that I have described elsewhere (Newton 1972a, chap. XVIII), the magnitude of a solar eclipse depends less upon the individual accelerations $\dot{n}_{\mathrm{M}}$ and $\dot{\omega}_{\mathrm{E}}$ than upon the parameter D'' to be defined in § 9 below. Estimates of D'' for various epochs within the historical period are shown in figure 1. Here we are interested in the accuracy with which we know D''. From

[†] Ginzel did this with the so-called 'eclipse of Archilochus', which is a passage of poetry. I see no reason to assume that this is more than a literary eclipse. Even if we assume that the passage is a genuine record, we do not know the place of observation. See table 4 and the accompanying discussion below.

figure 1, it seems reasonable to conclude that we know D'' with an accuracy of about $2''/\text{century}^2$ for times since about -700.

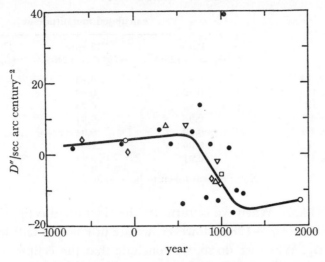

FIGURE 1. The acceleration parameter D'' as a function of time from about -700 to the present. The solid line has been drawn by eye to fit the plotted points. The points are based upon various kinds of observation, as follows: ●, the occurrence of solar eclipses; ◇, measured times of lunar eclipses; ○, measured times of lunar conjunctions or occultations; △, measured times of solar eclipses; ▽, measured magnitudes of solar eclipses; and □, the position of the Moon derived from an astronomical table.

In calculating the circumstances of a solar eclipse, it is convenient to assign values to the individual accelerations. Since the circumstances depend mostly upon D'' rather than upon the individual values, we must use individual values that are consistent with the estimates of D''. For convenience, we may pick any reasonable value of $\dot{n}_M$, say, but we must then calculate the value of $\dot{\omega}_E$ from its definition in §9.

As an example in calculating ancient eclipses, I shall use the various possibilities for the 'eclipse of Archilochus' that has already been mentioned. (See Fotheringham (1920), or Newton (1970), pp. 91–93.) The approximate historical period for this 'eclipse' is the first half of the -7th century. From figure 1, we see that an appropriate estimate of D'' is $2 \pm 2''/\text{century}^2$. For $\dot{n}_M$, I shall use -41.6 (Newton 1970, p. 272). Thus we must use

$$\dot{\omega}_E = -1288 \pm 59.$$

For convenience, I take the uncertainty in $\dot{\omega}_E$ to be 64, or 5 %.

The calculations are summarized in table 4. The first column lists seven possible dates that have been proposed for the 'eclipse'. Most discussions take Paros and Thasos as equally likely possibilities for the 'place of observation'. Thus the next two columns give the magnitudes calculated for these places, using $\dot{\omega}_E = -1288$. A magnitude greater than 1 means that the eclipse was total and remained so for a measurable length of time. The last column gives the magnitude calculated for Thasos when we change $\dot{\omega}_E$ by 1 part in 20 to -1224.

Table 4 shows explicitly the fallacy involved in the identification game. Suppose we had not previously used this 'eclipse' in estimating the accelerations (and indeed I have not and shall not use it), but suppose that we want to use it. For purpose of illustration, let us further suppose that we know $\dot{n}_M$ and that we are concerned only with estimating $\dot{\omega}_E$. If we take -1288 as the best estimate of $\dot{\omega}_E$ on the basis of other evidence, we conclude from table 4 that the eclipse was

TABLE 4. CALCULATED MAGNITUDES OF VARIOUS POSSIBILITIES FOR THE
'ECLIPSE OF ARCHILOCHUS'

	calculated magnitude at		
date	Paros ($\dot{\omega}_E = -1288$)	Thasos ($\dot{\omega}_E = -1288$)	Thasos ($\dot{\omega}_E = -1224$)
−710 Mar. 14	0.88	0.93	0.84
−688 Jan. 11	0.92	0.96	0.94
−661 Jan. 12	0.83	0.75	0.81
−660 June 27	0.95	0.94	0.89
−656 Apr. 15	0.99	0.94	1.02
−647 Apr. 6	0.96	1.01	0.97
−634 Feb. 12	0.91	0.88	0.92

Note: $\dot{n}_M$ was taken to be −41.6.

−647 April 6 seen at Thasos. When we in turn use the 'identified eclipse' in estimating $\dot{\omega}_E$, we necessarily find a value near −1288. However, we are just as well entitled to use −1224 as our initial estimate of $\dot{\omega}_E$. When we do so, we conclude that the eclipse was −656 April 15 seen at Thasos. When we then use this 'identification' in estimating $\dot{\omega}_E$, we find a value near −1224.

In other words, 'eclipses' identified by means of the identification game always tend to confirm the values of the accelerations used in making the identifications, and such 'identifications' contain no new information.

If we consider tables 3 and 4 together, I do not see how one can reject any date listed in table 4 as a possibility for the 'eclipse of Archilochus'.

The changes in magnitude produced by the change in $\dot{\omega}_E$ range from 0.02 to 0.09. The reason, aside from rounding error, is that the change depends upon the geometry of the eclipse path and the relation of a point to that path. The root-mean-square change is about 0.06 and we are justified in using this value as the typical uncertainty in calculating the magnitude of an eclipse near −700.

The base values of the astronomical parameters used in calculating the magnitude (other than the accelerations) derive from data obtained near 1900. Thus, for a given uncertainty in D'', the uncertainty in the magnitude varies with the square of the time measured from 1900. In addition, figure 1 shows that D'' is far from being constant in time. Therefore, as we extrapolate backward in time from about −700, we must also increase our estimate of the uncertainty in D''.

TABLE 5. THE UNCERTAINTY IN CALCULATING THE MAGNITUDE
OF AN ANCIENT ECLIPSE

year	uncertainty in magnitude
0	0.03
−700	0.06
−1300	0.18
−2000	0.54

It seems conservative to double the uncertainty in D'' if we go back six centuries before −700, and it seems conservative to double it again if we go back to, say, −2000. For still earlier dates, it seems pointless to speculate on the basis of present information. The resulting uncertainty in the magnitude, as a function of time before the beginning of the common era, is shown in table 5. The values in this table differ considerably from the conclusions of Stephenson (1970), who says that we can calculate the magnitude for eclipses near − 1400 with an uncertainty that does not exceed 0.01.

8. The accuracy of ancient measurements of time

We are now almost ready to turn to the consideration of the accelerations. Before we do so, we must decide upon the types of data that we will use in estimating them.

Many workers have used only observations that a particular solar eclipse was seen at a particular place, and they have usually tried to confine themselves to eclipses that, in their opinions, were total or nearly so. (In doing this, they may have overestimated the degree of totality, as we discussed in §6 above.) Two beliefs seem to underlie this course of action: (1) An observation of a total eclipse is considered to be a highly precise observation. (2) An ancient observation that involves the measurement of time is considered to have very low precision. These beliefs will now be discussed.

The first belief may arise from the fact that the duration of totality at a particular place is at most a few minutes. However, this would matter only if the time of the eclipse were recorded and used, and the fact would be useful only if the time measurement had the accuracy that is denied it by the second belief. Therefore, the belief requires that we use the eclipse without reference to the time of occurrence, even when the time is available.

A particular eclipse observation gives us a functional relation between the acceleration of the Earth and that of the Moon. In order to determine this function, we can start by assuming a value for the lunar acceleration and calculating the path of totality on a sphere that rotates uniformly with the present angular velocity of the Earth. We usually find, for an ancient eclipse, that the place of observation does not lie within the path limits on this sphere. In order to reconcile the calculation with the observation, we calculate the extra rotation needed to bring the point within the path, and we impute the extra rotation to the acceleration of the Earth. The precision of the observation, when used in this way, depends upon the angle through which the Earth can be turned and still have the point lie within the path. That is, it depends upon the width of the path in longitude, measured along the parallel of latitude that passes through the place of observation.

The width of an eclipse path, measured perpendicular to its length, is rather narrow. However, most paths extend nearly in an east–west direction and their width in longitude may be considerable. I measured this width for about forty eclipse paths used in an earlier work (Newton 1970) and found that its root-mean-square value is about 8°. This is equivalent to 32 min of time. If we estimate the acceleration by using the centre line of the path, the maximum error is equivalent to about 16 min, and the standard deviation of the error is almost exactly equivalent to 10 min. This, and not the half-duration of totality, should be taken as the measure of precision of an observation that an eclipse was total at a known point.

In order to find the functional relation that the observation imposes upon the accelerations, we repeat the calculations with other values of the lunar acceleration. It has always been sufficient to use two such values.

The second belief is probably connected with the general idea that ancient and medieval time keeping was poor in accuracy. This may be right so far as the general public was concerned, but astronomers are a different matter. Ancient astronomers had a time keeper whose accuracy was not exceeded until this century. This time keeper is the celestial sphere. The important question is: Did ancient and medieval astronomers use it?

We know that Islamic astronomers used celestial observations in order to measure time because, in a considerable fraction of the Islamic records that have been analysed (Newcomb 1875; Newton 1970), the data needed to calculate the time have been preserved along with the rest of the record. We have somewhat less information about Hellenistic time measurements. Ammonius and his brother Heliodorus say explicitly that they measured the time astronomically in an observation† made in 503. Ptolemy found the time astronomically in an observation made in 138, and in fact he recorded the data used. We do not know whether this was the common Hellenistic practice or not. So far as I am aware, we have no clues as to how the Babylonian astronomers measured time, except for the fact that their time units correspond to rotation of the heavens by 1° and by 30° (one sign of the Zodiac). This suggests but does not prove that they measured time astronomically.

TABLE 6. ERRORS IN ANCIENT AND MEDIEVAL MEASUREMENTS OF TIME

description of the measurements	historical period years	standard deviation of the measurement min
Babylonian measurements of lunar eclipse times	−720 to −490	31
Hellenistic measurements of lunar and solar eclipse times	−200 to +364	26
Hellenistic measurements of lunar occultations	−294 to +98	32
Islamic measurements of lunar eclipse times	854 to 1001	15
Islamic measurements of solar eclipse times	829 to 1004	9

In any event, we have several samples from which we can estimate the accuracy of old time measurements. These samples are summarized in table 6. The standard deviation for a sample in table 6 is not calculated from the mean for that sample. Instead, all standard deviations are calculated using the variation of D'' shown by the solid line in figure 1. We see that the timing accuracy of the Islamic astronomers is about 10 min. This is the same as the time-equivalent precision of a solar eclipse, and hence an Islamic time observation is entitled to the same statistical weight as an observation of a total solar eclipse at a known point. The timing accuracy of the Babylonian and Hellenistic astronomers is about 30 min and hence one of their observations should receive about one-ninth of the weight given to a total solar eclipse, other things being equal. This certainly does not say that we are entitled to ignore their time measurements. In fact, we should use them even if they had much lower precision yet because, as everyone knows who has tried to measure a physical quantity, it is important to measure it in a variety of ways in order to lessen the effects of bias that are inherent in almost every method of measurement.

† This observation does not appear in any formal work. Apparently Ammonius and Heliodorus noted some observations in their copy of Ptolemy's *Almagest*. Their copy was then used as the base for further copies, into which their notes were also copied. See the introduction to the cited edition of Ptolemy (*ca.* 152).

There is an element of irony in the belief that an observation of a total solar eclipse is more precise than other types of observation. The irony arises from the fact that the belief, though often advanced with vigour, is irrelevant. It would be relevant only if early records of total solar eclipses actually existed. There is a record (Newton 1972a, p. 399) from which we may conclude, with reasonable confidence, that the eclipse of 840 May 5 was total at the German town of Xanten. This is the earliest record I have found from which we may reasonably infer that an identifiable eclipse was probably total (or else annular and central) at a known place. By 840, contemporaneous Islamic astronomers were making time measurements of the same precision as an observation of a total eclipse, as we see from table 6. Thus, at every stage of history, measurements of time are at least comparable in precision to the observations of solar eclipses that are known to us from that stage.

This statement may surprise the reader, because the literature is full of claims about records of identifiable total eclipses seen at known places. I have investigated every such claim that I have found. In every case, the claim rests upon assumptions that are easily disproved by counter-example; an example of such an assumption is contained in the statement that Fotheringham (1920) made about the 'eponym canon' eclipse, which I quoted above. Even the Xanten record of the eclipse of 840 May 5 is not entirely satisfactory.

Of course, the eclipses in some earlier records may have been total at a known point, and a person may assume as a matter of opinion that they were. However, it seems to me that he is not justified in elevating the opinion into a heavily weighted scientific datum. In particular, he is not justified in doing so when the report is a 'literary' or 'magical' eclipse, as many of the early 'total eclipses' are. Certainly table 6 provides no justification for giving more weight to a single literary eclipse that has been identified by the identification game than is given to all the known measurements of time made by ancient and medieval astronomers.

Time measurements, as well as occurrences of solar eclipses, should be used in the estimation of the accelerations.

9. THE ACCELERATION PARAMETER D''

The acceleration $\dot{\omega}_E$ is subject to considerable variation on time scales from a few months to a few centuries, and $\dot{n}_M$ is subject to variation on at least the scale of a century. As a result, it is necessary to distinguish carefully at least three types of averaging or smoothing of the accelerations. The nature of the variations, and suggested definitions of useful kinds of averages, are given elsewhere (Newton 1972b). Here I shall present only the kind of average that is of most interest in the study of ancient civilizations. This is the kind that has been called the 'epochal average' and that has been denoted by putting angle brackets around the symbol for the acceleration. Here, for simplicity, the term 'epochal average' and the brackets are omitted.

The linear combination
$$D'' = \dot{n}_M - 0.033862\dot{\omega}_E$$

means the second derivative of the lunar elongation D, taken with respect to solar time rather than to ephemeris time. D'' is well determined by a large body of data with dates ranging from about -700 to the present. We know considerably less about any acceleration parameter independent of D'' than we do about D'' itself. In the interest of brevity, I shall discuss only D'' in this section.

Twelve estimates of D'' based upon about 370 observations of the occurrence of solar eclipses appear in Table XVIII.11 of Newton (1972a). Each estimate is based upon eclipses recorded within a particular half-century or century ranging from the fifth through the thirteenth century.

Thirteen more estimates of D'' are given in Table XIV.4 of Newton (1970). Each of these estimates is based upon a particular kind of observation made within a fairly short time interval and within a limited geographical region. The estimates are divided with regard to the kind of observation as follows: 25 observations of the occurrence of solar eclipses divided into three sets, 40 measurements of the times of lunar eclipses divided into four sets, 21 measurements of the times of solar eclipses divided into two sets, 10 measurements of the magnitudes of solar eclipses divided into two sets, and one set containing eight observations of the times of lunar conjunctions or occultations. In addition, there is a value deduced from the tabular mean longitude of the Moon at the epoch 1000 November 30, mean solar noon at Cairo; we know rather well the period of the observations upon which this value (from the Hakémite tables) was based. There is a 14th value of D'' for the epoch 1050 that is based upon the occurrence of solar eclipses. Most of the observations that went into this value were later used in a more detailed study (Newton 1972 b). Hence this value should not be used as an independent one.

The reader should notice that the quantity tabulated in Newton (1970) is not D'' but rather the quantity $-D''/1.6073$.

Finally, Martin (1969) has analysed about 2000 telescopic observations of lunar occultations made between 1627 and 1860. This analysis allows us to estimate D'' at an epoch near the present.

These 26 estimates of D'' are plotted against time in figure 1, which identifies each value according to the kind of measurement upon which it is based. We see that all kinds of measurement agree as well as could be expected, and that the points based upon the occurrence of solar eclipses have more scatter than the other points. The agreement is particularly interesting since the occurrences of solar eclipses are taken entirely from records made by people who were not professional astronomers, while the other values are based upon measurements of time or magnitude made by professionals.

The solid line shown in figure 1 represents about the best estimate of the time history of D'' that we can make in the present state of knowledge. Since we have neither a theoretical model nor a phenomenological basis for making a formal statistical study of the variation of D'', I have merely drawn the line by eye. It is plausible that the line gives D'' as a function of time since -700 with an uncertainty of about $2''/\text{century}^2$, and I used this uncertainty in the considerations of §7 above. However, we have no estimates for epochs between 1300 and the present, and the error for that period may be greater than $2''/\text{century}^2$.

The most striking feature of figure 1 is the rapid decline in D'' from about 700 to about 1300. When we remember that the values plotted in figure 1 represent the average between any epoch and 1900, this decline means (Newton 1972 b) that there was a 'square wave' in the osculating value of D'', and that the osculating D'' during the period 700–1300 had a value around $40''/\text{century}^2$ or more. Such changes in D'', and such values, are incapable of explanation by present geophysical theories.

The small value of D'' during the period of classical antiquity (before about 500) should also be noted. From -700 to $+500$, the mean D'' was probably smaller in magnitude than it has been at any time during the past 1000 years. This fact may account for the relative success of nineteenth-century calculations of ancient eclipses and lunar phenomena, which were made in almost complete ignorance of the accelerations.

10. Summary

Ancient and medieval astronomical data allow us to form 25 independent estimates of the important acceleration parameter D'', at various epochs from about -700 to $+1300$. These estimates, combined with modern data, show that D'' has had surprisingly large values and that it has undergone large and sudden changes within the past 2000 years. It even changed sign about the year 800. The uncertainty in the value of D'' at any epoch from -700 to 1300 is about $2''$/century^2.

We can also use ancient astronomical data for chronological purposes, but with some limitations. An uncomfortably large number of ancient records are either untrue or are in error by amounts larger than those expected from the technical ability of the times. Further, the recorded dates are often in error by serious amounts, even in terms of the calendrical system used by the observer. In a sample of nearly 700 records of solar eclipses, the year is wrong in about one record out of four. The errors range up to 550 years. In addition, the dating is so badly garbled in at least ten cases that we cannot tell which eclipse is meant. These records are a particular hazard. We happen to know that they are unidentifiable only because we know the chronological system used in dating them. If we did not know the system, we would not necessarily know that the data have been garbled and we could easily be led to a false identification.

Solar eclipses can be useful under restricted circumstances. As an approximate rule for the usefulness of eclipses, we can use them only if we already know the date within about a decade; occasionally we can tolerate a larger uncertainty. If the *a priori* uncertainty in date is appreciably larger, we frequently find that more than one eclipse meets the conditions of the record. Multiple possibilities in identification have often been resolved by choosing the largest eclipse, but this procedure is usually unjustified, for two reasons. First, we cannot calculate the magnitude of an ancient eclipse with the necessary accuracy. The error in the calculated magnitude of an eclipse near -1400, for example, can be 0.20 or more, even when we use all available information about the accelerations of the Earth and Moon. Secondly, we cannot usually assume that the eclipse was large. Only about one recorded eclipse out of nine was total. The probability that an eclipse was recorded is reasonably independent of the magnitude for all magnitudes greater than about 0.8 or 0.85. Even if a record says that an eclipse was total, we cannot assume true totality. Apparently astronomers did not learn to distinguish annular eclipses from total eclipses until about the year 1000.

I wish to thank R. E. Jenkins of the Applied Physics Laboratory, Johns Hopkins University, for many helpful discussions and for a critical review of this paper. I also wish to acknowledge the support of this work by the Department of the Navy, through its contract N00017-62-C-0604 with the Johns Hopkins University.

References (Newton)

al-Biruni, Abu al-Raihan Muhammad bin Ahmad 1025 *Kitab tahdid nihayat al-amakin litashih masafat al-masakin*; there is a translation into English, using the title *The determination of the coordinates of positions for the correction of distances between cities*, by Jamil Ali, American University of Beirut, Beirut, Lebanon, 1967.

Anglo-Saxon Chronicle ca. 1154 There is an edition of the six main Anglo-Saxon texts by Benjamin Thorpe in *Rerum Britannicarum medii aevi scriptores*, 1861, no. 23, v. 1, London: Longman, Green, Longman, and Roberts.

Anna Comnena *ca.* 1120 *Syntagma rerum ab Imperatore Alexio Comneno gestarum*; there is an edition by L. Schopen, in 2 volumes, in *Corpus scriptorum historiae Byzantinae*, 1839, (ed. B. G. Niebuhr), Bonn: Weber's.

Beer, A., Ho Ping-Yu, Lu Gwei-Djen, Needham, J., Pulleyblank, E. G. & Thompson, G. I. 1961 *Vistas in astronomy* **4**, 3–28. Oxford: Pergamon Press.

Clemens, S. L. 1889 *A Connecticut Yankee in King Arthur's Court*, chap. 6. New York: Harper and Bros.

Delambre, M. 1819 *Histoire de l'astronomie du moyen age*. Paris, Chez Mme. Ve. Courcier.

Dubs, H. H. 1938 *Osiris* **5**, 499–522.

Fotheringham, J. K. 1920 *Mon. Not. r. astr. Soc.* **81**, 104–126.

Ginzel, F. K. 1899 *Spezieller Kanon der Sonnen- und Mondfinsternisse*. Berlin: Mayer and Muller.

Gregoras, N. *ca.* 1359 *Romaikes istorias*; there is an edition by L. Schopen, in 3 vol., in *Corpus scriptorum historiae Byzantinae*, 1829 (ed. B. G. Niebuhr). Bonn: Weber's.

Haggard, Sir H. Rider 1886 *King Solomon's mines* chap. 11. New York: Harper and Bros.

Herodotus *ca.* −446 *History*. There is a translation by George Rawlinson, first published in 1858, reprinted by Tudor Publishing Co. New York, 1947.

Ibn Iyas *ca.* 1522 *Chronique*. There is a translation into French by Gaston Wiet, in 4 vol., under the titles *Histoire des Mamlouks Circassiens* (2 vol.) Imprimerie de l'Institut Français d'Archéologie Orientale, Cairo, 1945, and *Journal d'un bourgeois du Caire* (2 vol.), École Pratique des Hautes Études, Paris, 1960.

Jeffreys, Sir Harold 1970 *The Earth*, 5th ed., Cambridge University Press.

Jones, C. W. 1943 *Bedae opera de temporibus*, Menasha, Wisconsin: George Banta Publishing Co.

Martin, C. F. 1969 A study of the rate of rotation of the Earth from occultations of stars by the Moon, 1627–1860, a dissertation presented to Yale University; intended for publication in *Astronomical Papers Prepared for the Use of the American Ephemeris and Nautical Almanac*.

Munk, W. H. & MacDonald, G. J. F. 1960 *The rotation of the Earth*. Cambridge University Press.

Neugebauer, O. 1955 *Astronomical cuneiform texts*, in 3 vol. London: Lund Humphries.

Newcomb, S. 1875 Researches on the motion of the Moon, in *Washington Observations*, U.S. Naval Observatory, pp. 1–280.

Newton, R. R. 1969 *Science, N.Y.* **166**, 825–831.

Newton, R. R. 1970 *Ancient astronomical observations and the accelerations of the Earth and Moon*. Baltimore, Md: Johns Hopkins Press.

Newton, R. R. 1972*a* *Medieval chronicles and the rotation of the Earth*. Baltimore, Md: Johns Hopkins Press.

Newton, R. R. 1972*b* Astronomical evidence concerning non-gravitational forces in the Earth-Moon system, *Astrophys. Space Sci.* **16**, 179–200.

Plutarch *ca.* 90 *De facie quae in orbe lunae apparet*; there is an edition by H. Cherniss in *Plutarch's Moralia*, v. 12. Cambridge, Mass.: Harvard Press, 1957.

Ptolemy, C. *ca.* 152 'E *mathematike syntaxis*; there is an edition by J. L. Heiberg in *C. Ptolemaei opera quae exstant omnia*, B. G. Teubner, Leipzig, 1898. There is a translation into German by K. Manitius, Leipzig: B. G. Teubner), 1913.

Snorri (Sturluson) *ca.* 1230 *Noregs konunga sogur*; there is an English translation by L. M. Hollander, University of Texas Press, Austin, Texas, 1964.

Stephenson, F. R. 1970 *Nature, Lond.* **228**, 651–652.

Contributions to the discussion on astronomy in ancient literate societies

A. DIGBY (*The Paddocks, Eastcombe, Stroud, Gloucestershire*) I think the most striking difference between Old World and New World calendrical systems which has emerged from the papers today is the use in America of a 260-day period combining the 20 day names with the numerals 1 to 13. There is a possibility that the number thirteen may have been determined by the characteristics and shortcomings of a peculiar sundial which could be used to determine annual as well as diurnal time by showing the declination as well as the hour angle of the sun.

The evidence for the existence of this instrument lies in examples of the year glyph which can be shown to be a drawing of two trapezes set at right angles on a ring. One lying north to south would cast a shadow which would move from west to east across the base of the instrument between the hours of about 7 a.m. and 5 p.m., while the other, a taller trapeze would cast a shadow that travels across the instrument in a direction from south to north and back reflecting the declination of the Sun. There is some evidence to show that the instrument was tilted with the base parallel to the axis of the Earth like a 'polar sundial' (paper read to a symposium on recent Mesoamerican research, Cambridge 1972). Under these conditions, the shadow would make four traverses of equal length in the course of the year: from the centre of the instrument to the northern extremity (autumnal equinox to winter solstice); north extremity to centre (winter solstice to vernal equinox); centre to south extremity (vernal equinox to summer solstice); south extremity to centre (summer solstice to autumnal equinox), each of $91\frac{1}{4}$ days duration.

There must have been some scale to the instrument, but the only drawing known is the drawing of the game of Patolli, shown in Codex Maggliabechiano. The board is cruciform, the arms of the cross have seven divisions, the two outermost of which are marked with a diagonal cross. The movement of beans, representing counters in the game is said to represent the movement of heavenly bodies, and Caso says the Patolli board was used to represent the 52-year cycle common in Mesoamerica.

If we can show that the Patolli board was derived from a scale from the instrument, we have a possible explanation of the origin of the thirteen-day period in the Mesoamerican calendar.

Experiments to determine the accuracy with which the difference in the position of the shadow at noon on two successive days could be detected gave surprisingly accurate results. It was impossible to determine the exact point at which the edge of the penumbra passed over a line on white paper, but once over the line, the penumbra appeared to be clearly separated from the umbra. The contrast between the black line on the paper and the light shadow of the penumbra seemed to emphasize the difference between the two parts of the shadow. It was then easy to determine the point at which the shadow touched the black line. There is no time to describe these experiments but I should be pleased to give fuller details to anybody interested. The results of 94 observations with different observers under conditions of English sunshine gave a mean of 0.29 mm with a standard deviation of 0.09 mm when the cast of the shadow was 20 cm, representing a movement of about 5'. Under conditions of Mexican sunlight the results might be better but pre-Columbian equipment poorer. If we take a less optimistic measure and assume an accuracy to about 10', the daily variation between the positions of the shadow would not be detectable for about 25 days before and after each solstice. This would explain the differentiation made in the divisions on the Patolli board, and give five divisions signifying fast movement and two signifying slow movement.

The hypothesis offers a plausible explanation of the thirteen day period but further examples of the Patolli board need to be studied, and above all a long series of experiments to determine the accuracy which can be obtained with the trapeze instrument under Mexican conditions of sunlight.

F. R. STEPHENSON (*Department of Geophysics and Planetary Physics, School of Physics, University of Newcastle upon Tyne*)

Late Babylonian observations of 'lunar sixes'

British Museum tablet 34075 (Sp. 171) is one of the few published late Babylonian texts devoted exclusively to observations of Lunar Sixes. The date of the tablet is clearly specified, and more than 100 complete observations covering a period of 4 years are recorded. Analysis of this tablet provides valuable information on the accuracy of timekeeping in Babylon, and leads to speculation on the design and mode of operation of Babylonian clepsydras.

(1). *Provenance and date*

Photographs of the tablet, together with a full transliteration, translation and commentary, will be published elsewhere. An excellent drawing by T. G. Pinches is shown by Sachs (1955, p. 225). Sachs (1955, p. vi) asserts that the collection to which the tablet belongs originated in Babylon. The very first entry allows us to date the tablet precisely. We read:

$$[m]u \; 1 \; kám \; m \; pi \; [\;],$$

i.e. 1st year of Philip. Alexander died on 29 Airu in the 14th year of his reign (10 June 323 B.C.– cf. Bickerman (1968)). The astronomical observations preserved to us begin on the following day, namely 1 Simanu.

At least five successive years were originally covered by the text, but the actual period may well have been much longer. Complete observations have only survived for portions of the years 323/322, 322/321, 321/320 and 320/319 B.C.

(2). *Summary of contents of text*

In each month the following six time intervals are recorded:
  (i) sunset to moonset on the 1st day of the month (designated *na*);
  (ii) moonrise to sunset on the last evening that the (almost full) Moon rose before sunset (*me*);
  (iii) moonset to sunrise on the last morning that the Moon set before sunrise (*šú*);
  (iv) sunset to moonrise on the first evening that the Moon rose after sunset (*ge₆*);
  (v) sunrise to moonset on the first morning that the Moon sets after sunrise (*na*);
  (vi) moonrise to sunrise on the last morning that the waning crescent was visible (*kúr*).

Each entry starts with the day of the month, the first day entry also naming the month. Intervals of time are quoted in *uš*, the time required for the heavens to turn through 1° (i.e. 4 min). Of the various remarks which may conclude a particular entry, most relate to weather conditions. The common expressions *a* (mist) and *dir* (cloud) are frequently accompanied by *nu pap* (not watched for). In such cases the required interval has been predicted by a procedure which is not yet fully understood, but which made use of the corresponding observational data from 18 years previously (A. J. Sachs, personal communication).

The ideogram *muš* (accurately measured) also commonly follows *a* or *dir*, and indicates a successful observation despite unfavourable weather.

I have confined my attention to those records which either make no mention of mist or cloud, or else are qualified by *muš* – i.e. exclusively observational material.

(3). *Computational procedure*

The majority of the measurements recorded on the tablet are quoted to the nearest *uš*. At first and last visibility (*na* and *kúr*), intervals are almost exclusively rounded to the nearest integer. Evidently it was not possible to achieve greater accuracy, for the narrow lunar crescent is difficult to observe. However, around full moon (*me*, *šú*, ge_6, *na*) the observers seem to have aimed at a significantly higher precision. More than one quarter of these measurements are quoted to the nearest $\frac{1}{2}$ *uš* (i.e. 30/60), with a few scattered results expressed (probably without justification) to the nearest 1/3 or even 1/6 *uš*.

For purposes of comparison, a computer program was designed which calculated the interval between the rising or setting of the Sun and Moon (as required) at Babylon, the geographical coordinates of which were taken as 32° 33′ N, 44° 25′ E. This program required only the input of a selected Julian calendar date and the approximate local time of the phenomenon (6 or 18 h), together with a code number (1 or 2) representing the phase of the Moon (i.e. new or full). Computed intervals were expressed in *uš* to enable direct comparison with the recorded intervals.

Allowance was made for lunar parallax, refraction and dip. In making the correction for dip, it was assumed that the observer was some 15–20 m above the ground (roughly the height of the walls of Babylon – Ravn 1942) but this effect is fairly small for any reasonable height. The moment of rising or setting of the Sun was taken to be the instant when the upper limb (apparently) reached the horizon. There seems no reason to doubt that the Babylonian astronomers, like their present day counterparts, adopted this definition. However, the possibility of an alternative definition will be discussed in §4 below.

In computing the various time intervals, the values of the rotational acceleration of the Earth ($\dot{\omega}/\omega$) and orbital acceleration of the Moon ($\dot{n}$) employed were those derived by the author from an analysis of ancient observations (Stephenson 1972). The results of this investigation were

$$10^9 \; \dot{\omega}/\omega \; = \; -25.6 \pm 1.2 \; \text{(s.e.) century}^{-1}$$

$$\dot{n} \; = \; -34.2 \pm 1.9 \; \text{(s.e.) ″/century}^2$$

These figures correspond to a mean rate of the lengthening of the day of 2.2 ms per century and a rate of recession of the Moon from the Earth of about 5 cm per year. The mean epoch of the ancient observations analysed was 100 B.C., not too different from the epoch of the observations recorded on the tablet under discussion. The above results for $\dot{\omega}/\omega$ and $\dot{n}$ agree well with those obtained by de Sitter (1927) from similar material. If the higher values $10^9 \; \dot{\omega}/\omega = -27.7 \pm 3.4$, $\dot{n} = -41.6 \pm 4.3$ derived by Newton (1970) from ancient observations of mean epoch 200 B.C. are assumed, my computed local apparent (i.e. sundial) times should be advanced by about 0.4 *uš*. The implications of such an amendment will be considered in §4.

Prior to the operation of the computer program, the Babylonian dates were first corrected to the Julian calendar. The tables of Parker & Dubberstein (1956) were used to obtain the approximate date (never more than a day early or late) and the correct date was established by a trial and error process. In general, errors of observation were found to be small (typically about 1 *uš*) compared with the mean interval between two successive risings or settings of the Moon (about 13 *uš*) so that the true Julian dates were readily obtained, even when the relevant portion of the tablet bearing the date of observation was missing

(4). *Results of analysis*

As expected, the results of analysis of the data recorded on the tablet show that the new Moon observations were much less precise than those at full Moon (by a factor of about 3). It can have been rare indeed that the Babylonians were able to observe the actual rising or setting of the thin crescent, for atmospheric absorption is severe at a low altitude and the crescent is silhouetted against a bright sky. Most measurements probably made use of interpolation. In what follows I have concentrated exclusively on the full Moon data. This is of a high standard.

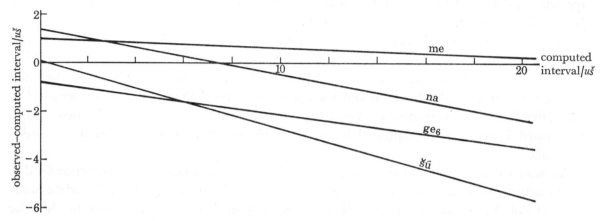

FIGURE 1. Detailed analysis of full Moon observations.

A total of 77 useful full Moon observations are recorded on the tablet. These observations are represented diagrammatically in figure 1, which is a graph of the observed–computed interval (in *uš*) as a function of the computed interval for each of the four categories *me*, *šú*, *ge_6*, *na*. Each line represents the best least squares fit to the appropriate set of data, after rejecting a few obviously faulty measurements (six in all). Individual points are not shown in order to avoid confusion. Rather surprisingly the four sets of data are quite distinct, both in intercept and gradient. The intercepts and gradients of the lines are as follows (assuming equations of form $y = a + bx$).

$$me: \quad a = +0.9 \pm 0.2, \quad b = -0.04 \pm 0.03,$$
$$šú: \quad a = 0.0 \pm 0.4, \quad b = -0.27 \pm 0.04,$$
$$ge_6: \quad a = -0.8 \pm 0.4, \quad b = -0.13 \pm 0.05,$$
$$na: \quad a = +1.4 \pm 0.3, \quad b = -0.18 \pm 0.03.$$

Let us consider the intercepts and gradients separately. Ideally, we should expect all four intercepts to be zero. Finite intercepts could arise from one of three causes: an incorrect definition of the rising or setting of the Sun and Moon, errors in the adopted values of $\dot{\omega}/\omega$ and $\dot{n}$, or the design of the clepsydra. In the latitude of Babylon, the Sun and Moon rise or set in about 2.5 min. Hence if the Babylonians defined the rising or setting of the Sun and Moon as the moment when the lower limb touched the horizon, my computed intervals *šú* and *ge_6* should be increased by about 1.3 $u\check{s}$ while the intervals *me* and *na* should be diminished by the same amount. The four intercepts in the diagram would thus become:

$$+ 2.2 \ (me), \ -1.3 \ (\check{s}\acute{u}), \ -2.1 \ (ge_6) \text{ and } + 2.7 \ (na).$$

Hence in order to minimize the intercepts we must accept the definition of rising or setting as the moment when the lower limb touches the horizon. This, of course, represents the very first or last sighting of the luminary.

The errors introduced by the uncertainties in the adopted values of $\dot{\omega}/\omega$ and $\dot{n}$ would seem to be comparatively small. The effect of alteration to these values is either to advance or delay both the apparent local time of moonrise and moonset. Thus for any random values of $\dot{\omega}/\omega$ and $\dot{n}$ we should expect the intercepts *me* and $\check{s}\acute{u}$ to be equal but opposite to the intercepts ge_6 and *na*. This is, of course, not apparent in the results of analysis. It would seem that the intercepts arise mainly from the design of the clepsydra.† This point will be considered below.

The gradients purely represent drifts in the clepsydra (they are in no way dependent on the choice of values of $\dot{\omega}/\omega$ and $\dot{n}$), but it is difficult to explain why the differences should be so marked and so systematic. These variations cannot be explained if an outflow type of clepsydra (whether a cylinder or a frustrum of a cone) is assumed.

The author is of the opinion that the existence of different intercepts, the remarkably small errors in individual measurements and the differences in gradient can most satisfactorily be explained by the assumption of four collecting vessels graduated in $u\check{s}$, each reserved for measuring a specific interval (possibly inscribed with the name of the interval). The intercepts can be then explained merely by an error in the marking of the first division (equivalent to a zero error). Assuming the graduations on all four vessels were equal, the varying gradients would be (wholly or partially, depending on whether or not a constant head device was used) the result of slightly differing diameters (the volume, of course, varying as the square of the diameter). From Proclus (*Hypotyposis Astronomicarum Positionum*, IV, 71–81, 87–99) we learn that in the time of Hero of Alexandria (perhaps 1st century B.C.) the apparent diameter of the Sun was measured by starting to collect water from a water clock at the moment of sunrise and continuing until the Sun had completely risen above the horizon. This quantity of water was then compared with the amount collected in a whole day. It is presumed that the Babylonians employed a similar mode of operation, every month waiting until all four samples had been collected before measuring the depth of liquid in each (hence the need for four vessels). We may suppose that, for some unknown reason, the glaring inconsistencies between the rates of filling of the collecting vessels passed unnoticed.

Analysis of other late Babylonian tablets recording lunar sixes may shed valuable light on this problem.

REFERENCES (Stephenson)

Bickerman, E. J. 1968 *Chronology of the ancient World*, p. 38. London: Thames and Hudson.
Newton, R. R. 1970 *Ancient astronomical observations and the accelerations of the Earth and Moon*. Baltimore: John Hopkins Press.
Parker, R. A. & Dubberstein, W. H. 1956 *Babylonian chronology 626 B.C. to A.D. 75*, Brown University Press.
Ravn, O. E. 1942 *Herodotus' description of Babylon*, trans. M. Torborg-Jensen, p. 20. Copenhagen: Arnold Busck.
Sachs, A. J. 1955 *Late Babylonian astronomical and related texts*. Providence, R.I.: Brown University Press.
Sitter, W. de 1927 *Bull. astron. Inst. Neth.* 4, 21–38.
Stephenson, F. R. 1972 Some geophysical, astrophysical and chronological deductions from early astronomical records. Ph.D. thesis, University of Newcastle upon Tyne.

† Some kind of clepsydra was definitely used, for during the intervals ge_6 and $\check{s}\acute{u}$, both the Sun and Moon are below the horizon, thus precluding other types of measuring time.

Phil. Trans. R. Soc. Lond. A. **276**, 123–131 (1974)   [123]

Printed in Great Britain

II. ANCIENT ASTRONOMY: UNWRITTEN EVIDENCE

Neolithic science and technology

By R. J. C. ATKINSON

Department of Archaeology, University College, Cardiff

For northern and western Europe, including Britain, the relevant evidence is much inferior to that available for the ancient East. In the absence of tomb-paintings, relief sculptures and documents, technological processes can be inferred only partially from their end-products, and there is no direct evidence of any general theory derived from observation of natural phenomena. Within these limits, however, it is clear that neolithic communities possessed great skill in 'civil engineering' and an adequate empirical knowledge of soil mechanics. Moreover, the corrected radiocarbon chronology implies the independent invention in the West of the techniques of mining and megalithic construction, and even perhaps of the wheel. In this context an interest in mensuration and in the properties of geometrical figures is not out of place, though no evidence survives for any system of numerical notation.

The purpose of this paper is to furnish a background of information about the state of neolithic technology, against which the evidence for the beginnings of astronomy in prehistoric Europe can be examined in greater detail. The area here considered is that part of Europe which lies to the north and west of a line drawn from Marseille to Helsinki, and south of the 60th parallel. This is the part of the continent which is farthest from the centres of early civilization and technology in the Ancient East and in which, therefore, innovations derived from those centres by a process of gradual cultural diffusion are likely to manifest themselves at a relatively late date, and generally in a modified or even vestigial form. It is accordingly also the area in which independent invention of techniques or ideas at an early date will most readily be detectable.

The period of time here surveyed necessarily has its upper limit at the date at which neolithic culture, economy and technology first became established in any significant part of the area defined above. This was nearly 7000 years ago, or about 4800 B.C. (Clark 1965). The lower limit is less obviously defined, if only because in modern usage the term 'neolithic' has essentially a technological connotation, and in the history of technology innovation can always be more closely dated than obsolescence. For present purposes the lower limit has been taken to be about 1700 B.C., because this appears to be the terminal date, excluding extreme cases, of the sites relevant to early astronomy to which attention has been drawn by Professor Thom (1967, 1971*a*).

At this latter date the stage of technological development reached in different parts of northern and western Europe varies from the Late Neolithic through the Early Bronze Age to the Middle Bronze Age; but it should be understood that the conventional (and largely arbitrary) stages so named conceal the continuing use of techniques of neolithic origin, on which the introduction and early development of copper and bronze metallurgy had relatively little effect.

The dates quoted above, and throughout this paper, are approximations to absolute dates, arrived at by applying corrections to the available radiocarbon dates. These corrections, as is now well known, are necessary in order to eliminate the observed discrepancies between the ages of specimens of ancient timber given by the counting of tree-rings and the ages of the same specimens given by the assay of their residual radioactivity (Suess 1970; Renfrew 1970).

Although the precise values of the corrections appropriate to given radiocarbon ages are still under discussion, it is sufficient for present purposes to say that at the beginning of the period here considered the radiocarbon dates have to be corrected by the addition of about nine centuries, and at the end by the addition of about three centuries.

It is worth noting that in astronomical terms the smaller of these corrections is still significant. For the rising and setting of the full Moon closest to the solstices, for instance, it represents at the period in question a difference in azimuth of rather more than one-fifth of the diameter of the Moon, which certainly exceeds the precision in alinement claimed for early prehistoric astronomers (Thom 1971 a). For this reason it is necessary to use *corrected* radiocarbon dates throughout the period here reviewed.

Within the wide limits of space and time defined above, considerable differences can be discerned between one region and another in material culture and in the level of technological advancement; but it is still possible to isolate certain basic elements of technological practice which are common to almost all the neolithic communities of western and northern Europe, and it is to these that special attention is paid below. Most of the examples cited, however, are drawn from Britain or from closely adjacent parts of Europe.

One further general point remains, however, to be made about the sources of our knowledge of the prehistoric technology of northwestern Europe, and about their status as evidence. We are accustomed to think of the ancient civilizations of the Near East as the cradle of science and technology; and the evidence available consists not merely of the surviving end-products of technological processes, but also of contemporary illustrations of the processes themselves. From Egypt, for instance, there are the actual remains of colossal stone statues, which are the end-products of processes of quarrying, carving and polishing, and of techniques of transport and erection; but there are also tomb-paintings and reliefs which show, often in considerable detail, how some of these processes were carried out.

By contrast, for analogous processes in neolithic Europe, which may be typified by the transport of the larger stones of Stonehenge (Atkinson 1960, pp. 116–122), there is no evidence at all. On geological grounds it seems safe to suppose that these stones *were* brought to Stonehenge from a distance of about 40 km; but how this was done is a matter not even of inference, but of pure speculation, based on a somewhat limited knowledge of the materials and mechanical devices which were available at the time in question.

It needs to be emphasized, therefore, how little our knowledge of neolithic technology owes to direct evidence, and how much rests upon a foundation of inference which is far from secure. Moreover, this caution applies with even greater force to our knowledge of the theory which lay behind practice. From Egypt, for example, there is documentary evidence, such as the celebrated Rhind mathematical papyrus, for the state of mathematical knowledge towards the end of the period here in question, about 1600 B.C. By contrast, nothing whatever is known from direct evidence about the state of mathematics in any part of northwestern Europe until after its conquest by the Romans; and what may be inferred from indirect evidence is little enough, and very uncertain. For this reason the term 'neolithic science' can be used, at best, in a very restricted sense.

It is with these provisos, therefore, that the basic common elements of European neolithic technology should be considered. The principal epithets used today by archaeologists to describe neolithic communities are 'stone-using' and 'agricultural'. The first of these reflects the older and literal meaning of the term 'neolithic', current until about 40 years ago and signifying that stone, and especially flint, was the best material available for hand tools used for cutting, chopping, boring and scraping; and that where appropriate such tools were shaped by grinding and polishing, whereas at earlier periods they had been shaped by flaking alone.

Already during the preceding Mesolithic stage, and indeed even earlier during the Upper Palaeolithic, a wide range of small hand tools of flint had already been evolved for special purposes, in the working of bone, antler and wood, and for the preparation of leather; and apart from the development of certain special forms of arrowhead, the generality of small neolithic flint tools show no marked improvement on their predecessors. The main technological innovation in this field, which is especially characteristic of the neolithic economy, is the development of heavy chopping tools – axes and adzes – for the felling and working of timber on a massive scale, for which the small light tools previously available were totally inadequate. These heavy tools, made of flint where it was available and of various igneous and metamorphic rocks elsewhere, mark the first use of shaped timber as a constructional and engineering material; and they symbolize a change from the relatively passive symbiosis with the natural environment, which characterizes the hunting, fishing and gathering economies of the Late Glacial and Post-Glacial periods in northern Europe, to a more active effort to exploit, to modify and ultimately to dominate the environment, which typifies the neolithic economy and those which have succeeded it (Cole 1959).

It may be added in passing that the efficiency of these prehistoric chopping tools is much greater than might commonly be supposed. Modern experiments have shown that oak trees of not more than 30 cm diameter can be felled with a flint axe in 30 min, and that for the clearance of forest the efficiency of stone and of iron tools differs by a factor of no more than 2 (Coles 1968, p. 7).

The demand for heavy woodworking tools itself gave rise to the development of two parallel industries producing axes and adzes, respectively in flint and in other types of stone. In both cases the products were distributed, probably in a semi-finished state, over distances of hundreds of kilometres from the source of the raw material. Moreover, in the case of flint the winning of suitable material from the ground involved a whole new technology of mining, very probably invented and developed independently in northern Europe, which is considered further below (p. 129).

In more recent usage, the term 'neolithic' also denotes the practice of agriculture; and it is this aspect of technology which is central to the change in the pattern of human economy called the Neolithic revolution (Cole 1959). The deliberate cultivation of cereals and other crops, and the controlled breeding and maintenance of herds and flocks of domesticated animals, provided either singly or in combination a surplus of stored foodstuffs, which acted as a buffer against the unpredictable vicissitudes of nature and allowed both a growth of population and an opportunity to develop new skills and new ideas.

Throughout the area of time and space here considered, the basis of neolithic life was a mixed farming economy, combining the growing of cereals with animal husbandry (Murray 1970). In its essentials this is a pattern of life at a peasant level which persisted without substantial change until the end of the medieval period in Europe, in which men and women share with animals the role of prime movers, and in which mechanical aids are restricted to simple tools and devices below the level of complexity deserving of the name of machine.

In much of the central part of the region defined, the dominant pattern of settlement was one of villages of substantial timber buildings, often with a tendency towards a common orientation that implies an accepted mechanism of social control (Piggott 1965, ch. 2). In many cases the buildings themselves exhibit a clear division of structure, which presumably reflects a corresponding division of function, between living-space and byre or barn.

The type of agriculture practised appears to have been the 'slash and burn' system, in which an area is cleared of scrub and undergrowth by cutting down and burning, leaving only the bigger timber standing, though probably killed either by deliberate bark-ringing or by the accidental effects of fire. The seed, usually some primitive variety of wheat, is then sown in the soil made powdery and sterile by fire, and enriched by the ashes. The same small plots are harvested, re-sown and re-harvested each year until, after about a decade, the depletion of the soil begins to have an appreciable effect on the yield. At this point the entire village community moves to another site, itself formerly occupied for about a decade by a previous generation and then abandoned. The precise duration of the cycle of occupation and abandonment is necessarily in some doubt; but it seems possible that a community would possess, say, half a dozen separate sites, each of which was occupied for about 10 years and then abandoned for half-a-century, to allow the regeneration of the soil and its plant nutrients.

Cultivation was initially carried out, probably, by hand with hoes and digging-sticks; but already before the end of the 4th millennium B.C. there is evidence for the use of some kind of traction-plough on fields with fixed boundaries and probably, therefore, of a more than temporary character (Fowler & Evans 1967; Evans & Burleigh 1969). This more advanced form of agriculture is appropriate to an environment in which natural forest has already given way in part to grassland, through the combined effects of deliberate clearance and of the prevention, by grazing, of the regeneration of woodland. Moreover, it is notable that in Britain the continental pattern of large nucleated villages is unknown. Instead, the few neolithic houses that have so far come to light are isolated buildings of a size much smaller than their continental counterparts (Piggott 1954, pp. 32–36). This difference must reflect some divergence in social structure and also, perhaps, in agricultural practice, with a greater emphasis on pastoralism in Britain.

The introduction of agriculture has a direct bearing, of course, on the early history of astronomy, since among all the technological innovations made by early man it is farming which most needs a reasonably accurate annual calendar. The need is greatest, perhaps, for the arable farmer, the ploughman; but the opportunities for making the astronomical observations on which any calendar has to be based are certainly far greater for the pastoral farmer, the shepherd who watches his flocks, and the sky, throughout the year and by night as well as by day.

For such primitive astronomers it is the horizon that provides the essential frame of reference – and, moreover, a *distant* horizon, the form of which would remain invariant under small local displacements of the observer. It should be understood, however, that in neolithic times a view of a far horizon was much more difficult to obtain than it is today, because forest, not open grassland or moorland, was the normal and natural habitat of the early agriculturalists. It needs a deliberate effort of will, for those accustomed to the wide open landscapes of today, to put these aside as man-made phenomena, and to think in terms of an environment in which forest was the rule on all but the highest ground, and open spaces of any size, below the tree-line, were the exception. It is easy, but mistaken, to project backwards into the prehistoric past the receding, undulating landscape of the chalk downs of southern England or the bare moorland, treeless and heather-covered, of the west and north; for these are man-made landscapes, many of them not developed before the Bronze Age (Dimbleby 1962, 1967, pp. 141–149).

What we must put in their place is forest in its natural state, in which the boles of oaks may rise to a height of 27 m before the first lateral branch of any size occurs, and to a total height of 43 m.

Clearance of natural forest of this kind is effected in part by the slow extension of areas of arable cultivation and in part, perhaps predominantly, by grazing. In the absence, initially, of more than small areas of grassland the leaves and twigs of forest trees, and perhaps of elms in particular, were fed to cattle, a process which it has been suggested is responsible for the marked decline in elm pollen which can be seen in pollen diagrams from a number of places in northern and western Europe, soon after the first arrival of neolithic farmers in the areas in question (Troels-Smith 1960). Sheep, goats and pigs, however, are even more efficient clearers of forest in the long run, not because they attack mature trees but because they nibble the tender bark of saplings and dig up their roots, and thus prevent the natural replacement of aged trees by young ones.

For these reasons the progress of forest clearance, and the opening up of astronomically significant horizons, must necessarily have been slow. It is therefore not surprising that almost all the archaeological sites in Britain and Brittany claimed as evidence for early astronomy date from the end of the neolithic period or indeed, for the majority, from the Early Bronze Age.

From the surviving remains it is clear that the most spectacular feats of neolithic technology lay in the field of what may be called 'civil engineering' – in the building, that is, of large earthworks and megalithic structures (Atkinson 1961). Among the earthworks three examples, all British, may be cited to illustrate the apparent limits of achievement.

The first is an extraordinary linear enclosure, about 90 m wide and 10 km long, which runs over the chalk downs of Cranborne Chase on the borders of Dorset, Wiltshire and Hampshire, and is known as the Dorset Cursus. Some twenty similar enclosures are known in Britain, though none of them exceeds 3.2 km in length. The construction of this one, in two stages, involved the digging of a ditch some 20 km in length and between 170 and 200×10^3 m³ in volume, and the piling of the excavated material to form a bank on its inner side (Atkinson 1955).

The second is the great circular earthwork at Avebury in north Wiltshire, which encloses nearly 12 hectares or about seven times the area of Trafalgar Square in London. Here the volume of solid chalk excavated from the ditch to a depth of some 10 m and piled in an outer bank, was about 100×10^3 m³ (Smith 1965).

The third example is the most remarkable of all neolithic earthworks, Silbury Hill, which stands 1½ km south of Avebury and is the largest mound of antiquity in Europe (Atkinson 1967). Its base covers 2.2 hectares and its flat top stands 40 m above the surface of the silt which has accumulated in the ditch surrounding the mound to a depth of 9 m. The date of construction is about 2750 B.C. (Atkinson 1969).

Recent excavations have shown that the builders of Silbury Hill had a very sound empirical knowledge of soil mechanics and were aware of the dangers of building a mound, in the manner of a modern mine waste tip, by dumping down the slope from the highest point. Though its exterior provides no hint of it, Silbury has in fact a very complex interior construction. It seems to have been built in successive stages, like a layer-cake, each stage being finished to a level top before the next was started. Moreover, each stage was apparently built in a series of relatively small dumps, working by accretion from the centre outwards; and in each constituent dump the material deposited was raked out into horizontal layers, with the free edges revetted by steeply rising retaining-walls of large chalk blocks. A horizontal section through the mound at any level would thus reveal a pattern of these walls, circumferential and radial, resembling a somewhat 'drunken' spider's web.

It should be added that the builders of Silbury Hill achieved a remarkable degree of accuracy in construction. The initial structures built at the centre of the base of the mound were strictly circular in plan, which allows the centre from which the work started to be determined accurately (Atkinson 1970). It is evident that this centre point was projected upwards as the work proceeded, and was used to ensure the concentricity of the succeeding stages. Careful survey has shown that the original centre at the base, the centre of the flat top, and the centre of the cone which is the best fit to the present surface of the mound all lie within a circle less than 1 m in diameter.

All the excavation involved in these massive earthworks was undertaken with no more than pick-axes of red deer antler, baskets, and the shoulder-blades of cattle used as shovels and scrapers. It is possible that wooden shovels were used as well, but none survives. Work studies carried out during the building of an experimental chalk bank and ditch near Avebury (Jewell 1963) have shown that these primitive tools have about half the efficiency of their modern counterparts – that is, steel picks, shovels and buckets. It is consequently possible to make rough estimates of the labour required for the building of specified neolithic earthworks. For Silbury Hill the figure is 500 men for 15 years. In view of the small size of the neolithic population, this represents a fraction of the 'gross national product' at least as great as that currently devoted by the United States of America to the whole of its space programme.

Engineering skills of a rather different kind were developed in the practice of megalithic architecture, in which large stones, weighing exceptionally more than 300 tonnes, had to be transported and erected. From the south coast of Spain to southern Sweden there is a widespread distribution of tombs for the collective burial of the dead, often used, apparently, for periods of several centuries (Daniel 1958). The walls and roofs of the burial chambers are formed of large stones, and in most cases the chamber was covered by a mound or cairn, although this has often subsequently been removed or eroded.

In a few instances there is petrological evidence that the building-materials were brought from a distance (Giot 1960, p. 94; Piggott 1962, p. 14); and in other cases the size of the structure and of its component stones is so large as to suggest a degree of megalomania. A single example – the Dolmen de Bagneux in a suburb of Saumur on the Loire – may serve as an illustration (Somerville 1928). Its internal dimensions are 18.5 m long, 4.9 m wide and 2.7 m high, so that it is hardly surprising that in modern times it has been used variously as a barn, a garage and a café. The four huge slabs forming the roof together weigh nearly 200 tonnes, and the largest of them is about 7 m square and weighs 86 tonnes. It is no exaggeration to say that even with modern equipment the building of this structure would be a notable achievement. With the primitive devices then available – levers, rollers, inclined planes and ropes (but not pulleys or capstans or windlasses) – it verges on the miraculous.

The handling of large stones was not confined, however, to the building of burial chambers. In both Britain and Brittany there are circles of standing stones, of which Stonehenge and Avebury are the best known, though hardly the most typical. In the same areas there are linear alinements of large stones, the largest being those near Carnac in south Brittany (Giot 1960, ch. 7) which are currently being studied by Professor Thom (1972). The principal example, the Menec alinement, contains 1100 stones arranged in 11 rows extending over a distance of nearly 1200 m.

In the same area, a short distance to the east, lie the broken fragments of the largest of all megaliths in Europe, the great menhir of Locmariaquer. It is 20.3 m in length, which is very close to the height of Cleopatra's Needle on the Embankment in London; and it weighs about 340 tonnes, or about 70 % more than the latter. There can be no doubt that originally it stood upright on its broader end; and it has recently been suggested (Thom 1971 b) that it served as a foresight for astronomical observations, some of them made over the open sea from a distance of about 16 km. The erection of this enormous stone must be counted among the greatest engineering achievements of prehistoric man in Europe.

With that in mind, it needs to be emphasized that the techniques of moving and raising such huge stones appear to have been invented and developed in northwestern Europe. No support can now be given to the older idea of an origin in the central Mediterranean or even farther to the east. The earliest radiocarbon dates for megalithic tombs come from Brittany, and it is accordingly there that we should seek the origin of this remarkable branch of technology.

Another branch of early technology which was also invented, probably, in northern or western Europe is the deep mining of flint. Neolithic flint mines are known from a number of localities, ranging from Portugal in the west to Poland in the east (Clark & Piggott 1933); but the earliest of them, so far as is known at present, occur at the same date of about 4300 B.C. in Belgium and in Sussex. In the Belgian sites (Clark 1952, pp. 174–175) the miners evidently understood very well the nature of the local geological stratification, and their deeper shafts were excavated through more than 9 m of unstable tertiary and quaternary sands and gravels before reaching the flint-bearing cretaceous beds beneath. In the Sussex mines (Curwen 1954, ch. 6) the shafts were spaced at close but irregular intervals, averaging about 15 m, so that the galleries radiating from their bases formed a continuous network, giving a very high ratio of extraction to waste. It is all the more remarkable, moreover, that this high efficiency was achieved without the use of any timbering, the roofs of the galleries being made deliberately low and arched, so as to be self-supporting. It should be noted too that in both areas the miners clearly knew the quality of flint that they were seeking, and were prepared to dig through up to ten inferior seams before reaching the one which they preferred.

One further invention may possibly have been made independently in northern Europe in neolithic times, namely the wheel; but it must be understood that the evidence on this point is still uncertain, and the hypothesis of independent invention is thus correspondingly speculative.

There can be no doubt that wheeled vehicles existed in ancient Mesopotamia in about 3000 B.C., if not earlier; and the evidence for the making of three-piece disk wheels in the area between the Black Sea and the Caspian has recently been put forward, for dates soon after the beginning of the third millennium (Piggott 1968). It should be noted, however, that the three sections of wheels of this kind are fixed together by very long dowels which pass right through the width of the central portion, and it is doubtful whether the necessary dowel-holes could be bored without the use of metal tools.

In peat bogs in the Netherlands and Denmark, however, one-piece disk wheels have been discovered, carved from the solid and therefore capable of being made without the use of metal tools. The radiocarbon dates of some of the Danish wheels, when corrected, come out soon after 3000 B.C., and those of the Dutch wheels somewhat later. It may well be, of course, that these are imitations, in the context of an inferior technology, of more advanced wheels developed somewhat earlier a long way further east; but the possibility of an independent invention cannot be summarily rejected.

The examples cited above have been intended to show that neolithic societies in northern and western Europe were neither as devoid of technical skills, nor as incapable of technological innovation, as has sometimes been supposed. It must be admitted, however, that all of them consist essentially of practical techniques which could have been evolved solely on an empirical or trial-and-error basis. It remains to be asked whether there is any evidence for the generalization of practice to the level of theory or, in short, for neolithic science as distinct from neolithic technology.

The short answer to that question is that there is very little evidence, but that it is correspondingly important. It is clear, for instance, that the builders of neolithic earthworks and megalithic monuments had at least an empirical understanding of field geometry, and were able to lay out on the ground various rectilinear figures which included accurate right-angles (Atkinson 1961); but it is not until the Early Bronze Age that we have unequivocal evidence for something that surpasses the empirical.

Professor Thom (1967, 1972) has drawn attention to numerous examples in western and northern Britain, and rare instances in Brittany, of stone settings of this period which are not circles in the strict sense, but form geometric figures of a more complex kind (ellipses, eggs and 'flattened circles') which cannot possibly be explained away as bad shots at a true circle (which is, after all, the simplest of all regular plane figures to set out with precision). It should be noted, moreover, that the figures in question are not merely complicated in a geometrical sense. They also exhibit properties of shape (e.g. the ratio of the perimeter to the principal diameter) which suggest a special respect for whole numbers and an interest in geometry which goes well beyond the requirements of practical surveyors. In sum, there is clear evidence for some knowledge of pure mathematics. Moreover, the recurrence of the same shapes, but of different sizes, implies the use of some unit of measurement, whether or not this was the megalithic yard deduced by Professor Thom.

In spite of the very compelling evidence that the builders of Early Bronze Age stone 'circles' were pure mathematicians in embryo, it must none the less be admitted that archaeology provides *no* evidence for the existence at this time for any system of numerical notation. In pure geometry the apparent lack of any such notation constitutes only a peripheral difficulty; but in astronomy, where the intervals between observed events have to be recorded, and stored in retrievable form, over periods that may exceed the working life of individual observers, this difficulty becomes much more central. It may be argued, of course, that a system of notation did exist, but that it was used exclusively on perishable materials such as wooden tally-sticks or parchment, so that no trace of it now survives. Such an argument, however, is not merely at variance with the fairly obvious need for long-term, and therefore permanent, record. It also cuts away the foundations of the whole archaeological and historical method of inquiry, which requires that one should deal with the evidence as it is, and not as one would prefer it to be. It is this contradiction, between the positive evidence for prehistoric mathematics and astronomy on the one hand, and the negative evidence for recorded numeracy on the other, which now most urgently needs to be resolved through the combined attentions of prehistorians and astronomers.

REFERENCES (Atkinson)

Atkinson, R. J. C. 1955 *Antiquity* **29**, 4–9.
Atkinson, R. J. C. 1960 *Stonehenge*. London: Penguin Books.
Atkinson, R. J. C. 1961 *Antiquity* **35**, 292–299.
Atkinson, R. J. C. 1967 *Antiquity* **41**, 259–262.
Atkinson, R. J. C. 1969 *Antiquity* **43**, 216.
Atkinson, R. J. C. 1970 *Antiquity* **44**, 313–314.
Clark, J. G. D. 1952 *Prehistoric Europe: the economic basis*. London: Methuen.
Clark, J. G. D. 1965 *Proc. Prehist. Soc.* **31**, 58–73.
Clark, J. G. D. & Piggott, S. 1933 *Antiquity* **7**, 166–183.
Cole, S. 1959 *The Neolithic revolution*. London: British Museum (Nat. Hist.).
Coles, J. M. 1968 *Proc. Soc. Ant. Scotland* **99**, 1–20.
Curwen, E. C. 1954 *The archaeology of Sussex*, 2nd ed. London: Methuen.
Daniel, G. E. 1958 *The megalith builders of Western Europe*. London: Hutchinson.
Dimbleby, G. W. 1962 *The development of British heathlands and their soils*. Oxford: University Press.
Dimbleby, G. W. 1967 *Plants and archaeology*. London: John Baker.
Evans, J. G. & Burleigh, R. 1969 *Antiquity* **43**, 144–145.
Fowler, P. J. & Evans, J. G. 1967 *Antiquity* **41**, 289–291.
Giot, P. R. 1960 *Brittany*. London: Thames and Hudson.
Jewell, P. A. (ed.) 1963 *The experimental earthwork on Overton Down, Wiltshire*, 1960. London: British Association for the Advancement of Science.
Murray, J. 1970 *The first European agriculture*. Edinburgh: University Press.
Piggott, S. 1954 *Neolithic cultures of the British Isles*. Cambridge University Press.
Piggott, S. 1962 *The West Kennet long barrow: excavations 1955–56*. London: H.M.S.O.
Piggott, S. 1965 *Ancient Europe*. Edinburgh: University Press.
Piggott, S. 1968 *Proc. Prehist. Soc.* **34**, 266–318.
Renfrew, C. 1970 *Proc. Prehist. Soc.* **31**, 280–311.
Smith, I. F. 1965 *Windmill Hill and Avebury*. Oxford: Clarendon Press.
Somerville, A. B. 1928 *Antiquity* **2**, 147–160.
Suess, H. E. 1970 *Proc. XII Nobel Symposium*. London & New York: Wiley.
Thom, A. 1967 *Megalithic sites in Britain*. Oxford: Clarendon Press.
Thom, A. 1971*a* *Megalithic lunar observatories*. Oxford: Clarendon Press.
Thom, A. 1971*b* *J. Hist. Astron.* **2**, 147–160.
Thom, A. 1972 *J. Hist. Astron.* **3**, 11–26, 151–164.
Troels-Smith, J. 1960 *Ivy, mistletoe and elm: climate indicators – fodder plants*. Copenhagen: Danmarks Geolog. Undersøgelse.

Phil. Trans. R. Soc. Lond. A. **276**, 133–148 (1974) [133]

Printed in Great Britain

Voyaging stars: aspects of Polynesian and Micronesian astronomy

By D. Lewis

School of Pacific Studies, Australian National University, Canberra, Australia

[Plate 21]

In Polynesia and Micronesia, where concepts are virtually identical, astronomy and navigation form one inseparable science. Sky 'domes', stellar zones, the seasons, yam growth and the Pleiades, as well as time spans of settlement, are mentioned in the paper, and attention is drawn to apparent maritime technological parallels elsewhere.

An explanation is advanced to account for survivals into the ethnographic present, *vide* the author's 1700 miles as navigators' apprentice, observations from a Sun-oriented trilithon and a stone instructional device, hitherto unrecorded, that is still in use. Solar observational platforms, navigational sighting stones and astro-navigation are touched upon. Pan-Pacific beliefs in stellar control of weather, astronomical lore inappropriate to its present location and the excellence of astro-navigation lead to speculation on possible one-time widespread diffusion through the neolithic world of related astronomical concepts.

'There is no specific word for "astronomer" in the Gilbertese tongue', Grimble wrote. 'If you would find an expert on stars, you must ask for a *tiaborau* or navigator' (Grimble 1931, p. 197). The study of astronomy was treated by the Tongans as a branch of navigation, according to Collocott (1922, p. 157). Similarly, in eighteenth-century Tahiti, Forster (1778, p. 501) remarked upon the subservience to navigation of the sciences of astronomy and geography. Such a relationship is hardly surprising considering that the Polynesian and Micronesian habitat, if we exclude New Zealand, is of the order of two parts of land for every thousand of water.

A few words about possible origins, time spans and technologies, of necessity grossly over-simplified, are desirable if we are to relate the astronomy of Oceania in any way at all to that of the ancient world in general. The ancestors of the Micronesians and Polynesians moved into the island world from Asia, the former coming to occupy the northwestern sector and the latter the west-central and central Pacific. Figure 1 shows these two areas and also Melanesia, which has unfortunately had to be excluded from this paper for lack of space. Archaeologists now tend to identify the makers of Lapita pottery as the most important Polynesian precursors (Groube 1971; Green 1972). Their coastal settlements appear mostly in the second millennium B.C. scattered over a huge area of Melanesia from New Britain to Fiji, in the Polynesian Santa Cruz Reef Island, and in Tonga, on the western margin of Polynesia. The rapidity of the dispersal of the Lapita pottery makers suggests that they were possessed of an advanced marine technology. This latter supposition gains indirect support from Haddon and Hornell's conclusion to their monumental work *Canoes of Oceania*, that the 'vessels used by the proto-Polynesians had frames and were plank-built, rather than ordinary dugout hulls with strakes' (1938, p. 40), in other words, were highly sophisticated craft. The size of double canoes in early historic times is attested by Cook, who saw some in Tahiti that were longer than the *Endeavour* (Haddon & Hornell 1938, p. 43). The inserted frames themselves and their attachments to planking are similar to those found in pre-Viking Scandinavian craft, leading Hornell (1935) to raise the possibility of the Pacific designs being distant offshoots of, or influenced by, some general neolithic or post-neolithic boat building tradition.

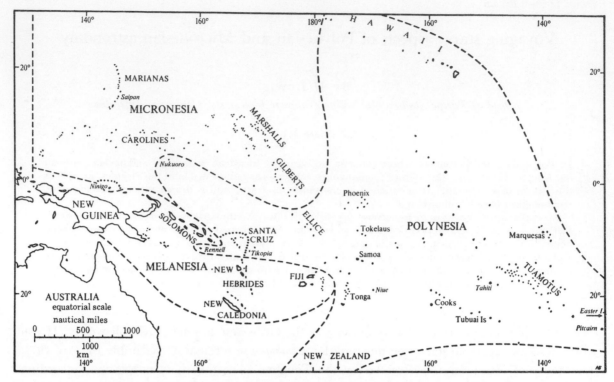

FIGURE 1. Map of Pacific regions showing Polynesia, Micronesia and Melanesia.

The entity we know as 'Polynesian' seems to have evolved in Tonga during the six or so centuries following settlement (Groube 1971), though it was from Samoa early in the Christian era that the dramatic move into the vast spaces of the central Pacific took place, the last eastern Polynesian outpost, New Zealand, being colonized from tropical eastern Polynesia by about A.D. 800 (Pawley 1966; Green 1966).

The economy of the neolithic Polynesians and Micronesians was in general based on yam, *taro*, sweet potato and breadfruit cultivation, harvesting the ubiquitous coconut, reef gathering and fishing. (This generalization ignores many anomalies, the role of pandanus in the Gilberts and Marshalls, for instance, or the remarkable absence of shell middens from Samoa.) Voyaging in Oceania seems to have been motivated more by quest for adventure, plunder and prestige, among other causes, than by trade cycles, which were more typical of Melanesia.

ASTRONOMICAL CONCEPTS

Astronomy, including nautical astronomy, is practically the same in Polynesia as in Micronesia, every significant concept being, in some degree, duplicated. The considerable differences that exist between individual archipelagos do not, as one might have anticipated, polarize across the linguistic-cultural 'frontiers' between the Polynesian Ellice and the Micronesian Gilberts. On such evidence as is available to us today, it does not appear justifiable to speak of separate Polynesian and Micronesian *systems*, but only of an Oceanic one with local variations (Lewis 1972, pp. 11, 160–162). A dissenting view, that the systems were distinct, is advanced by Akerblom (1968, p. 12), on the basis of a documentary study.

The usual cosmological concept is of sky domes, single or multiple and often solid, centred

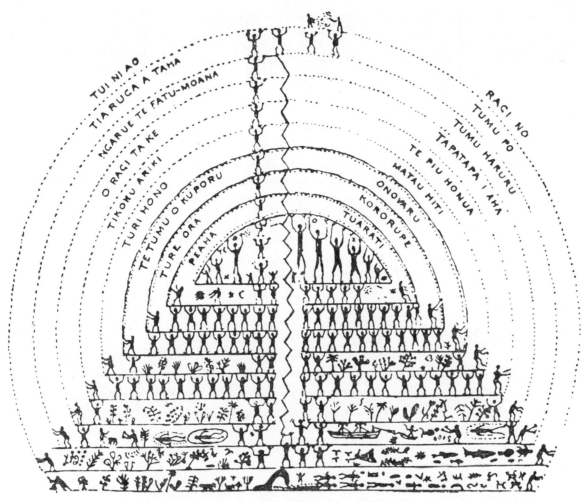

FIGURE 2. A Polynesian (Tuamotuan) heaven.

upon the home island or group, other groups having their own celestial cupolas, openings around the horizon allowing communication between them. Figure 2 shows a Polynesian heaven from the Tuamotu (Henry 1928); and figure 3 a Micronesian sky dome from the Gilberts (Akerblom 1968, p. 136). The similarities are obvious.

Navigator-priest-astronomers were familiar with the solstices and sometimes the equinoxes, the celestial equator, the zenith (which had special navigational significance), the distinction between stars and planets, and the precise azimuths of all navigationally useful stars, together with the approximate dates of their appearance in the night sky. The Maori of New Zealand were not atypical in having names for close on two hundred stars (Akerblom 1968, p. 22; Best 1922; Grimble 1931; Goodenough 1953; Makemson 1941). The magico-religious aspects of astronomy cannot be discussed adequately here, but religious beliefs and Polynesian star and sun myths are well documented (Williamson 1933; Fornander 1878, 1880). The rarity of Micronesian star myths, compared with their Polynesian profusion, remarked upon by Goodenough (1953, p. 4), is not readily explicable.

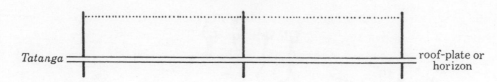

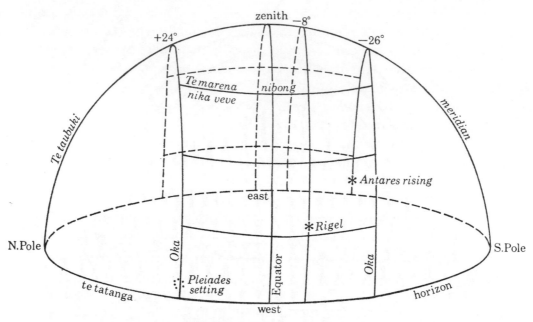

FIGURE 3. A Gilbertese sky dome.

WHO WERE THE ASTRONOMERS?

They were specialists. As J. R. Forster noted in Tahiti (1778, p. 528), 'geography, navigation and astronomy are known only to few'. Most of these were of high rank, like Tupaia, a dispossessed high chief and priest-navigator from Raiatea near Tahiti, who was Cook's most knowledgeable geographical informant (Beaglehole 1955, p. 117n.). Tupaia's known world in terms of islands radiating from Tahiti, embraced every major group in Polynesia and Fiji, except Hawaii, New Zealand, and the isolated Easter Island (Hale 1846, p. 122). Yet we know virtually nothing about his actual astro-navigational concepts.

The secrets of Tongan astronomy were held by a hierarchy of navigator families of varying, but always chiefly rank. So jealously guarded were their more esoteric concepts that it was not until 1969, well after the overwhelming bulk of their lore had been lost, that leaders of the Tuita navigator clan consented to divulge to me, an outsider, the half-forgotten residuum of what a nineteenth-century blind Tuita once claimed, according to Ve'ehala (personal communication 1969),† to be secrets that only he and the devil knew. We will discuss some of these secrets under *fanakenga* sky zones and under zenith stars.

Similar examples of secrecy are legion. Thus in the Marshalls, Winkler (1901, p. 505) tells us it was 'strongly and religiously forbidden to divulge anything concerning this art [navigation] to the people'. Such sanctions, coupled with the tendency, inherent in oral disciplines,

† See appendix to the references on contemporary indigenous authorities.

for even the most learned individuals to be variously informed, must in large part be to blame for the paucity of our knowledge of Pacific astronomy.

Sun observations from stone structures

Accounts come from all parts of Oceania. The observation places seem usually to have been temple platforms. Fornander (1878, p. 127) mentions one in Hawaii that was located between two cliffs at the point where the rising sun at the summer and winter solstices just tipped the northern and southern cliffs respectively. Several similar platforms were seen by Laval (1938, pp. 213–214) in Mangareva. Where suitable natural marks were lacking, sighting stones were erected. According to Heyerdahl (1961, pp. 189, 222), one Easter Island temple platform appeared to have been used for noting the equinoxes as well as the solstices, an unusual refinement for Polynesia (Akerblom 1968, pp. 17, 22).

The Gilbertese, on Butaritari at any rate, distinguished much finer gradations in the Sun's annual progression. It entered a named station, determined by the altitude of the Pleiades, every tenth day. Antares was used to fix the autumnal equinox. Observations were made from platforms on top of tapering structures that varied in height from 0.6 to 3.7 m. The dawn observation ritual, *te kauti*, was believed to give strength in love and war (Grimble 1931, pp. 205–211).

The evidence for Tongan sun observation is circumstantial. A solitary coral rock trilithon called *Ha'amonga a Maui*, a structure unique in Oceania, stands near the north-east coast of Tongatapu (see figure 4, plate 21). Tradition holds it to have been erected by Tu'itatui, the eleventh *tu'i tonga* or sacred ruler, who reigned around A.D. 1200 (Gifford 1929, p. 52). While no memory of its purpose survives, the present king, who is an amateur astronomer, discovered its long axis to be on the precise alinement of the summer solstice sunrise and thinks it probable that a trough on the horizontal coping stone was similarly related to the winter solstice.

If a personal digression may be permitted, it was to help confirm this latter possibility that I was invited to Tonga in 1969. King Tupou being indisposed on the morning of 21 June, I ascended the massive monument in the pre-dawn darkness (by way of a prosaic ladder) as his deputy. The sequence of events was predictable enough and, of course, there was no continuity with, nor attempt to re-create prehistoric observance. So I was unprepared for the evocative impact of the moment when the sun burst up over the sea horizon, bathing the lintel stone in blinding light, and the low chanting of the people massed below in the darkness swelled to a thunderous crescendo. However, the depression on the horizontal stone was too worn to permit of a definite conclusion.

Stones for navigation

Certain Micronesian and Polynesian stone structures undoubtedly subserved the purposes of nautical astronomy and one still does. Three sets are sighting stones or groups of stones and the other is a training device. It should be mentioned in passing that there is no evidence of any significant navigational artefact ever having been used at sea in Oceania. The oft-cited Hawaiian 'magic calabash' is no more than a chief's travelling trunk (Bryan 1936; Buck 1938).

The Hanga'i'Uvea

This solitary basalt stone set on edge, whose name means 'facing Uvea', is situated on the Tongan island of Niuafo'ou. Compass bearings taken along its 1.5 m length by the yachtsman-anthropologist Rogers (personal communication 1969) indicated a course that would take a voyager to a point 16 km upwind of Uvea, some 210 km distant. Landfalls in the Pacific were and are invariably made a little to windward and upcurrent of the objective.

The other two sets of sighting stones are both in the Gilberts. The first comprises the Butaritari stones, a group on the northern rim of the big Butaritari atoll (E. V. Ward, personal communication 1969) so situated that it can only indicate courses to the Marshalls, 264 km further north. Voluminous Butaritari traditions collected by Grimble (MS. n.d.) of contact with the Marshalls, the unique 18 m voyaging canoes seen on Butaritari by Wilkes (1845, vol. 5, pp. 74, 94), and a valid extant star path course to the Marshalls given me by a classically trained *tia borau* (Lewis 1972, p. 160), reinforce this assumption.

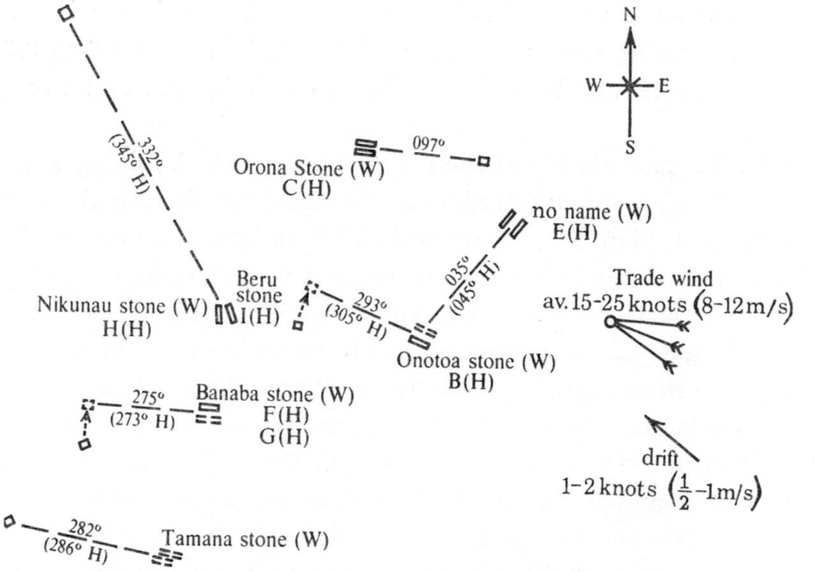

FIGURE 5. Arorae 'stones for voyaging'.

The second is called *Te Atibu ni Borau*, the 'stones for voyaging', on Arorae, the southernmost island of the Gilberts, described by E. V. Ward in 1946 (MS. 1946) and by Hilder a decade later (1959, 1963, 1972). Ward saw thirteen of these coral slabs in place and the former positions of four more could be determined (figure 5). Most were in groups of three that were named for the islands towards which their sight lines pointed (or perhaps indicated the transits of the azimuths of stars on the islands' bearings). All the targets were in the Gilberts except for Banaba or Ocean Island and Hull Island or Orana. The former was indeed once in contact with the Gilberts (H. E. Maude & H. C. Maude 1932), but the latter is one of the Phoenix group, uninhabited and unknown to the pre-contact Gilbertese. In fact H. E. Maude, the former administrator, was a member of the party that first decided to name Hull Island 'Orana' – in 1938 (personal communication 1969). The perils of uncritical reliance upon unsupported local tradition are obvious. All the other names, including Banaba, can be accepted and all indicate close approximations to the bearings of the islands concerned. The incompatibility of the

FIGURE 4. *Ha'amonga a Maui*, Tonga. A summer solstice-oriented trilithon.

FIGURE 6. Rewi instructing his daughter on 'stone canoe'.

numerous extant traditions as to the stones' origin (agreeing only on their navigational purpose) suggests their considerable antiquity and lends weight to Hilder's supposition (1972, p. 88), that the site represents an ancient school of navigation.

The stone canoe

The so called 'stone canoe' of Beru in the Gilberts has not been, as far as I know, previously recorded (Lewis 1972, pp. 184–187). It stands behind the house of the *tia borau* Rewi, where it was built by his father on the model of one made by his own father, Tebotua, the present navigator's grandfather. Rewi knows of only three other examples, though this is but negative evidence. The device is instructional and is currently being used by Rewi to train his own children (figure 6, plate 21 and figure 7). The longer axis is east–west and the shorter north–south. Seated upon it as on the thwarts of a canoe, the boy and girl learn beneath the night sky the directional stars of the islands.

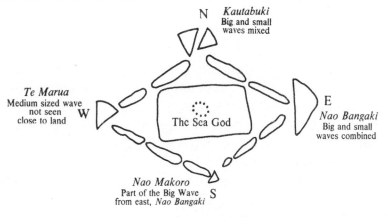

N *Kautabuki* Big and small waves mixed

Te Marua Medium sized wave not seen close to land W

The Sea God

E *Nao Bangaki* Big and small waves combined

Nao Makoro Part of the Big Wave from east, *Nao Bangaki* S

FIGURE 7. 'Stone canoe' diagram.

The structure is at other times conceived of as an island, when the triangular stones set at the corners represent by their sizes and angles the swell distortions detectable at sea beyond an atoll's sight range. The largest corner stone, for instance, corresponds to *nao bangaki*, the great swell from the east. Hidden from view in the photograph by the seat stone is a lump of brain coral personifying the sea god who, in Rewi's words, 'is most important of all. He helps us sail over the sea because he rules the sea'. Here, in contrast to the Arorae voyaging stones traditions, we have useful ethnographic data, in whose absence the 'stone canoe's' purpose would have been, to say the least, obscure. It might equally have been a sacrificial altar as a navigational artefact.

THE SEASONS

In Tonga the same word, *ta,u* stands for 'year' and for 'yam season', while a large part of the nomenclature of the 12 or 13 lunar months is based on this plant's seasonal changes (Collocott 1922, pp. 164, 166). A year varying between 12 and 13 months is also used in Samoa and in Tahiti, again related to the yam seasons and, in Tahiti, to the ripening of breadfruit and to fishing as well (Williamson 1933, vol. 1, pp. 126, 154–155, 167, 172).

The Gilbertese calendar, more accurately described, in Grimble's opinion, as a 'nautical almanac', is divided into the 'time for voyaging', *Aumaiki*, and the inclement *Aumeang* season. The former is associated with Antares in the ascendant and the latter with the Pleiades, the star's or the constellation's height, expressed in terms of meeting house (*maneaba*) purlins and roof beams, determining the division of the corresponding season into eight 'months' (*bong*). The names of most *bong* refer to the circumstances of navigation, for instance *Baro* is a contraction of *buta te ro*, 'cast off the cable' (Grimble 1931, p. 202).

Weather and currents are to this day firmly held to be under the sway of Antares and the Pleiades. The four learned *tani borau* Iotiebata, Teeta, Abera and Rewi, the most accomplished astronomer-navigators in the Gilberts today, all went to great lengths to reconcile this lore, in which they implicitly believed, with actual observed phenomena. Thus an aberrant current encountered when at sea in Iotiebata's sailing canoe was explained by Iotiebata as being the result of 'a struggle for mastery' between the controlling stars, 'which left the current free to move in any direction' (Lewis 1972, pp. 113–114). There is reasonable agreement between pilot book date (Ward 1967) and the seasonal current and weather patterns described by the *tani borau* – though the causes ascribed to the phenomena are somewhat different.

THE PLEIADES

This constellation is important in other calendars where we saw under Sun observations from stone structures that it determined the stages of the Sun's progression besides the Gilbertese. It plays a similar role in Mangaia (Gill 1876, p. 317), possibly in Tonga (Collocott 1922, p. 166), in Pukapuka, where it indicates the month when turtles land to lay eggs (E. Beaglehole & P. Beaglehole 1938, p. 351), and in Tahiti, whence comes the following account from an 1806 mission diary (*Transactions L.M.S.* 1806):

For some months there had been 'much disputing' as to whether the Pleiades (*mata-ree*) would set in daylight before or after the 'death of the Moon'. When the constellation was seen to be still above the horizon after the Moon had changed those who had been proved right 'shouted and triumphed over the other party'.

The amount of popular feeling aroused by the disagreement is noteworthy. The incident is also a reminder that naked eye observations, when calendars too are approximate, can give only approximate results. The point is also made for the Gilberts by Grimble (1931, p. 201).

DIVISION OF THE SKY INTO ZONES

Here we have a widely held concept that often has navigational connotations, particularly in the role asigned to zenith stars. As mentioned earlier the Gilbertese divide the sky, *uma ni borau*, 'the roof of voyaging', into latitudinal zones by analogy with the rafters and purlins of their meeting houses. The celestial equator is fixed by the declination of Rigel, so is 8° south of our own (Grimble 1931, p. 197–200), a point whose possible significance will be discussed later. The Carolinian heavens are similarly divided into latitudinal bands or *jaan* (paths) (Goodenough 1953, p. 4).

Hawaiian cosmogony, as expounded by Kamakau (1891), lumps together as navigation and land-ruling stars all those between the northern and southern limits of the Sun – the two highways of the navigation stars. Beyond are the strange or outside stars, among which, however, the Southern Cross and Polaris receive mention. (This concept does not appear to bear any European imprint. Some other of Kamakau's statements may have been influenced by his high school education and his general Christian upbringing.) The other main authority, Kepelino

(1932, p. 19), gives a more elaborate classification into fixed, moving, ruling and protecting stars that guide towards land. The passage about the latter having been severally interpreted by Beckwith (Kepelino 1932), Makemson (1941, p. 13) and Akerblom (1968, p. 39–40), Professor Elbert, at the request of Dr Finney (personal communication 1969), kindly re-translated it from the original Hawaiian. Protecting stars are 'suspended (*kau*) severally over various lands, such as Hoku-lea in the Hawaiian islands, and the Southern Cross over the lands of Tahiti, etc'. *Hoku-lea* appears to be Arcturus which, in A.D. 1000 culminated just north of the Hawaiian chain (V. Radhakrishnan, personal communication 1970). The Southern Cross, on the other hand, is very far indeed from being in Tahiti's zenith, though it indicates the island's approximate bearing. However, the word 'Tahiti' instead of 'Kahiki' is used in the Hawaiian text of the second phrase, an indication of European 'borrowing'.

The Tongans likewise divide the heavens into latitudinal zones, the northern, middle and southern *fanakenga* (Collocott 1922, p. 158). These divisions are reflected in the sea below, the northernmost zone being the warmest and vice versa.

'My father said any true sailor knew when he had crossed any of these *fanakenga* because of the temperature', the 88 year-old Sione Fe'iloakitao Kaho, the oldest surviving Tuita, told me. Indeed the clan owes its prominence to the celebrated feat of Kaho's great grandfather, the blind Tuita, who succeeded in re-orientating a lost royal flotilla by putting his hand in the water and announcing that Fiji was just below the horizon and indicating its direction. (He had also asked his son Po'oi to tell him the positions of certain stars.)

Kaho asserted, probably with truth, that his great grandfather's action was not to test the sea temperatures alone. He was also making contact with 'the devil' – the old sea god *Tangaloa*.

STARS IN NAVIGATION

The succession of stars that rise or set on the same azimuth makes up the star path or *kavienga* (Tonga), *kavenga* (Tikopia). By this one steers. The Carolinian 'star compass', discussed below, is but a specialized variant of the horizon star path, whose advantage is to reduce the number of stars that need to be memorized. The same degree of precision obtains with either, as nearly 2700 km steered by master navigators using one or other system demonstrated (Lewis 1972). The accuracy was such as to justify the Tongan aphorism, 'the compass may go wrong, the stars never'. Daytime steering is by the Sun and the angle of selected ocean swells.

A pertinent question is whether worthwhile data can be collected in the ethnographic present after one or two centuries of acculturation. Traditional nautical astronomy is still the favoured system in the Central Carolines and, to varying degrees, in the Santa Cruz Outer Reef Islands, Tikopia, Ninigo, the Gilberts and Tonga. Even in the Carolines there are lacunae and much more has been forgotten elsewhere. But while the old methods continue in use at sea, unlooked for reefs and unexpectedly empty ocean exert sanctions against distortions; the navigator's data and his conclusions therefrom are subject to confirmation by landfall. The validity of extant Carolinian nautical astronomy was demonstrated in 1969 by Hipour's return Saipan voyages, when he twice made accurate landfalls after crossing 720 km of open sea devoid of intervening land, navigating solely by eye in accordance with 'star compass' sailing directions, orally transmitted for some three generations since last used (Lewis 1971).

Survivals seem to have been of two kinds – relatively straightforward practices like horizon star steering, and elaborate concepts that resist modification because of their incompatibility with their Western equivalents. Examples of the latter include such stellar orientation systems as reference islands visualized as 'moving' beneath the star points (*etak*) (Alkire 1970, pp. 51–55; Gladwin 1970, pp. 181–195), the navigational application of zenith stars (Lewis 1972, pp. 223–243), and the Carolinian 'star compass'.

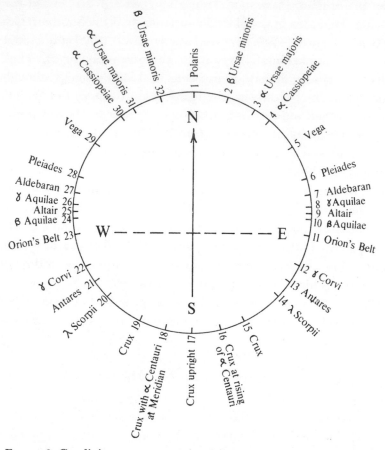

FIGURE 8. Carolinian star compass (modified from Goodenough 1953).

This last mental construct owes nothing to Europeans, its directional references being the 32 to 36 points on the horizon where given stars rise or set, regardless of whether they be currently visible or not (Goodenough 1953; Gladwin 1970, p. 148). It can be readily seen from the diagram (figure 8) that the star points are not equidistant, but crowded together in the east–west plane and spread out in the north and south. Herein lies the incompatibility: no practical correction factor could be devised to convert sidereal points into true ones. The 'star compass', therefore, may be expected to remain intact until its final replacement by the magnetic instrument.

Zenith stars

Horizon, star path, or compass stars stand in sharp contrast to overhead stars, that culminate in or near the observer's zenith. By the former one steers, by the latter one cannot; their function is to indicate what we term latitude.

My informant Ve'ehala explained in 1969 that the Tongan word *fanakenga* or sky zone has an additional meaning in the private usages of the Tuita navigator clan. A *fanakenga* star in this sense is 'the star that points down to an island'. When it is right overhead it indicates that 'you are nearing the island'. Experienced Tikopian navigators likewise insisted that the appropriate 'on top star' being overhead showed that they were near the island, and went on to stress the distinction between 'on top' and steering stars (Rafe, personal communication 1968; Tupuai & Samoa, personal communication 1969).

Of course, neither zenith star observations nor any other non-instrumental techniques can give the slightest clue as to longitude and so, at first sight, the Tikopian and Tongan formulations of being 'near land' when only latitude is known, seem strange. They are quite logical, however, in the light of a universal Polynesian and Micronesian navigational principle of always making a 'windward landfall', that is, of imparting a deliberate bias to the course in order to arrive a little upwind of the objective. The practice of true latitude sailing has never been reported from Oceania.

There is a documented instance of zenith stars having once been used for navigation in the Carolines (Sanchez 1866, p. 236), a more circumstantial one from the Gilberts (Sabatier 1939, pp. 94–95), and less direct evidence from Hawaii (Kepelino 1932) and Tahiti (Henry 1907, p. 121–124), as well as the contemporary Tongan and Tikopian sources we have already mentioned. Neyret suggests a zenith star interpretation of the Fijians' association of a particular star with the corresponding island (1967, p. 24).

The Tahitian concept is of star pillars supporting the sky dome. That it had navigational significance is mere inference from the circumstance that most 'pillars' were also zenith stars of geographically significant islands (Lewis 1972, pp. 239–241). Regardless of the correctness of this interpretation, the fact that one of these important props of the cosmos was Polaris, which is invisible until 1600 km north of Tahiti, points towards long consciously navigated voyages having once been made half way at least towards Hawaii.

How accurately can latitude be estimated by naked eye observation of zenith stars? A controlled trial during a 3200 km non-instrumental voyage from Tahiti to New Zealand gave results, in average weather, generally accurate to within 30'. The latitude error of the New Zealand landfall was 26'. The double-hulled yacht used in this experiment provided a stable platform, from which to make observations, equivalent to a double or outrigger voyaging canoe (Lewis 1967, pp. 281–285).

Altitude of the Pole Star

Estimation of latitude from the altitude of Polaris is only possible north of the equator, in Hawaii, the Gilberts, Marshalls and Carolines. The record is silent apart from the Carolines, where the question remains open. When Hipour was approaching Saipan, and the Pole Star had doubled its height from 7° 30' at Puluwat in the Carolines to 15°, he insisted that the phenomenon was devoid of navigational significance. The following year the same navigator told Edwin Doran (personal communication 1970) that he *had* drawn conclusions from the observation. I fear that the explanation may lie in Hipour having noticed me surreptitiously checking the star's height by loosely extending my hand at arm's length. However, the equally celebrated Satawal navigator Repunglug, who also made a return voyage to Saipan – by sailing canoe in 1970 – expressed the Pole Star's altitude in a local unit, *ee-yass*, one and a half at Satawal, two at Saipan (McCoy, personal communication 1970). The correct ratio, whatever the unit of measurement, should have been 1:2. Here the matter rests.

THE SIGNIFICANCE OF CERTAIN CONCEPTS AND OF THEIR DISTRIBUTION

Weather prediction by the stars

Atmospheric signs apart, star-weather correlations are necessarily seasonal. Predictions of day-to-day changes can have no more validity than 'stars foretell' columns. What is so significant is the near-universal distribution throughout Oceania of erroneous beliefs to the contrary.

Captains Cook (Cook & King 1784, p. 144) and Andia (Corney 1913–19, vol. 2, p. 284) refer to star-weather divination in Tahiti, where one sign was the 'bending' of the Milky Way by the wind. The three stars Canopus, Sirius and Procea (together called *Maan*) were pointed out to me as controllers of wind direction and weather on the 'para-Micronesian' atoll of Ninigo. My informant, Itilon, had confidently relied upon them during numerous inter-island sailing canoe voyages of up to 96 km in length. Gilbertese beliefs about sidereal control of currents and weather were mentioned earlier. Gladwin (1970, p. 212) found stellar weather forecasting to be an essential ingredient in trained Carolinian navigators' repertoires. He makes the intriguing observation that the forecasts of the two navigational schools, *warieng* and *fanur*, more often contradict each other than are congruent, the fine weather stars of one school being frequently the ones that foretell bad weather in the other. It strains one's credulity to imagine that the same kind of erroneous deduction about stars influencing weather could have been made quite independently in corners of Oceania so remote from each other as are Tahiti, Ninigo and the Carolines.

Concepts possibly derived elsewhere

There is a Maori legend of the Sun travelling southward to join *Hine-takurua*, the Winter Maid and, after the June solstice, returning northward to *Hine-raumati*, the Summer maid (Best 1922, p. 14). The myth seems logical enough to us here in the Northern Hemisphere, but for New Zealand, in the 30s and 40s of south latitude, it reverses summer and winter. Makemson (1941, p. 85–6) suggests that the story represents memories of a sojourn in the Northern Hemisphere. An alternative explanation would be borrowing from Hawaiians or Micronesians.

A concept that the Maori may well have carried southward from tropical Polynesia is their identification of Rigel (*Puanga*) with the zenith (Makemson 1941, p. 30). This star's A.D. 500 declination, as worked out by Dr Radhakrishnan, was 11° 06′ and New Zealand does not anywhere extend north of 34° S. But neither do Tahiti, about 17° S, nor Samoa, 13° 30′, though closer, really fit the bill.

Nevertheless, we cannot entirely dismiss the possibility of Rigel-zenith identifications reflecting one-time real observations because, as we have seen, Rigel is the apex of the sky dome in the equatorial Gilberts (Grimble 1931, p. 198). Did the Gilbertese, then, originate further south? A major Samoan immigration actually took place around A.D. 1400 (Maude 1963, p. 7). But Rigel's declination in A.D. 1000 was 9° 52′ S, so even then it did not culminate over Samoa, whose northernmost point reaches only up to 13° 30′ S. It was a zenith star for Samoa, however, in 500 B.C., which was about the time the archipelago was settled.

As so often in the field of Pacific astronomy, we are left with contradictory evidence, pointing to no definite conclusion.

A concept seemingly locally arrived at

This is the Tahitian and Samoan belief that the sun descends into the sea each evening and traverses a submarine passage during the night to arise in the east next morning (Williamson

1933, p. 113). The Tahitians alleged that people on Borabora, farther west, had heard the hissing as the sun plunged into the ocean at sunset (Ellis 1839, vol. 3, p. 170). Such an idea could hardly have entered the minds of anyone except islanders conscious of their sea horizons.

Discussion

The pan-oceanic distribution of essentially similar astronomical, astro-navigational and sidereo-meteorological concepts is remarkable, considering the probably diverse origins, both temporal and geographical, of the elements that went into the formation of Micronesian and Polynesian populations and cultures, and the relatively modest scope of voyaging at European contact. Something like 800 km without intervening land was about the limit of recorded deliberate journeys, though marine technology and nautical astronomy were perfectly adequate, when the target was an archipelago for 3200 km (Lewis 1972, pp. 21, 101–102). There is, of course, ample evidence of the importance of accidental drifts in settlement and inter-group encounters (Sharp 1963; Golson 1972; Riesenberg 1965), but recent computer studies have shown (within the limits of the model) that pure drifts could not have been responsible for the crossing of a number of the long seaways that we know must have been traversed. The sailing of consistent cross-wind courses would have been required (Levison, Ward & Webb 1972).

We are left with the alternatives of an unlikely degree of pre-Pacific cultural homogeneity among Polynesian and Micronesian precursors, or of astro-navigationally orientated voyages – whether planned or forced is immaterial – having been longer and more frequent in earlier times. That the latter is the correct explanation is suggested by evidence from various areas of voyaging which suffered a pre-European decline. The reasons are quite obscure, though increased agricultural efficiency could well be among them.

There is time for but one example. Tupaia told Cook that his father had known of islands farther south than he did (Beaglehole 1955, p. 157). More significantly, Tupaia's own enormous geographical horizons far exceeded the range of eighteenth-century Tahitian voyaging. His information could not have been derived, in any large measure, from drifters, since the overwhelming majority of trade wind drifts have been from east to west, that is, from Tahiti towards Tonga and Samoa (Golson 1972). But the Tongans, who were the leading seafarers in contact times, and stood to gain most from involuntary arrivals, had but vague ideas about Eastern Polynesia, whereas Tupaia could name and point towards any number of Tongan, Samoan and Fijian islands. Myths of islands having drifted apart would appear to symbolize the severence of once-close relations. One such story relates how Rarotonga became separated (by 800 km) from Raiatea on account of sacrilege (Williams 1846, pp. 47–48, 88). Another is the tale of a (400 km) 'land bridge' once having existed between Niue and Tonga and having long since sunk (Loeb 1926, p. 12).

But the bold range of Pacific voyaging, even in contact times, and the excellence of the indigenous navigational systems leaves us to speculate with Doran, Kehoe and Jett (in Riley 1971), on a more general question: whether the oceans of the world could not once have been pathways to a far greater extent than we generally allow. Of course, the fallacy of assuming that our fragmentary data about the astronomy and astro-navigation of eighteenth to twentieth century Oceania can open some kind of ethnographic window into the neolithic past of continental cultures, earlier by millennia and up to half the Earth's circumference away, is too obvious to labour. But observations from the Pacific may be indirectly relevant.

They demonstrate, for instance, the surprising efficacy of non-instrumental navigational techniques for finding even small targets, thus removing one major objection to trans-oceanic pre-instrumental voyaging in general. Secondly, they draw attention to the possibility that certain astronomically oriented megalithic sites elsewhere may have had navigational as well as other correlates. Should confirmation of such association be forthcoming, neolithic Oceania may be shown not to have been alone in conceiving as one single entity the 'pre-sciences' of navigation, geography and astronomy.

The author is most grateful to Mr P. W. Gathercole for discussing the paper with him and presenting it at the meeting in his absence.

REFERENCES (Lewis)

Akerblom, K. 1968 *Astronomy and navigation in Polynesia and Micronesia.* Stockholm Ethnographical Museum Monograph no. 4.

Alkire, W. H. 1970 Systems of measurement on Woleai Atoll, Caroline Islands. *Anthropos* **65**, 40.

Beaglehole, E. & Beaglehole, P. 1938 Ethnology of Pukapuka. *Bull. Bernice P. Bishop Mus.* no. 150. Honolulu: Hawaii.

Beaglehole, J. C. (ed.) 1955 *The journals of Captain James Cook on his voyages of discovery,* vol. 1. Cambridge University Press for the Hakluyt Society. Addenda and corrigenda 1968.

Best, E. 1922 *The astronomical knowledge of the Maori.* Wellington: Dominion Museum Monograph no. 3.

Bryan, E. H. Jr. 1936 Letter to Captain Haug, personal communication 1970.

Buck, P. 1938 Letter to Captain Haug, personal communication from Bryan 1970.

Collocott, E. E. V. 1922 Tongan astronomy and calendar. *Occ. Pap. Bernice P. Bishop Mus.* no. 8, 157. Honolulu: Hawaii.

Cook, J. & King, J. 1784 *A voyage to the Pacific Ocean ... in the years* 1776–80. (3 vols.) London: Nicol & Cadel.

Corney, B. G. (ed.) 1913–19 *The quest and occupation of Tahiti by emissaries of Spain during the years* 1772–6. (3 vols.) London: Hakluyt Society.

Ellis, W. 1839 *Polynesian researches ... in the Society and Sandwich Islands.* (2nd edn.) (4 vols.) London: Fisher, Son and Jackson.

Fornander, A. 1878 (vol. 1) 1880 (vol. 2) *An account of the Polynesian race.* London: Trübner.

Forster, J. R. 1778 *Observations made during a voyage round the World (in the 'Resolution'* 1772–5). London: G. Robertson.

Gifford, E. W. 1929 Tongan society. *Bull. Bernice P. Bishop Mus.* no. 61. Honolulu: Hawaii.

Gill, W. W. 1876 *Myths and songs from the South Pacific.* London: King.

Gladwin, T. 1970 *East is a big bird.* Cambridge, Mass.: Harvard University Press.

Golson, J. (ed.) 1972 *Polynesian navigation. A Symposium on Andrew Sharp's theory of accidental voyages,* 3rd edn. Wellington: Reed for the Polynesian Society.

Goodenough, W. H. 1953 *Native astronomy in the central Carolines.* Philadelphia: University Museum, University of Pennsylvania.

Green, R. C. 1966 Linguistic subgrouping within Polynesia: the implications for prehistoric settlement. *J. Polynes. Soc.* **75**, 7.

Green, R. C. 1972 Revision of the Tongan sequence. *J. Polynes. Soc.* **81**, 79.

Grimble, A. 1931 Gilbertese astronomy and astronomical observances. *J. Polynes. Soc.* **40**, 197.

Grimble. A. n.d. MS. notes, Kings of Butaritari and Makin as narrated by informant 1938 (by courtesy of H. E. Maude).

Groube, L. M. 1971 Tonga, Lapita pottery and Polynesian origins. *J. Polynes. Soc.* **80**, 278.

Haddon, A. C. & Hornell, J. 1938 Canoes of Oceania. *Spec. Publs. Bernice P. Bishop Mus.* nos. 27–29 (1936–1938). Honolulu: Hawaii.

Hale, H. 1846 *Ethnology and philology of the Wilkes exploring expedition.* Philadelphia: Lee and Blanchard.

Henry, T. 1907 Tahitian astronomy. *J. Polynes. Soc.* **16**, 101.

Henry, T. 1928 Ancient Tahiti. *Bull. Bernice P. Bishop Mus.* no. 48. Honolulu: Hawaii.

Heyerdahl, T. 1961 *Easter Island and the East Pacific* (vol. 1), *Archaeology of Easter Island.* Santa Fé: Monographs of the School of American Research and The Museum of New Mexico, no. 24, pt. 1.

Hilder, B. 1959 Polynesian navigational stones. *J. Inst. Nav. (Lond.)* **12**, 90.

Hilder, B. 1963 Polynesian navigation. *Navigation (J. Inst. Nav. Wash.)* **10**, 188.

Hilder, B. 1972 Primitive navigation in the Pacific, in Golson, J. (ed.) *Polynesian navigation,* 3rd edn. Wellington: Reed for the Polynesian Society.

Hornell, J. 1935 Constructional parallels in Scandinavian and Oceanic boat construction. *Mariner's Mirror* **21.**

Kamakau, S. M. 1891 Instructions in ancient Hawaiian astronomy as taught by Kaneakahoowaha . . . (translated from *Nupepa Kuoka* of 5 Aug. 1865, for the *Maile Wreath* by W. D. Alexander). Honolulu: Thrum's Hawaiian Annual.

Kepelino. 1932 Traditions of Hawaii, Martha Warren Beckwith (ed.). *Bull. Bernice P. Bishop Mus.* no. 95. Honolulu: Hawaii.

Laval, H. 1938 *Mangareva, L'histoire ancienne d'un peuple polynésien.* Brain-le-Compte: Maison des Pères des Sacrés Coeurs de Picpus.

Levison, M., Ward, R. G. & Webb, J. W. 1972 *The settlement of Polynesia: a computer simulation.* Minneapolis: University of Minnesota Press.

Lewis, D. H. 1967 *Daughters of the wind.* London: Gollancz; Wellington: Reed.

Lewis, D. H. 1971 A return voyage between Puluwat and Saipan using Micronesian navigational techniques. *J. Polynes. Soc.* **80,** 437.

Lewis, D. H. 1972 *We, the navigators.* Canberra: Australian National University Press; London: Angus and Robertson; Honolulu: Hawaii University Press; Wellington: Reed.

Loeb, E. 1926 History and traditions of Niue. *Bull. Bernice P. Bishop Mus.* no. 32. Honolulu: Hawaii.

Makemson, M. 1941 *The morning star rises.* New Haven, Conn.: Yale University Press.

Maude, H. E. 1963 The evolution of the Gilbertese Boti. *Mem. Polynes. Soc.* no. 35. Wellington.

Maude, H. E. & Maude, H. C. 1932 The social organisation of Banaba or Ocean Island. *J. Polynes. Soc.* **41,** 265.

Nyeret, J. M. 1967 Piroques Océaniennes. *Neptunia* no. 85, (supplement *Triton,* no. 81).

Pawley, A. 1966 Polynesian languages: a subgrouping based on shared innovations in morphology. *J. Polynes. Soc.* **75,** 39.

Riesenberg, S. H. 1965 Table of voyages affecting Micronesian islands. Supplement to Simmons, R. *et al.* Blood group genetic variations . . . Micronesia. *Oceania* **36,** 156.

Riley, C. L. *et al.* (ed.) 1971 *Man across the sea: problems of pre-Columbian contacts.* Doran, E. Jr. The sailing raft as a great tradition, 115; Kehoe, A. B. Small boats upon the North Atlantic, 275; Jett, S. C. Diffusion versus independent development, 16. Austin and London: University of Texas Press.

Sabatier, E. 1939 *Sous l'équateur du Pacifique. Les Iles Gilbert et la mission catholique* 1888–1938. Paris: Sacré-Coeur, ed. Dillen.

Sanchez y Zayas, E. 1866 The Marianas Islands. *Nautical Mag.* no. 35.

Sharp, A. 1963 *Ancient voyagers in Polynesia.* Auckland and Hamilton: Paul's Book Arcade; Sydney: Angus and Robertson.

Transactions of the London Missionary Society (Tahiti). Mission diary 19 April 1806. Canberra: A.N.U. Microfilm PMB 72.

Ward, E. V. 1967 *Sailing directions, the Gilbert and Ellice Islands Colony.* Tarawa: Secretariat G. and E.I.C.

Wilkes, C. 1845 *Narrative of the United States exploring expedition during the years* 1838–42 (5 vols.). Philadelphia: Lea and Blanchard.

Williams, J. 1846 *A narrative of missionary enterprises in the South Seas.* London: Snow.

Williamson, R. W. 1933 *Religious and cosmic beliefs of Central Polynesia* (2 vols.). Cambridge University Press.

Winkler, Capt. 1901 On sea charts formerly used in the Marshall Islands, with notices on the navigation of the islands in general. *Rep. Smithson. Instn.* no. 487. Washington 1889.

APPENDIX: CONTEMPORARY INDIGENOUS AUTHORITIES

It would be anomalous not to recognize the precepts and demonstrations of these trained masters of astronomy and astro-navigation as primary sources, and important ones at that. They are acknowledged in the text as 'personal communications' and are listed in this appendix to the references, together with something of their qualifications and status, in lieu of references to published work.

Abera of Nikunau, Gilbert Islands, is a trained navigator (*tia borau*). He makes frequent sailing canoe voyages among the southern and central Gilberts.

Hipour of Puluwat, Carolines, is a highly trained initiated navigator (*ppalu*). He has been voyaging for years in his canoe over an east–west range of something like 1300 km.

Iotiebata of Maiana, Gilberts, is a *tia borau* who makes frequent canoe passages between Tarawa and Maiana. Once 5 weeks storm-drifted, he kept his bearings by stars, Sun and ocean swells and, eventually, recognized land clouds.

Itilon of Ninigo, Manus Group, is a canoe captain and star navigator, who has visited islands up to 100 km away in his 16 m vessel.

Kaho, Hon Sione Fe'iloakitau of Tonga is the 88 year-old great grandson of the famous blind Tuita navigator, Kaho Mo Vailahi. He is the uncle of the present Tuita title holder.

Rafe of Tikopia learned navigation by surreptitiously listening to the elders, who were withholding instruction because of losses of youths at sea. He captained canoes to Vanikoro (180 km) and to New Hebrides (175 km).

Rewi of Beru, Gilberts, is the *tia borau* who was instructed by his father with the aid of the 'stone canoe'.

Samoa of Tikopia was trained by the elders. He has made the canoe voyage to Vanikoro.

Teeta of Kuria, Gilbert Islands, is a *tia borau* trained in the full classical *maneaba* tradition. He made secret wartime voyages without chart or compass throughout most of the archipelago.

Tupui of Tikopia was trained by the elders. He has navigated the Tikopia–Anuta return canoe voyage (114 km each way).

Ve'ehala, Hon., Governor of Ha'apai, Tonga, formerly of the Tradition Department, is the most knowledgeable member of the Tuita navigator clan. His chief informants were an old woman, Makelesi Lakoti, who had been taught the chants by the grandfather of the present Tuita, and his own grandfather Ve'eto, who was over 90 when he died in 1959.

Phil. Trans. R. Soc. Lond. A. **276**, 149–156 (1974) [149]

Printed in Great Britain

Astronomical significance of prehistoric monuments in Western Europe

By A. Thom

'Thalassa', The Hill, Dunlop, Ayrshire, Scotland

[Plate 22]

The accuracy of megalithic man's linear measurements is shown by recent work in Orkney and Brittany. Everywhere his geometry is based on integral right-angled triangles. The Sun was observed at the solstices and equinoxes. The fact that the foresights for the latter show a declination of $+0.5°$ and not $0°$ proves that the year was divided into two equal parts. By similar analysis of other foresights the megalithic calendar has been established.

The evidence that the Moon was observed in its extreme positions is extensive but the most interesting sites are those which show the small perturbation of the inclination of the lunar orbit. To establish the necessary sight lines some method of extrapolation was necessary. The actual sectors used for this are found in Caithness and in Brittany. Alinements at Le Ménec perhaps provide an adjustment for the varying speed of the Moon in its orbit.

1. The megalithic yard in Brittany and Orkney

The results of our researches up to 1970 have been described in two books (Thom 1967, 1971). Here we propose to give briefly some results of the recent work done by the groups who worked with us in 1970, 1971 and 1972 in Brittany and Orkney.

It has been said that the megalithic yard (0.829 m) is simply the length of a man's step and that megalithic circles were set out by pacing. The megalithic culture lasted for a long time and it may be that the earlier circles were set out by pacing but undoubtedly at some time or another an accurate standard yard was established, and universally adopted. Anyone who takes the trouble to study and master the geometry of Avebury will see that it is inconceivable that the Avebury ring was set out by pacing (Thom 1967, p. 89).

Let us look at what we find in Brittany. The Ménec alinements consisted originally of 12 rows about a kilometre long. Most of the 1600 stones carry the re-erection mark and as the people who re-erected them had no knowledge of the original geometry, or indeed of any geometry, the stones are often badly displaced. In 1970 the parties working with us made the first accurate large, 1:500, scale survey of the site. A statistical analysis of the west end shows that the rows were set out originally with a quantum of $2\frac{1}{2}$ megalithic yards which will be called the megalithic rod. Each row started with a node on a line crossing the rows at 1 in 2. After the gap, where the stones have been removed from the agricultural land, we re-determined the quantum and the nodes. These nodes are shown in figure 1 relative to a transverse line. The figure also shows the nodes as brought forward by calculation (using a rod of 6.800 ft or 2.073 m) from the starting line at the west. Considering that the rows are over 900 m long the agreement is remarkable and cannot be due to chance. It disposes completely of the idea that pacing was used and it gives us a value for the megalithic yard of 2.721 ± 0.001 ft or 0.8294 ± 0.0004 m.

1600 km to the north in Orkney the Ring of Brogar gives us an almost identical value. The ring consists of tall flat stones forming a true circle obviously intended to be 50 rods in diameter. Using all the stones and stumps as they are today we find a mean diameter of 340.0 ± 0.6 ft (~ 103.7 m), or neglecting two stones known to have been re-erected 340.7 ± 0.4 ft (~ 103.9 m) so that the megalithic yard from the Brogar Ring is probably between 2.720 and 2.725 ft (0.829–0.831 m). For Britain as a whole I gave 2.720 ± 0.003 ft in Thom (1967). The above values speak for themselves. They could not have been produced by pacing.

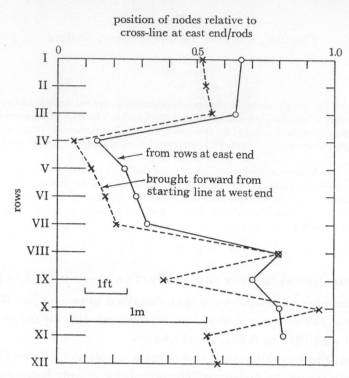

FIGURE 1. Displacement of the nodes in the Ménec rows after 900 m.

2. THE ASTRONOMY OF MEGALITHIC MAN

(a) General

It is necessary to be quite clear about the observing procedure, used by megalithic man at his lunar observatories, which was to observe the setting and rising Moon on two (or three) successive nights straddling the monthly declination maximum. On each night a stake was put in the ground showing the position from which the Moon's limb appeared to graze the foresight. Figure 2 shows on the left the stake positions A and B for 2 nights' observations and on the right at A, B and C for 3 nights' observations. To make the explanations simpler we shall assume that the observer, instead of working along a straight line across the line of sight, moved forward a constant distance $2a$ each night. His stake positions are then as shown at A' and B' (or A', B' and C'). These points lie on curves which are, to some scale, a graph of the lunar declination. The sagitta G is not constant but the change in 2 days will not effect the result. For three stakes it can be shown that the amount by which the stake B has to be advanced to make it correspond to the maximum declination is $\eta = p^2/16G$. This can be compared with the expression $\eta = p^2/4G$ for the two-stake case fully discussed in (Thom 1971). In both cases $2p$ is the distance between the first and last stakes. If G is assumed to be constant then only 2 nights' observations are required, and the value of $p^2/4G$ can be found using a grid or sector such as we find at six observatories, four in Caithness and two in Brittany. If R is the radius of the sector then,

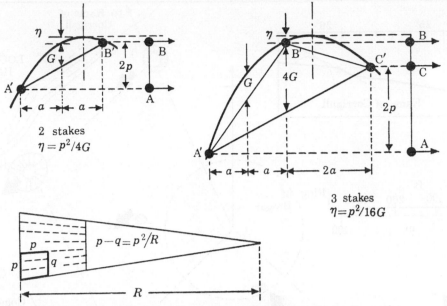

FIGURE 2. Extrapolating to the maximum from two or three observations.

referring to figure 2, by similar triangles $p - q = p^2/R$. Hence the radius R should be $4G$ for the two-stake and $16G$ for the three-stake case. Today when we know the distance to the foresight we can calculate G and so we can show that all the known sectors have approximately the correct radius for the two-stake case. Evidently at these observatories G was assumed constant from lunation to lunation. The three quantities, parallax, semidiameter and the sagitta G all rise or fall as the distance of the Moon from the Earth rises or falls. This distance goes through a cycle in an anomalistic month (perigee to perigee). If at a particular standstill perigee happened to coincide with the declination maximum (period = 1 tropical month), at the next standstill, 18.6 years later, perigee would be only 3 days behind the monthly declination maximum. This means that for several standstills megalithic man would make his observations at a time when the sagitta was nearly a maximum and only after 90 years would he find a minimum. If, for example, the Mid-Clyth sector was built between 1670 and 1570 B.C. then (Thom 1971, figure 7.2) we see that we ought to expect it to have a sector radius less than that calculated with a mean G. In fact the radius *is* low. This happens to lend some confirmation to the date of about 1600 B.C. found at the only reliable solstitial site so far examined in Caithness, namely that at Cnoc na Maranaith (map reference 132332).

(b) Brogar, Orkney

Figure 3, plate 22, shows the Brogar site from the air. Three of the cairns close to the Ring can just be seen. We made a tacheometric survey (with levels) of the whole site and carefully measured the profiles of the hills in the directions indicated by the cairns. A small-scale copy of this survey will be found in Thom (1973). The azimuths were carefully measured from the centre so that the declination indicated by the notches on the profiles could be estimated from any point on the survey. An examination of figure 4 shows that we are here dealing with a lunar observatory of a most remarkable nature. No less than three natural foresights were used, one for the major standstill and two for the minor (rising and setting). Backsights still remain, showing for each foresight three of the values of $\pm(\epsilon \pm i \pm s \pm \varDelta)$ where ϵ is the obliquity of the

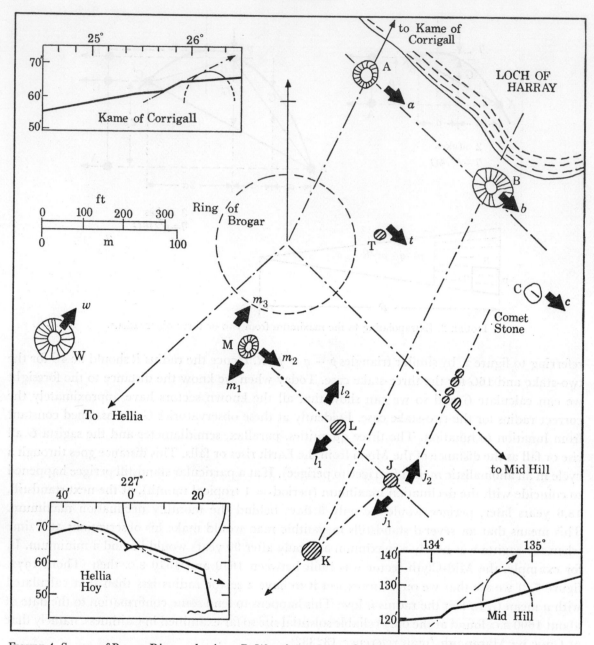

FIGURE 4. Survey of Brogar Ring and cairns: B, W, cairns or mounds about 4.25 m high; A, M, despoiled cairns
now about 1.8 m high; C, low mound with Comet Stone oriented on Mid Hill; T, mound recorded in 1849
(Thomas 1851) but not noticed in 1972. The other eight mounds are all low but definite. Arrow heads show
the positions which an observer would ideally have occupied to obtain the declinations listed below:

$$
\begin{array}{lll}
a \text{ and } b & \text{declination} = -(\epsilon - i + s) & \left.\begin{array}{l}\\ \\ \\ \end{array}\right\} \text{ on Mid Hill} \\
m_2 & \qquad\qquad\quad\; -(\epsilon - i - s - \varDelta) & \\
t & \qquad\qquad\quad\; -(\epsilon - i) & \\[6pt]
j_1 & \qquad\qquad\quad\; -(\epsilon - i) & \left.\begin{array}{l}\\ \\ \\ \end{array}\right\} \text{ on Hellia notch} \\
l_1 & \qquad\qquad\quad\; -(\epsilon - i + s - \varDelta) & \\
m_1 & \qquad\qquad\quad\; -(\epsilon - i + s) & \\[6pt]
m_3 & \qquad\qquad\quad\; +(\epsilon + i + s - \varDelta) & \left.\begin{array}{l}\\ \\ \\ \end{array}\right\} \text{ on Kame of Corrigall} \\
l_2 & \qquad\qquad\quad\; +(\epsilon + i + s) & \\
j_2 & \qquad\qquad\quad\; +(\epsilon + i + s + \varDelta) &
\end{array}
$$

FIGURE 3. The Ring of Brogar from the air.

ecliptic, i is the inclination of the lunar orbit, s is the Moon's semidiameter and Δ is the small perturbation found by Tycho Brahe. It is not yet possible to be certain which point on the Hellia profile was intended to be used, but fortunately there are no alternatives at Kame of Corrigall. Taking mean values for i, s and Δ we are able to deduce that the obliquity of the ecliptic which best fits the foresights is about $23°$ $52.9'$ (1570 ± 100 B.C.). This enables us to estimate where the observer should have ideally placed the backsights for various combinations of ϵ, i, s and Δ. These positions are shown on the figure by broad arrow heads. It will be seen that in all cases the arrows lie close to one or other of the cairns. The arrows for the cases not shown lie either in the cultivated land to the north, where the smaller cairns would have been ploughed out, or else in Loch Harray.†

The ground at the site rises somewhat irregularly from the shore to the ridge on which the three cairns M, L and J are situated. The level ground at the top provided the necessary freedom of movement for the observer or observers ranging themselves into positions so that the Moon grazed the foresight at Hellia, and 9.3 years later at Kame of Corrigall. When Mid Hill was being used the observer perhaps moved along a line from near M to near A. The Comet Stone is orientated on the foresight but it may have been a warning stone and it is not clear yet why it was not placed further back. The foresights are from 6 to 13 km distant so that the errors shown in the position of the backsights are, in declination, very small (mostly less than a minute). It is remarkable how much of this site has survived. A modern tractor plough could remove all trace of the smaller cairns in a single season. It is little wonder that we have so much difficulty in understanding many of the other Western European sites situated on arable land or near buildings.

(c) Brittany

In figure 5 we see the immediate neighbourhood of the Carnac alinements and some of the nearby menhirs. The 6.1 m high menhir M on the hill top at Manio would, in the absence of the trees which now surround it, have been visible from all round. We have now run long accurate traverses from M through the woods in various directions and so determined that accurate lunar and solar declinations are given by M as viewed from a number of marked sites details of which will be found in figure 5. On the large-scale map there are stones at Keriaval and these also appear to give lunar declinations with M.

But evidently this observatory was not considered satisfactory. Perhaps some of the lines were too short and the terrain prevented them being lengthened. Whatever the reason the Manio site was replaced by the huge lunar observatory centred on Er Grah or Le Grand Menhir Brisé, the largest artificially cut and dressed stone in Europe. The position of this stone, now fallen and broken in four parts totalling some 340 tons, is shown in figure 6. It is placed so that there were level positions for observing at rising and setting all four cases of $\pm (\epsilon \pm i)$, but whether these were all completed we do not know. There are stones in the right place at Kerran, Kervilor, Le Moustoir and Quiberon, and suitable extrapolating sectors at Petit Ménec and St Pierre. We failed to find anything like a backsight near St Pierre until in 1972 we came upon a large menhir lying on the beach near high tide mark. The gradually rising level of the sea has cut into the land leaving the menhir where it slid down from its original position. Careful

† Some accurate check measurements made in Orkney in 1973 show that the foresights at Brogar do not behave as two-dimensional silhouettes when the observer changes his position. Consequently the arrows in figure 4 may be somewhat displaced. It is proposed to make measurements from each backsight independently.

We also found traces of another cairn as shown on the Ordnance Survey at the intersection of the lines KB and MJ.

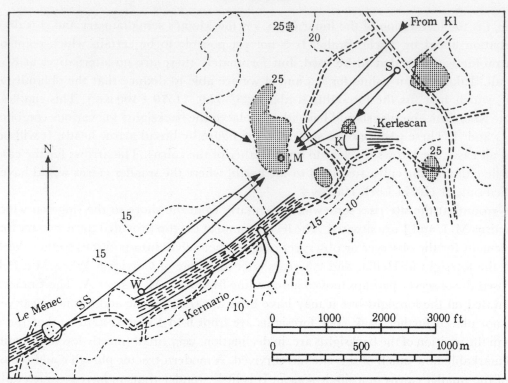

FIGURE 5. Position of menhir at Manio relative to the Carnac alinements. M, menhir at Manio 6.1 m high; S, menhir (2.75 m) to M gives solar declination at winter solstice; K1, small menhir near Kerlagad to M gives lunar declination; W, group of fallen menhirs close to the alinements to M gives lunar declination; SS, inside Le Ménec east cromlech to M gives solar declination at summer solstice; K, menhir (3.7 m) to M gives a calendar date; and S to K, gives a lunar declination.

calculation based on the large scale map shows it to be in the correct position but this remains to be verified by actual astronomical observations in 1973.

Megalithic man could not have built and operated for many years an observatory of this size and accuracy without noticing the effects of the changes in G. We suggest that his natural reaction was to build a sector longer than normally necessary, such as we find at Petit Ménec. If G were determined at each lunation there would always have been a sufficient length of sector. But to find G necessitated observations on three nights and so presumably the next step was to make use of these three observations directly. The length of the Ménec alinements up to the bend is $16G$ for two of the nearby backsights and the base width at the west end is close to $4G$, the greatest value of $2p$ normally needed. Perhaps at first there was a single sector here, of radius $16G$, built of the larger stones which now form the west end of the lines. But it was almost certainly standard procedure to use for some of the observations the technique of having an observer for each limb and it is possible that the fact that the distance between the two, i.e. the ground equivalent of the Moon's diameter, was approximately equal to $4G$, led the observers to combine these two lengths by taking an average. We know today that this was theoretically wrong but it seems that megalithic man did evolve a peculiar empirical method using the average and in Thom (1972 b) it is shown that Le Ménec alinements can be used with this method to give the required extrapolation distance with sufficient accuracy. Here, without going into detail, we shall simply show the remarkable relation which emerges from the analysis of the Ménec rows.

Let the distances of a row from AB (figure 7) at the west end and at the bend be a and b. It is

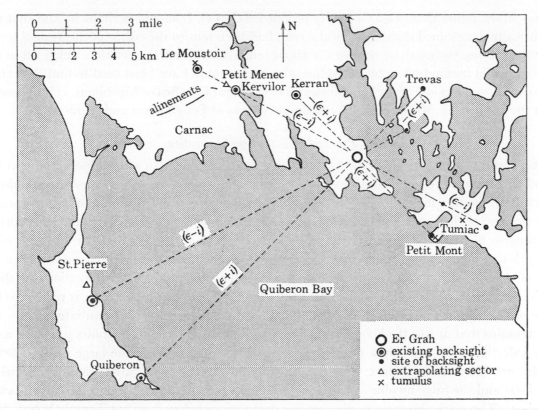

FIGURE 6. Le Grand Menhir Brisé (Er Grah) as a universal lunar foresight.

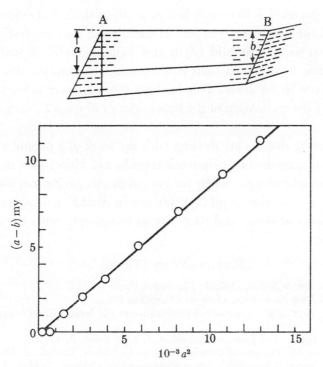

FIGURE 7. The quadratic relation in the Ménec alinements.

shown in the figure that $(a-b)$ is closely proportional to a^2. It should perhaps be said that the statistically determined dimensions of the rows had been sent to the editor of the *Journal for the History of Astronomy* before we noticed the above remarkable quadratic relation which led to the discovery of a method whereby the Ménec alinements could have been used to find the extrapolation distance of any of the four backsights to the west of the Sea of Morbihan. It is suggested that the rows at Le Ménec replaced the simple sectors at Petit Ménec and St Pierre.

(d) Pre- and post-standstill observations

It is useful to speculate about the methods used by megalithic man at the observatories as the standstill approached. He must have started observing many months before the standstill. At each declination maximum he would establish a position on the ground marked by a stake. The position of these stakes would oscillate with a period of 173 days and an amplitude corresponding to 9′ (Δ). This would of course be superimposed on the gradual rise in declination towards the standstill maximum. I have elsewhere suggested how the observers may have made use of several oscillations occurring before and after the standstill. Observations of these oscillations could have indicated the date of the standstill and how this date was related to the time of the nearest oscillation maximum. Whether this suggestion be substantiated or not the fact remains that at 5 of the known lunar observatories there are extra menhirs so placed as to give a declination about half a degree less than the maximum. Thus the large menhir 200 m to the northwest of Le Ménec village gives with the Manio stone, a declination 29′ below $(\epsilon - i - s)$ and the huge menhir called Goulvarh near Quiberon, shows a declination 27′ below $(\epsilon + i - s)$. There are also extra stones showing declinations below $(\epsilon + i)$ at Temple Wood, Mid Clyth and Yarrows (Thom 1971, pp. 50, 94 and 99).

3. CONCLUSION

A criticism made of our work is that we failed to produce statistical evidence that megalithic man actually observed the lunar perturbation. It seems to us that the testimony of the Brogar cairns, backing up what we find at Mid Clyth and Temple Wood, is such that it cannot be gainsaid. If this evidence be accepted then it is legitimate to analyse the combined material from the 25 sites as given in my book. This analysis shows among other things that the sites yield a mean value for i the inclination of the lunar orbit of 5° 08′ 52″, very close to the modern value of 5° 08′ 43″.

The conclusions must be that we are dealing with the work of a people who had come far in the application of scientific method to astronomical problems. How many more of their achievements remain to be brought to light when we get rid of our prejudices and make a scientific attack on the evidence scattered over all NW Europe in tumuli, cairns, menhirs, petroglyphs, etc.? There is so much to be done, and there are so few people with the ability to make the necessary measurements.

REFERENCES (Thom)

Thom, A. 1967 *Megalithic sites in Britain*. Oxford: Clarendon Press.
Thom. A. 1971 *Megalithic lunar observatories*. Oxford: Clarendon Press.
Thom, A. & Thom, A. S. 1971 The astronomical significance of the large Carnac menhirs. *J. Hist. Astron.* **2**, 147–160.
Thom, A. & Thom, A. S. 1972*a* The Carnac alignments. *J. Hist. Astron.* **3**, 11–26.
Thom, A. & Thom, A. S. 1972*b* The uses of the alignments at Le Ménec, Carnac. *J. Hist. Astron.* **3**, 151–164.
Thom, A. & Thom, A. S. 1973 A megalithic lunar observatory in Orkney. *J. Hist. Astron.* **4**, 111–123.
Thomas, F. W. L. 1851 *Archaeologia* **34**, 88–136.

Phil. Trans. R. Soc. Lond. A. **276**, 157–167 (1973)   [157]

Printed in Great Britain

Astronomical alinements in Britain, Egypt and Peru

By G. S. Hawkins

Smithsonian Astrophysical Observatory, Cambridge, Massachusetts, U.S.A.

[Plates 23 and 24]

Photogrammetric air surveys have been made of Stonehenge, the Great Temple of Amon-Re, Karnak, and the desert lines near Nasca, Peru. The latter designs have been correlated in time with pottery of type Nasca 3 and 4. New astronomical alinements have been found for the trilithons and station stones at Stonehenge. The Amon-Re temple alined precisely, within the limits of present-day measurements, with the rising of the Sun at midwinter at the time of rebuilding of the structure by Tuthmosis III. No significant astronomical alinements were found for the desert lines. The possible significance of the British and Egyptian alinements are discussed together with the implications of the negative result in Peru. Some comments are made concerning the precision of the megalithic yard, and the geometries and stellar alinements of prehistoric British structures.

1. Introduction

(a) Prehistory and astro-archaeology

If prehistory is defined to cover that period in various localities before the use of written records, then – neglecting those contemporary cultures that, although non-literate, can be studied by verbal communication – the problem of reconstructing the past becomes exceedingly difficult. As Childe (1947) and Daniel (1962) asked: 'How far was it legitimate to infer non-material facts about prehistoric man from the material remains that survive'? Those non-material facts are the perishable qualities of a culture – ideas and ideals, philosophy and religion, knowledge and awareness of the environment. Here we are forced to make guesses, as Daniel puts it; nevertheless, we must make intelligent guesses, while guarding against the dangers of excessive speculation. In the worst extreme, the inherent vagueness of prehistory has been taken advantage of to defend a particular scheme or theory beyond warrantable logic – the hyperdiffusionist theory of a 'heliolithic' race, the theory of unilinear cultural evolution, and the reading of prognostications from the physical dimensions of the Great Pyramid. Yet despite such episodes, which lie so heavily in the academic literature, it is precisely those non-material questions relating to the thought processes of ancient man that are of interest to the modern scholar. Indeed, it is the direction of research indicated by the title of this discussion meeting.

Clark (1970) suggests that, by combining various disciplines within prehistoric studies, we should be able to reconstruct some assessment of the levels of perception and self-awareness, the behaviour patterns, and perhaps the ideology of early man. In reconstructing man's perception of the environment formed by the sky – the observation and marking of the movements of Sun, Moon, and stars – Hoyle (1966a) considers it an essential preliminary to ask 'How would *we* do it?' Of course, in this approach, one recognizes the enormous gap in time and culture, the so-called cognitive gulf, where the only common denominator is the calculable movement of the celestial bodies, and perhaps some parallels of response within the human brain.

To a large extent, astro-archaeology relies on material facts – the alinement of menhirs, post-holes, and other prehistoric structures with objects visible in the sky. These alinements, when verified by accurate survey and calculated to the given archaeological epoch, are material facts, as durable as arrowheads, burins, and burials. The alinements are artefacts in their own right, to be considered and assessed with other information in the study of a particular culture. There is no preconceived scheme or theory to be proved in astro-archaeology. Acceptance and evaluation are a matter for researchers in their various areas of expertise. However, acceptance of the alinements *per se* has a tacit but important corollary: the culture in question had knowledge of and an interest in the movement of astronomical objects as a function of time. It is beside the point to argue about the description of this knowledge. We are conditioned by our own vocabulary when we tend to describe the ancient endeavours as 'scientific', though for some prehistorians, as Daniel (1962) points out, the latter word is synonymous with 'modern'. Yet an equally valid description might well be 'magical understanding' (Schwartz 1965). The critical matter lies within the corollary that follows from acceptance of the alinements.

The author (1968) has suggested criteria to be applied in establishing the validity of alinements:

(i) *Construction dates should not be determined from astronomical alinements.*

(ii) *Alinements should be restricted to man-made markers.*

(iii) *Alinements should be postulated only for a homogeneous group of markers.*

(iv) *All related celestial positions should be included in the analysis.*

(v) *All possible alinements at a site must be considered.*

These standards are difficult to apply rigorously in all cases; nevertheless, they provide desirable prerequisites for credibility. The reasons for the choice of these criteria are given at length in the original publication. Early work had been made suspect by lack of attention to these points. Lockyer (1901), in his paper 'An attempt to ascertain the date of the original construction of Stonehenge from its orientation' was at odds with criterion (i). In his suppositions he set himself a series of traps. *If* (so the argument goes) the builders observed the Sun when it was 2′ above the horizon, and *if* they built the avenue to aline with this direction at the solstice, and *if* the engineering accuracy was at the limit of the resolving power of the human eye, then the construction date was 1680 B.C. ± 200 years. Each of these suppositions can be legitimately doubted, and the construction date can now be determined by unambiguous methods. In later work, Lockyer (1906) mixed tumulus and menhir, castle and ditch (see (iii)), used natural features such as hills (see (ii)), and tended to constrain the findings to a particular scheme (see (iv), (v)). His foray into egyptology (Lockyer 1894) collapsed almost entirely on the basis of ignoring criterion (i). Only a small part of Lockyer's work was valid, and archaeological opinion was prudent in rejecting this part with the whole. Earlier work on Maeshowe by Spence (1894) and on Stonehenge by Stukeley (1740), Smith (1771), and others, lacks the precision of survey and detail of computation and falls into the category of speculation.

There are number systems preserved in rings of holes and stones, in rows of objects, and in the notational markings in cave and mobiliary art. These again are material facts and are of significance in astro-archaeology when the numbers correlate with the natural periods of the Moon, seasons and eclipses. Acceptance of the correlation implies long-term observation and assessment by prehistoric man. At the lowest level, acceptance of the numbers has an important corollary: There was a number system, an understanding of numerical values, before the invention of writing.

Stonehenge presented the first comprehensive pattern of astronomical alinements and correlation of number sets with astronomical periods to be discovered (Hawkins 1963). These alinements have been accepted with varying degrees of reservation by Daniel (1964), Newham

(1964), Hoyle (1966b), Atkinson (1967), Newall (1967), Thom (1967a), and Clark (1970). The perception of astronomical periodicities by the builders of Stonehenge has been accepted by Hoyle, and the significance of the existence of numbers at the site by Daniel (1965).

(b) Sun and Moon extrema on the horizon

The pattern of alinements at Stonehenge consisted of the solstice extremes of the Sun (four positions) and the high and low positions of the Moon on either side of the solstice lines (eight), as shown schematically in figure 1. There was also evidence for east–west lines corresponding to

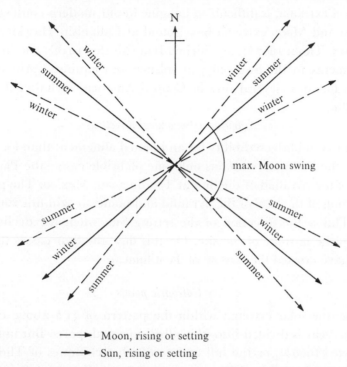

– – –▸ Moon, rising or setting
———▸ Sun, rising or setting

FIGURE 1. The four azimuthal extrema of the Sun at the solstices, and the eight extrema of the Moon.

equinoctial positions. The pattern would, of course, be modified by irregularities in the altitude of the skyline and vary with latitude. The observational phenomena have been fully described elsewhere (Hawkins 1965a). Briefly, the Sun appears to move from one arm of the pattern to another in a period of 6 months, whereas the Moon 'swings' in half a month. In addition, the swing of the moon is modulated by the regression of the nodes (direction of intersection of the plane of the Moon's orbit with that of the Earth) so that any particular phase (the full Moon nearest in time to the winter solstice, say) returns approximately to an extreme position after 18.61 years has elapsed. The seasonal Moon returns to an extreme more precisely after three cycles of almost exactly 56 years.

In passing, it must be said that this pattern of directions was found in the analysis of the structure. It was not a sought-after, preconceived scheme. Also the pattern was not laid out at the site from a common centre, as shown in figure 1. This is a representation more compatible with present-day thinking. Rather, the lines were offset, and, as will be demonstrated later in this paper, very few or perhaps none of the lines originated at the geometrical centre. This peculiarity, if we might call it such, together with the disregard of the conventional north–south meridian, may have tended unduly to delay the discovery of the underlying astronomical basis. Granting that the builders observed, noted, and set the stones and archways to point to natural phenomena, an astronomer comes across a further enigma, a humanistic one – the extrema have no urgent, practical value. The periodicities are, to be sure, connected with eclipses, ocean tides, and the calendric passage of time, but none of these factors would seem to justify materialistically the enormous amount of effort. The reasons seem to be creations of the prehistoric mind. It is sufficient to note that the Sun and Moon are the brightest luminaries, exhibit the greatest movement with respect to the stellar background, and, uniquely for the Moon, show an apparent change of shape from night to night. What meaning was placed on these extrema in prehistory, particularly the Moon extrema, is difficult to imagine in our modern context.

Alinements to Sun and Moon extrema have found at Callanish (Hawkins 1965 b), at other British megalithic sites (Thom 1967 b), at Chichen Itza (Morley 1956), and in France (Thom & Thom 1971). Alinements to the Sun only, at solstice and equinox, primarily for calendric purposes, have been found at several sites in Central America, including Uaxactun (Morley 1956) and Monte Alto.

(c) Singular astronomical directions

It is more difficult to establish credibility for an isolated alinement than for a pattern. To the author's knowledge, the literature contains only one plausible case – the Plataforma Adosada on the western side of the pyramid of the Sun at Teotihuacan, Mexico. The platform is skewed by 6° from the direction of the base of the pyramid and points to azimuth 290.95° east of north (Marquina 1951). This is the direction of the setting Sun when the declination is $+19.7°$, numerically equal to the latitude of the site. On this day, the Sun passes through the zenith and can be said to have crossed the 'tropic of Teotihuacan'.

(d) Calendric points

In a general sense, the solar extrema within the pattern of §1 b above can be said to be calendric because the year is divided into two, albeit unequal, parts. But unless we accept the offset theory of Hoyle (1966 b), or the hill and rim-flash techniques of Thom (1967 b), these solstice dates cannot be observed to the precision of a day. Spence (1894) suggested a year divided into quarters and marked by solar azimuths; Lockyer (1906) extended this megalithic calendar to eighths with prehistoric solar alinements in Britain, and Thom (1967 b), to sixteenths. Heliacal rising of certain stars, although calendric, is not within the present consideration, because the phenomenon is not alined with the horizon.

(e) Star alinements

A star rises at the same point on the horizon each night of the year. There is a slow drift in azimuth with a pattern of extrema somewhat similar to that shown in figure 1 and with a period of 26 000 years. This drift would become detectable to unaided vision after about 20 years, and a man-made alinement would thereafter become inoperative.

In the author's opinion, the rising of a star is an uninspiring phenomenon. Even the brightest star in the sky, Sirius, can be seen only on less than 10 % of the occasions when rising over a sea horizon (personal observations, Costa del Sol, Spain, 1966) and even then it is a faint object, scintillating into invisibility. Nor can the horizon be seen on a dark, moonless night; and when the Moon is up, visibility of the star is impaired. If star alinements exist, one would expect them to be toward stars brighter than magnitude zero at sites with elevated skylines.

Lockyer (1894, 1906) found star alinements for British megalithic monuments and for most Egyptian temples. His work was unsound as judged by criteria (i) to (v), and his attitude toward archaeological facts was pretentious. It is doubtful if any part of this early work is valid. Dow (1967) has suggested an alinement of the streets of Teotihuacan with Sirius, the Pleiades, and certain northern stars. Thom (1967 b) reported megalithic alinements with 13 stars ranging in brightness down to magnitude +2.† In Lockyer's and Thom's work, construction dates were assumed.

(f) Planets

Owing to the inclination of the orbit to the plane of the ecliptic, a planet takes up positions similar to those of the Moon, shown in figure 1, but the azimuthal swing and the periodicity are different for each planet. At the time of writing, no prehistoric planetary alinements have been found.

2. STONEHENGE, A NEW SURVEY

The alinements at Stonehenge are to Sun and Moon extrema (§ 1 b), as shown in figure 1, with the apparent exception of the omission of summer moonset lines, and with the possible addition of equinox lines. The original set (Hawkins 1963), derived from small-scale plans, was criticized by Atkinson (1966) and Hoyle (1966a) as showing misalinements greater than might be expected from the megalithic builders. Hoyle considered the misalinements to be deliberate offsets made to aid in observations of the extrema. At least part of the error was inherent in the charts themselves, in the reading of the charts, and in the determination of skyline altitudes from contour maps. A photogrammetric air survey was carried out, horizons were measured, and the alinements were redetermined (Hawkins 1971). Not all the lines could be checked, because some of the original data were derived from holes excavated in the past and returfed. The error was quoted in terms of the angular distance of the object above or below the skyline in the direction of the alinement, the azimuthal error being approximately a factor of 1.5 greater. In general, the new values showed greater accuracy with the root-mean-square error decreasing marginally from 1.1 to 0.9 %. In particular, the trilithon alinement to midsummer sunset (23 to 24 seen through 59 to 60)‡ increased in accuracy from an error of 3.2 to 1.7, midsummer moonrise (8 to 9 seen through 53 to 54) from 1.5 to 0.2, and the extreme moonrise (9 to 10 seen through 53 to 54) from 2.0 to 0.5. The error in the diagonal 91 to 93 remained excessively high (3.6) and was therefore deleted from the set. It would now appear that for station stone 91 to mark a moonrise extreme, some viewing line other than that through the centre of the monument would have to be postulated. Also, the extremely large error for midwinter moonset, declination $+18.7°$, if it were to be marked by sarcen 20–21 as seen through trilithon 57–58, was confirmed, the error in the original determination being $+5.1°$ and that in the new, $+4.7°$. In fact, the error is such that the Moon would set within but to the extreme right of the slot. One is not sure whether to identify this as a misalinement accepted by the builders, or a building error. To correct the 'error', the builders would have had to distort the symmetry of the architecture, because the observation lines corresponding to figure 1 are not exactly symmetrical. During the second analysis, it was noticed that station stone 93 was just visible through the archways, and in exactly the correct position, as estimated from the plan, to act as a marker for the Moon. This prompted a new analysis of the trilithons taken as a group standing separate in the absence of the sarcen circle. The results are given in table 1.

† There are 45 stars brighter than magnitude +2.
‡ The numbers follow the key of the Department of the Environment.

TABLE 1

point	seen from	azimuth	object	declination	error
93	57–58	299.3°	winter Moon	+18.8°	−0.1
91	51–52	118.3	summer Moon	−18.8	−2.9
53–54	57–58	140.7	summer Moon	−29.1	−1.1
57–58	53–54	320.7	winter Moon	+29.1	−1.1
51–52	59–60	138.7	summer Moon	−29.1	−2.2
59–60	51–52	318.7	winter Moon	+29.1	−0.1
55–56	heel	231.4	winter Sun	−23.9	−1.4
heel	55–56	51.4	summer Sun	+23.9	+0.5
H	51–52	142.8	summer Moon	−29.1	−0.1
H	53–54	130.4	winter Sun	−23.9	−0.4

It can be seen that alinements existed *ca.* 1800 B.C. between the trilithons and the outer stones, and between the trilithons themselves. Archaeological evidence (Stone 1924; Atkinson 1960) indicates that the trilithons were erected before the time of the sarcen circle that surrounds them. The astronomical finding therefore raises the question of whether the five trilithons had a period of observational use in the early stages of Stonehenge III, standing as an independently designed structure. Related is the question of whether the construction of Stonehenge III was according to a single, integrated plan or followed a series of developments akin to second thoughts.

Before accepting these alinements as intentional, one must note the excessive error in 91 as seen from 51 to 52. This line can be ignored without affecting the possible validity of the remainder of the table. After the Stonehenge alinements had been published, hole H was seriously questioned as to whether it was a man-made feature (Atkinson 1967), but the point has appeared again with significance in the computer output. (Could there be a corresponding point, H′, in the undug western sector?) Arguing on the positive side for acceptance of table 1, it should be noted that (i) all extrema of the pattern are marked (± 18.8°, ± 23.9° and ± 29.1°) either at rising or at setting or at both; (ii) the accuracy of the previously suggested Moon diagonal in the station stone rectangle has been considerably increased because the trilithons are to the north and south of this line (indeed, there may have been some type of central structure at the time of the rectangle whereby the observing points were offset from the geometrical centre); (iii) trilithon 57 to 58 appears from the stereo plan (Hawkins 1971) to be rotated noticeably in the counterclockwise direction when compared with its corresponding partner 53 to 54, and this rotation places it squarely to face 93, the accurate far-sight to the Moon. These items are, of course, no more than conjectures and are offered for consideration within the total archaeological study. Table 1 has been included for completeness of data, and the astro-archaeological hypothesis does not stand or fall on its acceptance.

3. LINES IN THE NASCA DESERT

In the dry foothills of the Andes, many areas have been found – from the Cantogrande valley near Lima to the pampas of Jumana and Colorada near Nasca – where huge geometrical and figurative markings are inscribed on the desert pavement. The areas were formed by the erosion of thick alluvial fans, leaving a natural pavement of irregular pebbles, reddish in colour and blackened on top by desert varnish. When the pebbles are lifted or scraped away, the light yellow subsoil is exposed. A typical high-altitude photograph is shown in figure 2, plate 23. The ancient lines have been described as a gigantic astronomical calendar by Kosok (1949) and Reiche (1949, 1968), with features pointing to the solstice Sun, the Moon extrema, and stars.

Two areas were chosen for study by the procedures established for Stonehenge and other sites (Hawkins 1968). One was near the Ingenio valley, long. 75° 08′ W, lat. 14° 42′ S; the other, near Nasca, 74° 59′ W, 14° 49′ S. The results were negative on all counts – the lines did not show significant alinement with Sun and Moon extrema, stars, planets, or any currently unidentifiable astronomical object of fixed declination.

In an attempt to derive an archaeological date, a search for pottery shards was made in a traverse 2 km in length and 5 m in width. The most frequent occurrence was of styles 3 and 4 in early Nasca ware,† which corresponds to dates between 100 B.C. and A.D. 100 (Rowe & Menzel 1967). Although it is likely that the lines are contemporary with this pottery (since the fragments are scattered over the surface and whole pots have been found on mounds within the lines), it strictly indicates no more than that the lines predate the pottery. The analysis was therefore performed for the first centuries A.D. and B.C. and for a few previous millennia.

To summarize: All linear features in the area were measured (criterion (v)), and these were tested for Sun and Moon extrema and for all stars brighter than magnitude +2 (criterion (iv)). Reiche, on the other hand, computed for only a few selected lines. A total of 30 overlapping, high-resolution air photographs were taken and a photogrammetric plan was drawn to the scale of 1:2000. Altogether, 93 linear features were measured, and since each could be viewed along two directions, this gave a total of 186 azimuths for consideration. Some of the lines extended beyond the edge of the charts to a distance of several kilometres.

Allowing an error of up to 1° in declination, we would expect about one direction in 10 to correspond, purely by chance, with one of the 10 Sun or Moon extrema or equinox positions, as given in § 1b above. Actually, 39 of the alinements matched the Sun or Moon positions. A closer inspection of the computer output showed that only a few of the 39 correlations were made with significantly differentiated lines. Thus, the hypothesis of a Sun–Moon pattern fails to account for the majority of the lines, even if we agree to reserve our judgement on the few that appear to fit. In the latter category, it is pertinent to note that the large rectangle, perhaps the most prominent feature in the area studied (figure 2), points to the winter Moon extreme at declination 18.5°. However, there are no matching rectangles for any of the other seven extrema taken up by the Moon in its 18.6-year cycle.

Alinements were computed for the 45 stars brighter than magnitude +2 and the Pleiades cluster (an asterism of interest in Pre-Columbian cultures). Each star changed approximately 0.36° in declination during the course of a century. Because of the circumpolar areas in the sky, the declination for a line is limited to a range of 150°, between ± 75°. Thus, we would expect by chance approximately $(46 \times 0.36)/150$ stars per century along a given direction, or 0.11. In actuality, there were on the average 0.09 stars per line per century over the period 5000 B.C. to A.D. 1900, thus confirming the probability estimate. For the period of archaeological interest, 100 B.C. to A.D. 100, the averages were 0.05 and 0.06 stars per line per century, values again indicative of zero correlation.

To test for alinement with currently unidentified objects of fixed declination (such as the centre of a prehistoric constellation, nova, or planetary configuration), we adopted the follow-

† Sometimes the word is spelled 'Nazca', following an older rendering of the name of the nearby town. The modern spelling on maps and on location is 'Nasca'.

ing procedure. The lines at the Ingenio site were divided into two groups (east and west of the Panamerican Highway, which is the dark, oblique line in figure 2). We computed the number of duplicate declinations in the sets. This would show whether a particular astronomical object had been used to fix the direction of certain of the lines in both groups. We expected 27 coincidences on the basis of chance; 33 were found. This was an insignificant excess. Similarly, there was no evidence of duplication between the Ingenio site and Nasca farther south.

Thus, the line complex could not be accounted for by astronomical alinements to the stars, Sun, Moon, or planets. This somewhat unfruitful exercise in astro-archaeology tends to negate the lurking feeling that alinements might perhaps be found with any ancient structure. Astronomically speaking, the lines in the Nasca desert are random. Speculations as to their purpose have been given elsewhere (Reiche 1968; Hawkins 1969, 1973).

4. The temple of Amon-Re at Karnak

The majority of Egyptian temples are river-oriented – that is, they face the Nile whether they are on the east or on the west bank. Furthermore, Posenor (1965) has cited evidence to show that the river was regarded as a fundamental direction, as a general north–south meridian. It would therefore seem superfluous to examine these river temples for astronomical alinement. However, in the case of the temple of Amon-Re, Barguet, in his definitive work (1962), concluded on the basis of architecture and inscriptions that the structure had a cosmic or astronomical significance. On the same evidence, he inferred that the temple pointed not toward but away from the river, to the desert to the southeast. It is therefore of interest to compute the declination of this particular axis.

A preliminary measurement was obtained from the air-survey charts of the Franco-Egyptian Centre at Luxor, as well as from German maps of the terrain made in World War II. There was a major rebuilding of the temple, particularly the Hall of Festivals, by Tuthmosis III. Hayes (1961) gives the reign of this pharaoh as established by various authorities, from which one can estimate the date of rebuilding to be *ca.* 1480 B.C. At that epoch, the obliquity of the ecliptic was 23.87°. Calculations show that the declination of the Sun, with disk tangent to the skyline and centred on the axis, was $-23.9°$, with an estimated error of 0.2° arising primarily from the reading of the maps. Thus, with a concordant declination, the temple pointed to the sunrise at the time of the winter solstice. For Amon-Re, this alinement was precise to within the limits of measurement ($\pm 0.2°$); the accuracy might be found to be further improved if ground surveys were taken. Because of the slow change in obliquity, the pointing to the midwinter sunrise would be an observable phenomenon for many centuries before and after the assigned epoch. The alinement is a property of the structure – a material fact. Whether it was an intention of the builders so to mark the southern extreme of yearly sunrises is difficult to establish. One cannot readily turn to criterion (v), because, as has been pointed out, the temples of the Nile seem to aline, in the first instance, perpendicular to the river. Nor can we expect the evidence of support to be clearly shown in the hieroglyphics, because the inscriptions are rich in metaphors, allusions, and ambiguities; nor do the writings describe the pharaonic knowledge of what we call physical science with exactitude. The calendar system and the method of reckoning time, for example (Neugebauer & Parker 1960; Parker 1950), were deduced from fragmentary data, not from a definitive text.

Barguet (1962) describes a small temple on the roof of the Hall of Festivals at the northeast

FIGURE 2. Ancient markings in the desert near the Ingenio valley, Peru. The large rectangle is approximately 800 m in length. The oblique, dark line is the Panamerican Highway.

Although there were a scattering of astronomical lines in the Peruvian analysis, the major portion of the desert pattern seems not to be so alined. The alinement at Karnak is for a single temple and must therefore be regarded as a tentative result, although it is the most important temple complex in Egypt and there seems to be internal evidence for the astronomical significance of the temple in the hieroglyphics. The new alinements, found at Stonehenge, based on the trilithons and perimeter stones, may not be important in themselves, but become significant when added to the number of alinements found previously for the site.

If we concede a similarity in the critical faculties of modern and prehistoric man, it is not surprising to find evidence for a knowledge of astronomical alinements and periodicities. The work required no more than watching for and marking a natural event – counting, remembering, and/or recording. As Stahlman said (Stahlman & Gingerich 1963): 'Celestial phenomena – ranging from stellar risings and configurations, to novae, comets, and solar, lunar, and planetary risings and positions – seem always to have appealed to the human psyche as at once incontestable and, in a sense, unique events. Time is measured in and by the heavens. Man has always sought to reduce celestial uniqueness to repetitive patterns and thus to "understand" them, but he has also learned that the patterns are never quite complete.' Stahlman was writing primarily within a historic context, but it is not unreasonable to extrapolate this statement, a short way at least, into prehistory. Hoyle transposed his own mind into prehistoric Britain with the question 'How would *we* do it?' He concluded that 'an excellent procedure for "us" to follow would be to build a structure of the pattern of Stonehenge, particularly Stonehenge I'.

On the other hand, it *is* surprising (Hawkes 1967) if one equates intellect with the arts and a certain style of graceful living, and if one expects an understanding of numbers to come with or after the invention of a written script.

The work in Peru and England was supported by the National Geographic Society. Photogrammetric surveys were made at Stonehenge by Hunting Surveys Ltd, and at Nasca by the Servicio Aerofotografico National of the Peruvian Airforce, with the cooperation of the Instituto Geofisico del Peru. Photogrammetric plans of Karnak were kindly supplied by the Franco–Egyptian Centre, and the work in Egypt was supported by a Smithsonian Institution Grant. The author has appreciated discussions with Professor F. L. Whipple, Dr Labib Habachi, Professor G. Clark, and staff members of the British Museum and the National Museum, Cairo.

REFERENCES (Hawkins)

Atkinson, R. J. C. 1960 *Stonehenge*. London: Pelican Books.
Atkinson, R. J. C. 1966 *Antiquity* **40**, 212.
Atkinson, R. J. C. 1967 *Antiquity* **41**, 92–95.
Barguet, P. 1962 *Temple d'Amon-Re a Karnak*. Cairo: French Oriental Institute.
Childe, V. G. 1947 Josiah Mason Lectures, Birmingham University.
Clark, G. 1970 *Aspects of prehistory*. Berkeley, California: University of California Press.
Daniel, G. 1962 *The idea of prehistory*. London: Pelican Books.
Daniel, G. 1964 *Nat. Hist.* (April), pp. 47–52.
Daniel, G. 1965 *Sunrise* **14**, 229–239.
Dow, J. W. 1967 *Am. Antiquity* **32**, 326–334.
Edwards, I. E. S. 1972 *Treasures of Tutankhamun*. London: British Museum.
Habachi, L. 1970 *Kemi, Rev. Philologie* **20**, 229–235.
Hawkes, J. 1967 *Antiquity* **41**, 174–179.
Hawkins, G. S. 1963 *Nature, Lond.* **200**, 306–308.
Hawkins, G. S. 1965a *Am. Sci.* **53**, 391–408.

Hawkins, G. S. 1965b *Science, N.Y.* **147**, 127–130.
Hawkins, G. S. 1968 *Vistas in Astron.* **10**, 45–88.
Hawkins, G. S. 1969 *Sci. Rep. natn. Geog. Soc.* (June), pp. 1–45.
Hawkins, G. S. 1971 *Nat. Geog. Soc. Res. Rep. for* 1965, Washington, D.C.: Nat. Geog. Soc.
Hawkins, G. S. 1973 *Nat. Geog. Soc. Res. Rep.* (in the Press).
Hayes, W. C. 1961 *Cambridge ancient history* **1**, pamphlet 10.
Hoyle, F. 1966a *Antiquity* **40**, 262–276.
Hoyle, F. 1966b *Nature, Lond.* **211**, 454–456.
Kosok, P. 1949 *Archaeology* **2**, 206–215.
Lockyer, J. N. 1894 *The dawn of astronomy.* London: Cassell.
Lockyer, J. N. 1901 *Nature, Lond.* **65**, 55–57.
Lockyer, J. N. 1906 *Stonehenge and other British stone monuments.* New York: Macmillan.
Marquina, I. 1951 *Memorias del Instituto Natl. de Antropologia e Historia* I. Mexico: Inst. Ant. e Hist.
Morley, S. G. 1956 *The ancient Maya.* Stanford, California: Stanford University Press.
Neugebauer, O. & Parker, R. A. 1960 *Egyptian Astron. Texts.* London: Brown University Press.
Newall, R. A. 1967 *Antiquity* **41**, 98.
Newham, C. A. 1964 *The enigma of Stonehenge*, private publication.
Parker, R. A. 1950 *Calendars of ancient Egypt.* Chicago: Univ. Chicago Press.
Posenor, G. 1965 *Nachr. Akad. wiss. Göttingen,* (Ph. -Hist. Kl.) **2**, 69–74.
Reiche, M. 1949 *Mystery on the desert*, private publication.
Reiche, M. 1968 *Mystery on the desert.* Nazca, Peru: private publication.
Rowe, J. H. & Menzel, D. 1967 *Peruvian archaeology.* Palo Alto.
Schwartz, Y. 1965 *Science, N.Y.* **148**, 444.
Smith, J. 1771 *Grand orrery of the ancient druids, Stonehenge.*
Spence, M. 1894 *Standing stones and Maeshowe of Stenness.* Glasgow: Carluke.
Stahlman, W. D. & Gingerich, O. 1963 *Solar and planetary longitudes from* −2500 *to* +2000. Madison, Wisconsin: University of Wisconsin Press.
Stone, E. H. 1924 *Stones of Stonehenge.* London: Robert Scott.
Stukeley, W. 1740 *Stonehenge.*
Thom, A. 1967a *Antiquity* **41**, 95–96.
Thom, A. 1967b *Megalithic sites in Britain.* Oxford: Oxford University Press.
Thom, A. & Thom, A. S. 1971 *J. Hist. Astron.* **2**, 147–160.

Phil. Trans. R. Soc. Lond. A. **276**, 169–194 (1974) [169]

Printed in Great Britain

Archaeological tests on supposed prehistoric astronomical sites in Scotland

By E. W. MacKie
The Hunterian Museum, The University, Glasgow

[Plate 25]

CONTENTS

An astronomical interpretation of the British standing stone sites has been developed in great detail by A. Thom using methods and data which have hitherto rarely, if ever, been used by archaeologists. The unfamiliarity of these methods, and the revolutionary nature of the conclusions drawn from them, have no doubt contributed to the difficulties which the profession is evidently encountering in coming to terms with Thom's ideas. However, the raw data on which the theories are based are, like all other archaeological data, susceptible to checking and testing in traditional archaeological ways – by field-work and excavation. If one is to do this, one must isolate the hard evidence on which the theories are built and these are the many long alinements – from standing stone to a mark on the horizon – which are claimed to have astronomical significance. The plausibility or otherwise of these alinements is something that all can assess by visiting the sites. Moreover, at several of these sites it is possible to devise tests by excavation for the astronomical interpretation, and the results of two such tests are described.

INTRODUCTION

In two recent books (Thom 1967, 1971), and in a number of papers going back over many years (Thom 1954, 1955, 1961*a*, *b*, 1962, 1964, 1966*a*, *b*, 1968; Thom & Thom 1971, 1972*a*, *b*, 1973) Professor Alexander Thom has offered prehistorians a detailed new interpretation of the origin and function of the standing stones and stone circles and some new insights into the astronomical, mathematical and geometrical skills possessed by their builders. Although suggestions have not been lacking in the past that the Neolithic and Early Bronze Age inhabitants of Britain, or a few of them, practised quite sophisticated astronomy, such earlier views have usually been founded on a large number of assumptions based on individual sites such as Stonehenge and Callernish (Lockyer 1906; Somerville 1912, 1923; Hawkins 1966). Thom is the first to have systematically surveyed large numbers of standing stone sites, to have looked for possible astronomical alinements in them and to have founded his theories on a mass of data drawn from many sites instead of a few.

THE THOM THEORIES

Thom's hypotheses – based on new information mostly collected by him in the field – fall conveniently into three groups, only one of which concerns this paper. The first major theory is that the stone erectors practised sophisticated geometry in laying out their circles and rings – having knowledge, for example, of Pythagorean triangles – and used a precise and invariable unit of length in so doing. This is the 'megalithic yard', equal to 0.829 m (2.72 ft) and strikingly similar to the modern Iberian *vara* of between 0.843 and 0.838 m (2.766 and 2.7495 ft) (Thom 1967, p. 34). The second theory concerns the cup-and-ring rock carvings of southwest Scotland (and can presumably be extended to those of other regions); these were, Thom suggests, drawn out with the same elaborate geometry as in the circles and rings and were based on a unit of length independently inferred from the carvings to be 20.5 mm (0.808 in) or $\frac{1}{40}$th of the megalithic yard (Thom 1968).

The third major theory suggests that many standing stones and stone circles formed part of alinements to the horizon which were intended accurately to mark the rising and setting points of various celestial bodies at significant times (Thom 1967). The solar sites among these are particularly important and were presumably designed to make possible the keeping of an accurate calendar by pinpointing the days when the Sun was at its extreme (solstitial) and central (equinoctial) positions on the horizon. Another major part of this astronomical theory is that many other alinements were designed to record the much more complicated motions of the Moon, and that this was done in order to predict eclipses (Thom 1971). In this paper I shall consider only some aspects of the astronomical theory.

Methods of approach

When assessing the value of a new and controversial theory – the acceptance of which would require some drastic changes in long established ideas – it is important to be quite clear on the nature of the evidence on which the novel hypothesis is based. In the case of those of Thom it is evidently not proving easy for the archaeological profession to come to terms with Thom's ideas which would credit the prehistoric British population of the late Neolithic and early Bronze periods with skills in practical surveying, advanced geometry and observational astronomy which are far above any hinted at by the more traditional archaeological evidence (Childe 1955; Hawkes 1967; Hogg 1967). So unexpected indeed are the Thom interpretations when set against previous archaeological ideas that it is only fair to ask that the evidence on which they are based be subject to careful scrutiny and the theories themselves tested where possible. Obviously it would be wrong to reject these theories on the facile ground that they do not accord with the previously generally accepted picture of prehistoric Britain. Our failure to find evidence of sophisticated intellectual activity among the barrows, cairns, standing stones, stone circles and henge monuments of 4000 years ago cannot mean that such evidence does not exist. It need only mean that most of the archaeological profession was not equipped either by training or temperament to discover it.

Equally, however, the theories should not be accepted uncritically and it would be just as scientifically naïve to assume that they are correct simply because the data collected has been subjected to impeccably accurate and skilled mathematical analysis. It is of course axiomatic in science that such analyses can only be as reliable as the data they use. Therefore it is essential to isolate and test the factual basis of the Thom astronomical theories. What is this evidence? In essence it consists of the identification of a large number of long alinements from standing stones to prominent points on the horizon and of the discovery that these sight-lines cluster round significant prehistoric astronomical declinations such as those of the Sun at the solstices and equinoxes, the Moon at its four extreme positions, the rising and setting points of various stars and so on (Thom 1967, Fig. 8.1). If one assumes that the alinements are genuine the histogram of their declinations is by itself a highly significant body of evidence since these alinements only cluster in this way when converted into declinations. A histogram of their azimuths would show a more random distribution. However, it is clearly on the genuineness of these alinements on the ground – or in other words on the objectivity of the means by which they have been identified – that the whole theory depends.

Possible tests for the Thom theories

There are a variety of tests which can be applied to both the evidence that Thom has assembled and to the various theories which he has devised to explain it. They can best be identified as the answers to the following questions: (1) Have the alinements been identified objectively? (2) Are the horizon notches and mountain peaks, which have been chosen as the foresights visible from and used at any given stone, self-indicated or inferred? (3) Are they the most likely ones, and the only ones, to be seen from the sites concerned? (4) Have any of the alinements been chosen because they were expected in a particular place? (5) Does the archaeological dating of the structures inferred to be part of such alinements fit the fairly precise dates given to them on astronomical grounds? (6) Are there features at individual sites which the astronomical interpretation requires to be present which can be checked by fieldwork and excavation? (7) Can the astronomical inferences be correlated with the cultural groupings seen in the stone circles and henge monuments and made on the basis of site plans and associated pottery and artefacts? (8) Does the astronomical theory involve equipment and techniques which a Neolithic technology is unlikely to have been able to produce? (9) Does it involve the storing of knowledge of a type and in a manner for which there are no known parallels among recorded non-literate societies?

I propose to consider here tests for three of these problems, which involve the examination of three different sites. The first, Ballochroy in Kintyre, relates to the plausibility and completeness of the chosen alinements (questions 1 to 3). The second site, which may supply an answer for question 6, is Kintraw, also in Argyll. The third is Duntreath in Stirlingshire which, with other evidence, is relevant to question 5. Finally I add a short section dealing with question 9, though here the evidence naturally comes from comparative material rather than from deliberately organized tests.

The importance of solstice sites

The first two sites mentioned are thought to be solar solstice observatories, and a few words are needed to explain why these are of crucial importance for the Thom astronomical theory as a whole. Because of the very small changes in the Sun's declination at these times the accurate detection of the summer and winter solstices is a formidable task for a people with a Neolithic technology even if one leaves aside the problems presented by refraction and temperature changes in the evening air near the horizon. In fact one may assume that, if they had mastered this, the *practical* problems of observing the Moon and stars for example would have been relatively simple by contrast. In the 24 h before and after the moment of the solstice the change

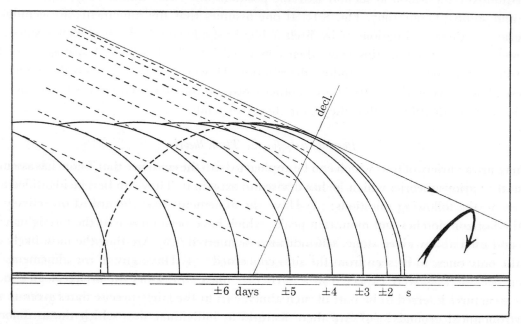

FIGURE 1. Representation of the changes in the solar declination during the 8 days before and after the summer solstice at about latitude 56° N. The Suns are shown half set on a sea horizon and the dotted lines and the black arrow mark the changes in its real position. The solar diameter is about 32′.

in the solar declination (which corresponds also with its position on the horizon at the equator) is 12″ and this daily movement increases with the square of the interval (figure 1). At the latitude of Scotland the actual movement of the Sun's disk that this represents along the horizon is greater, about 0.5′, because of the shallower angle that the path of the Sun's descent makes with the horizon in northerly latitudes: yet this still represents only about $\frac{1}{64}$th of the solar diameter. However, in the case of the most accurate kind of solar observatories, described below, the changes in the sun's declination are seen against a mark on the horizon which is about parallel to the diurnal path: here the declination change to be observed would be 12″, 0.2′, equivalent to $\frac{1}{160}$th of the solar diameter. Table 1 shows how the daily change in the Sun's real position decreases and increases before and after the solstices.

So if our hypothetical British Neolithic astronomer-priests did discover the exact length of the year, and were thereby able to construct an accurate calendar (an achievement which would have been essential if they carried out the more advanced observations of the Moon's movements), even up to the level of detecting cycles of eclipses suggested by Thom (1971) they must have had

TABLE 1

time from solstice days	difference from solstitial declination
s	0′ 0″ (= ± 23° 27′)
$s \pm 1$	0′ 12″
$s \pm 2$	0′ 48″
$s \pm 3$	1′ 48″
$s \pm 4$	3′ 12″
$s \pm 5$	5′ 0″
$s \pm 6$	7′ 12″
$s \pm 7$	9′ 48″

(diameter of solar disk 32′)

some technique for detecting this tiny solar movement of less than half a minute. There appear to be three ways of doing this available to societies which lack the means to make small, accurate instruments – one involving the sun-dial or gnomon and the other two long sight-lines to points on the horizon.

Techniques for solstice detection

If sufficiently large gnomons could be constructed it might be possible to tell not only the time of day from the angle of the shadow but also the time of year from the length of the shadow at midday. Possibly some of the tall, pointed-topped, standing stones in Arran and Orkney may have served such a purpose, though presumably flat, level, plastered platforms marked with scales would have been an essential part of the instrument. Such platforms have not been discovered and, in any case, the distance represented on the ground by the 12″ change in solar declination as seen in the position of the midday shadow of a 6.1 m high stone in Orkney (lat. 59° N) at midsummer is far too small to be detectable. Presumably the great 18.3 m high masonry gnomon built at Delhi in the eighteenth century by the Emperor Jai Singh II might have been able to detect the solstice; here the shadow fell on carefully built, graduated masonry arcs at right angles to the edge of the triangular gnomon (Bergamini 1971). A similarly tall masonry sun tower was built in the thirteenth century in China (Needham, this volume, p. 67). Both these structures clearly involve a skill and accuracy in building in mortared masonry which was far beyond the technological capabilities of Neolithic and Early Bronze Age Britons (Atkinson, this volume, p. 123).

The difficulty with the gnomons highlights the main problem of detecting the tiny amount of solar movement at the solstice – or, to put it another way, of detecting when this very slow and diminishing movement has ceased to go north and has started south again (at midsummer). It is clear that if a shadow could be cast over a long enough distance even a movement of 0.2′ would be represented by a substantial distance on a flat surface. However, because of the increasing width of the penumbra the exact position of the edge of the Sun's shadow becomes indistinguishable after a relatively short distance and this method is clearly impracticable. Yet if one in effect reverses the operation and watches the rim of the Sun's disk against some distant smooth edge such as a mountain slope, the optical problems become much simpler. The edge of the penumbra of the mountain's shadow is itself invisible of course, but it is there and the eye can detect its precise location when it sees the first flash of the brilliant, clearly defined rim of the sun appear from behind the mountain.

This is the basis of the method of solstice detection suggested by Thom (1967). If the observer arranges his position a few days before the solstice so that he sees the Sun set behind a mountain with a smooth, slightly convex right slope, and then manoeuvres himself so that the right edge of the disk momentarily reappears as a flash in the convexity as it sets, he will have exactly defined the solar position on the horizon for that evening. This will presumably be marked with a peg. On the following evening the Sun will set somewhat further to the right and the same procedure will find a peg a few metres to the *left* of the first one, the exact distance depending on the distance of the mountain. Even the tiny horizontal movement of 0.2′ in the 24 h before and after the solstice could be represented by a distance of a few metres on the ground if the line of sight is sufficiently long. At the peg farthest to the left would be set up a permanent marker, such as a standing stone, from which the day of the solstice could be checked in future years. Table 2 illustrates the distance on the ground subtended at the observer by an angular change of 0.2′ at various distances from him. It will be seen that an alinement at least 8 km long, and preferably 15 or 30 km, is needed for accurate solstice watching and, using these criteria, that it is improbable that any of the alinements so far detected at Stonehenge could have served as accurate solstitial instruments of this type.

TABLE 2. BASE LENGTHS OF RIGHT-ANGLED TRIANGLES WHOSE ANGLE
AT THE APEX IS 12″

length of triangle (i.e. distance to foresight)	example	length of base (i.e. horizontal distance needed to distinguish solstice)
80 m (260 ft)	centre of Stonehenge to Heel stone	4.8 mm (0.19 in)
915 m (1000 yards)	—	55 mm (2.16 in)
1.6 km (1 mile)	Duntreath to Strathblane hills	96 mm (3.7 in)
16 km (10 miles)	—	0.975 m (3.2 ft)
30.7 km (19.1 miles)	Ballochroy to Jura	1.84 m (6.1 ft)
43.6 km (27.2 miles)	Kintraw to Jura	2.66 m (8.62 ft)
48.1 km (30 miles)	—	2.89 m (9.5 ft)

The second method of using alinements was suggested by Hoyle (1966) to resolve this difficulty over the short sight-lines at Stonehenge. If the foresight is aimed a degree or two *inside* the estimated solstice position, the point of sunset or sunrise will pass it while the daily change in declination is large enough to be seen even with a short sight-line. The sun will rise or set again over the foresight marker some time after the solstice and the date of the latter should be obtainable with some precision by halving the interval between the two observations.

Credibility of the alinements

A long alinement for observing the solstice could simply consist of a backsight of a standing stone, marking where the observer was to stand, and a foresight which was a distant mountain peak or a notch on the horizon. If the stone was isolated and irregular there would be no built-in indication of where the foresight was and the identifier of such sight-lines might be open to the charge that he was selecting mountains in the appropriate place for a particular theory. Alinements with some sort of built-in direction indicator would clearly be more reliable and such might include a view from a stone in which there were really distant peaks in only one or two directions, or a flat-sided stone alined towards the peak concerned, or a second outlying stone pointing the way to the foresight. On the whole it seems improbable that an outlying standing stone could act as a solar foresight by itself unless it was used in the Hoyle method described above: if it were far enough away to be moderately accurate it would be too small against the bright disk of the sun and if it were near enough to hide most of the disk it would be insufficiently precise. These and other considerations must be kept in mind when assessing the plausibility of claimed astronomical alinements in the field.

Orientations and alinements

The archaeological literature contains several examples of claims being made about structures being alined towards astronomically significant directions. Lockyer (1906) and Somerville (1923) surveyed numerous barrows, cairns, stone circles and alinements of standing stones in Britain and concluded that large numbers had been laid out with astronomical considerations in mind. Hawkins (1966) claimed numerous astronomically significant alinements between the various stones and stone-holes at Stonehenge and Somerville (1912) found the same at Callernish. The temples and pyramids of Egypt have also provided material for the 'astro-archaeologists' (Hawkins, this volume, p. 157). The whole field is being surveyed by Baity (1973). However, it seems that Thom (1967) was the first to show how distant horizon marks might actually have been used to make precise astronomical observations in prehistoric times in the manner explained above and the inherent inaccuracy for solar observation of short sight-lines, commented upon earlier, seems to make it necessary to distinguish between on the one hand potentially useful, practical instruments and, on the other, structures which are simply built with their axes or sides orientated in an astronomical direction.

It seems very doubtful whether any useful, accurate observations of the Sun could be made simply by looking along the straight side of a masonry building, no matter how precisely alined. Even an arrangement whereby the observer stood on the steps of a pyramid at the solstices and saw the sun rise at opposite ends of a range of masonry buildings 60 m away – as perhaps at the Maya site of Uaxactun, Guatemala (Morley & Brainerd 1956, p. 300) – seems incapable of giving much more accuracy than the Heel stone at Stonehenge (MacKie 1968). In fact there is evidence that the Maya practised similar kinds of long-distance observations that Thom has suggested for Neolithic and Early Bronze Age Britain. In the *Codex Nuttall* there is a drawing of a man looking out through a temple doorway through a pair of crossed sticks – a simple device to ensure that his eye is in exactly the correct position – and presumably observing some phenomenon on the horizon from this high vantage point. Another drawing in the *Codex Bodleian* shows the eye behind the crossed sticks – this time seen from the front – and nearby a star or planet descending into a distant notch (Morley & Brainerd 1956, p. 258). This must surely imply that long alinements to the horizon were used by the Maya in their astronomical work.

It thus seems useful to make a distinction between *orientations* and *alinements* in primitive structures. It is reasonable to suppose that many buildings and constructions were *orientated* towards a particular direction for magical, religious or traditional reasons: the consistent directing of the long axes of Christian churches towards the east is a classic example and many prehistoric and ancient structures could well have been orientated towards the solstices and equinoxes from equally non-scientific motives. The term *alinement* by contrast could well be reserved for long sight-lines which are capable of being used as observing *instruments*. If this terminology is adopted it might help to reduce misunderstandings over what is being claimed about ancient astronomical practices in Europe, and about the motives which lay behind them.

An independent description and assessment of two sites in Argyllshire claimed as solstice observatories may help to answer the questions posed earlier (p. 171) about the credibility of the alinements. At Kintraw, at the head of Loch Craignish (grid ref. NM/831051) there is a standing stone from which the midwinter Sun could have been seen setting exactly behind Beinn Shiantaidh on Jura about 3800 years ago. At Ballochroy, on the west coat of Kintyre (grid ref. NM/730525) there are three standing stones from which the same phenomenon at midsummer would have been seen behind Corra Beinn, also on Jura. It is reasonable to found a preliminary assessment of the general plausibility of the Thom astronomical theories on these two sites, first because solstice alinements are both difficult to set up and essential as the basis for all other work by prehistoric astronomers (p. 172 above), and second because the author of the theories himself considers them important sites (Thom 1967, 1971).

Ballochroy

This site consists of three standing stones close together in a line running from northeast to southwest; there is a massive stone cist on the same line about 37 m to the southwest. The field in which the stones stand is behind a raised beach. Thirty km away across the sea in the northwest are the Paps of Jura, forming a series of four very striking peaks on the horizon (figure 2). Two of the stones have very flat right faces which are orientated unequivocally

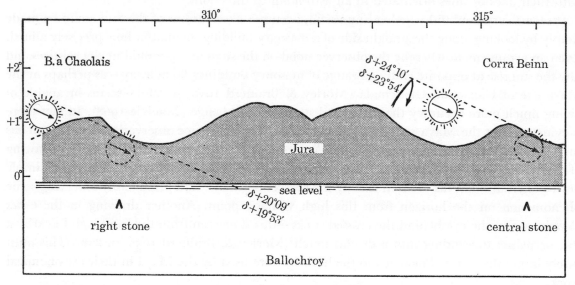

FIGURE 2. Scale drawing of the Paps of Jura as seen from the Ballochroy standing stones. The directions indicated by the flat sides of the two alined stones are shown with inverted V's and the declinations of the Sun setting behind the two hills concerned are given.

towards two of the Paps. The central stone, 3.3 m high, points at the right slope of Corra Beinn which has a declination of $+24° 10'$: if the Sun was setting behind this peak with its upper rim just grazing the edge of the right slope the centre of the disk would have a declination of $+23°$ $54'$, suitable, according to retrospective calculations, for a midsummer solstice of about 1800 B.C. The right-hand (northernmost) stone, however, is equally clearly alined towards the right slope of Beinn à Chaolais, the most southerly of the four Paps; the centre of the solar disk setting behind this mountain in the same way has a declination of about $+19°$ 53', well inside even the modern solstice.

The character and situation of the site do indeed make the astronomical interpretation of it most attractive. The Jura mountains conspicuous in the northwest (R.C.A.H.M.S. 1972, pp. 46–47, pl. 10 a), the fact that two of the slabs are alined precisely towards two of these peaks and the general absence of any other distant markers (except Cara island, described below), all combine to suggest forcibly that stones and mountains are interconnected in some way. Yet simultaneously there are some odd aspects to the site which remind one of the need to assess the objectivity of the chosen sight-lines on the ground by careful examination.

At first sight it is not easy to understand why Corra Beinn was selected as the foresight for the solstice: its right slope is about parallel to the angle of the Sun's descent but somewhat uneven (Thom 1971, Fig. 4.1). There is in fact a difference of about 2.7' in the declination of the edge of the solar disk according to whether one assumes that it grazed the top of the slope or the lower part (Thom 1971, Fig. 4.1). Both Beinn Shiantaidh, immediately to the left, and Beinn à Chaolais at the southern end of the row of peaks have smooth, concave right slopes which would be much more suitable as sunset markers: their declinations would be unambiguous and precise. To use these peaks the backsight markers would need to have been positioned further south-west down the coast.

As we have seen the right hand, shortest stone does in fact indicate the right slope of Beinn à Chaolais quite clearly but the declination of the centre of the convex slope, at $+20°$ 9', is difficult to interpret. The centre of the Sun's disk when it was just showing from behind this slope would be about $+19°$ 53'. If one considers the site without foreknowledge or preconceived ideas there is no reason from its layout to prefer either of the indicated alinements as the primary solstitial one. That to Beinn à Chaolais uses a more suitably shaped foresight but is marked by a smaller, flanking stone: that to Corra Beinn is marked by the tall central stone but uses a less precise foresight. The declination of Beinn à Chaolais falls uneasily between peaks of alinements on Thom's histogram at about $\delta + 16.89°$ and $+ 21.84°$ – which represent, he suggests, two points in the 16 'month' solar calendar (Thom 1967, Table 9.2) – and its significance remains obscure at the present.

It is possible, as Thom suggests, that the stones were sited in their present position in order that two alinements could be observed from the same spot, one for midsummer and one for midwinter. If one looks along the front edges of the stones, which are approximately in line, one sees Cara island 8 km away in the southwest and the western end of this gives a declination of $-23°$ 53' for the centre of the solar disk if its upper edge is just grazing the point. This of course is almost exactly the same as the declination of Corra Beinn and supports the view that the site was a dual purpose one. However, if one again looks at the alinement objectively one notices that the edges of the stones indicate a point some way east of the *eastern* end of the island; the chosen western end is not marked at all precisely. In addition, the small hump or peak on the eastern end is a much more conspicuous foresight marker than the featureless western end. These may be minor objections but they do illustrate that close analysis will often show that the astronomical interpretation of an archaeological site may not always be as simple as it seems at first sight. Moreover the megalithic cist (burial chamber), which is on the alinement to Cara island, was almost certainly once covered by a cairn and, if it is older than the standing stones – which is more than probable in view of the massiveness of its construction – then it would have completely blocked the view to the island. Nevertheless the plausibility of the claimed primary alinement at Ballochroy (to Corra Beinn) is greatly increased by its similarity to the second solstitial site to be described, at Kintraw.

Kintraw

This site consists of a single menhir and a large and a small cairn, the latter almost completely destroyed; they all stand in a level field on an otherwise steep slope rising from the head of Loch Craignish in Argyllshire, 19 km north of Lochgilphead (figure 3). The cairns were excavated in 1959 and 1960 (Simpson 1967) and a kerb of massive stones was found around the large one with a small, empty cist at this periphery. There was no central burial but part of the shaft-hole for an upright wooden post 7½ cm in diameter, and resting on the old ground surface, was traced in the centre of the cairn material.

Thom discussed this site (1967, p. 154) and suggested that the menhir marked the spot from which one looked down Loch Craignish to Beinn Shiantaidh one of the Paps of Jura. At about 1800 B.C. the midwinter Sun should have set behind the mountain so that as it disappeared its right edge flashed momentarily in the col between it and Corra Beinn to the west (figure 4). The declination of this notch as seen from the site is $-23°\ 38'$ so that the declination of the centre of the Sun's disk when its upper edge was just showing in the col would be $-23°\ 54'$ (Thom 1971, p. 42). This position is almost identical to that of Corra Beinn as seen from the Ballochroy stones ($+23°\ 54'$, p. 176 above) and it is hard to believe that this is the result of mere coincidence. One must recall too that these two declinations for hill slopes are not themselves identical ($+24°\ 10'$ as opposed to $-23°\ 38'$): they only become identical if one assumes that the real measurement is that of the centre of the Sun's disk when the edge is just showing. At the midwinter solstice the centre of the solar disk will be *further away* from $\delta\ 0°$ than the mountain slope of the foresight: hence one will add $16'$ (the Sun's semidiameter) to the measured figure. At the summer solstice, however, the centre of the disk is *nearer* $\delta\ 0°$ than the slope of the foresight: in this case $16'$ would be subtracted from the measured figure. In my view this is sufficient evidence by itself that the almost exact coincidence of the declinations of the centre of the Sun's disk at the two sites is a result of the sites having been laid out as complementary solstice observatories.

However, a problem was posed by the presence of a tree-covered ridge 1.6 km in front of the stones which hides the foresight mountain from the surface of the field (figure 4). How could the sunset phenomenon have been seen from the stone and how indeed could the position of the stone have been fixed originally if the claimed foresight is invisible from it? Thom suggested that the cairn had once been higher (there are sheep fanks nearby which could have been built from its material) and that it could have served as an observation platform from which one could see over the ridge. However this would not solve the problem of how the position of the site was first established. (In fact it is possible to see the Jura mountains from the ground beside the standing stone but as soon as one moves to the left or southeast – that is in the direction in which the essential sightings of the Sun setting on the col before and after the midwinter solstice would be taken – the col is obscured.)

Thom discussed the site again (1971, p. 37), mentioning Simpson's excavations and the discovery of the post-socket in the centre of the large cairn. The problem of how the cairn's position was fixed initially could have been solved by first locating the solstitial alinement with a series of sunset observations taken on the higher ground to the north of the field – a steep slope falling into a gorge with a stream. On exploring this slope Thom found a long narrow ledge at

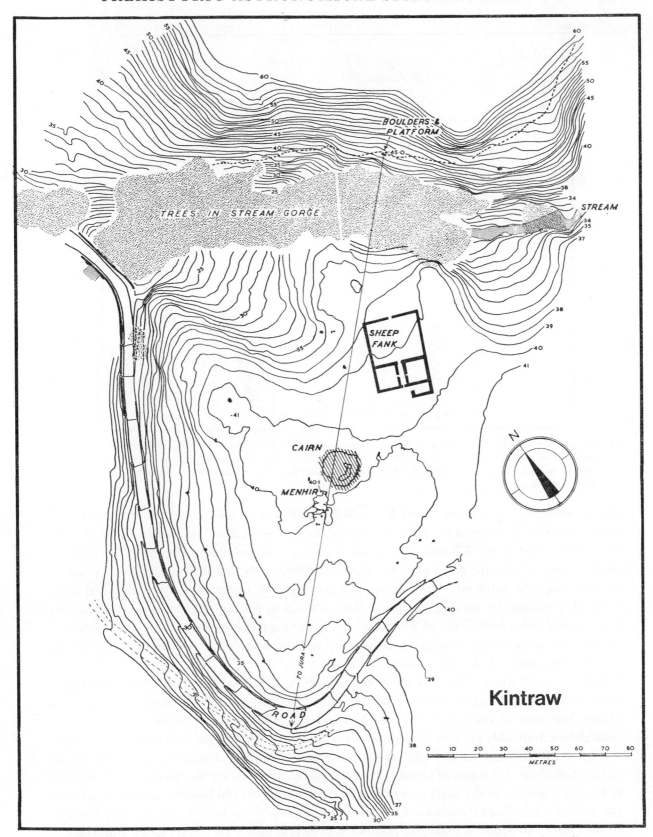

FIGURE 3. Photogrammetric contour plan of the Kintraw site showing the field with the cairns and standing stone ('menhir'), the stream gorge and the platform discovered beyond on the hillside. The contours are at 1 m intervals above mean sea level and spot heights at the stone and boulder platform are marked.

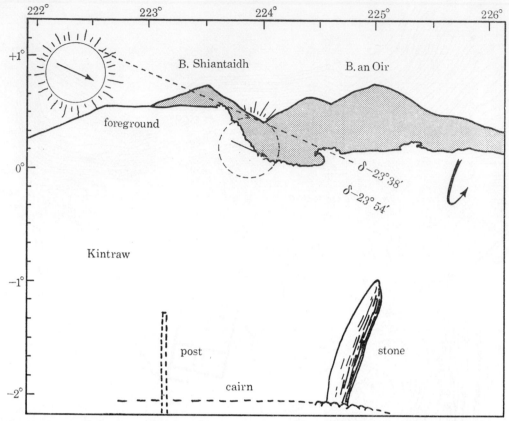

FIGURE 4. Scale drawing of the Paps of Jura as seen from the boulder platform on the hillside northeast of the standing stone at Kintraw, which is shown in the foreground. The Sun is shown as it should have appeared setting on midwinter's day about 3800 years ago, the direction of the change in its declination being marked by the black arrow.

the appropriate height from which Beinn Shiantaidh can be seen clearly over the ridge (in good weather). At the spot on this ledge cut by the alinement from the western slope of Beinn Shiantaidh through the Kintraw menhir was a large prostrate boulder (figure 3). He surmised that the menhir and the cairn/platform were positioned from this ledge or terrace – which he thought might be artificial – with the aid of a ranging post, traces of which were found in the cairn. The reasons for needing an observation platform in the flat field instead of on the steep slope were partly convenience of access and partly the suggested existence of a lunar alinement pointing to the north-northwest, which runs through the field. The cairn would serve for both but the hill ledge only for the solstice.

I discussed Kintraw with Professor Thom and Dr A. E. Roy in the autumn of 1969 and the former was kind enough to let me read the manuscript of his new book (1971) in which the above discussion of the site occurs and in which he first described the hill terrace and its possibilities. It quickly became clear that the site presented what might be a unique opportunity for an archaeologist to throw light on the astronomical theory by excavation. For the theory in effect predicts that traces of human activity ought to be found on the steep slope north of the field, and moreover at the exact spot from which the solstice should have been seen from behind the menhir (or perhaps from behind the cairn post, slightly to the left). If signs of such activity – in the form of an artificial platform, post-holes, potsherds and so on – were found this would be as near to proof for the astronomical theory as one could hope to obtain in archaeology since there is no other obvious reason why prehistoric activity of this nature should have occurred in just that spot on a steep slope and above a steeper gorge.

A preliminary reconnaissance of the site in June 1970 suggested to me that the terrace above the gorge was primarily a natural feature. It varies in width and runs for many metres along the slope, gradually descending towards the road bridge 130 m downstream; it extends for a much greater distance than was necessary or convenient for a cut platform to fix the time of the solstice. In any case the preliminary sightings for the winter solstice would have been taken to the left, or *upstream*, from the final alinement (see below). The ledge does run upstream and after a few metres it widens to a sloping platform several metres broad (figure 3); there is another large boulder at the lower edge which has apparently fallen down the slope to this point. A third large rock is on the ledge where it is much narrower, about 20 m downstream from the 'solstice' boulder. Thus there was no reason to select the central of the three boulders for investigation rather than one of the other two except that it is on the hypothetical line with the standing stone and Beinn Shiantaidh, 42 km away.

Excavations were carried out at the boulder for a week each in August of 1970 and of 1971. It was also intended to investigate the socket of the standing stone, and perhaps obtain a carbon-14 date for its erection, but it is a scheduled monument and the then Ministry of Public Building and Works in Edinburgh was unable to give permission. One long trench, extending up the slope from the ledge, was dug just to the west of the boulder and three more squares were opened behind the boulder and extended upstream from it and along the ledge (figure 5). The front face of the boulder was also exposed.

These trenches showed that the terrace was composed of a great thickness of bright red earth, containing many randomly scattered stones, which continued up the steep slope above the boulder: no trace of stratification was observed in it. So deep was this soil that bedrock was not reached anywhere, although a depth of a metre was achieved in places on the upper slope. The long trench running up this slope was designed to reveal any stone socket at the western end of the boulder and thereby to test whether this rock had once been standing upright. There was no socket but some stones were found jammed against its western end which might have been packing. The cutting also revealed the rounded end of what proved to be an extensive, nearly level layer of small stones which lay behind the boulder and ran further upstream.

The stone layer

Making a sharp contrast with the random scatters of stones and small boulders found in the soil farther up the slope was a compact layer of angular stones, some 15 cm thick and resting on the same red soil, which lay immediately behind the boulders (figure 6). The marker boulder was in fact found to be two adjacent stones set upright in the red earth and with pointed ends which met to form a large notch (below p. 183): into this notch the stone layer ran. As noted the layer appeared to end just to the west (downstream) of the boulders but was traced for more than 6 m upstream. The depth of the layer below the turf varied from less than 5 cm behind the boulders to 73 cm at the farthest point east exposed. The longitudinal section (figure 5) shows that the layer was almost exactly level along the slope and the cross-section shows a slope of about 10°, substantially less than that of the turf (figure 5). Thus in both directions the surface of the stone layer was much more level than the turf of the terrace above it.

The question of whether this stone layer was a standard feature below the turf on this terrace was considered. Two other nearby boulders were investigated, both of similar size to

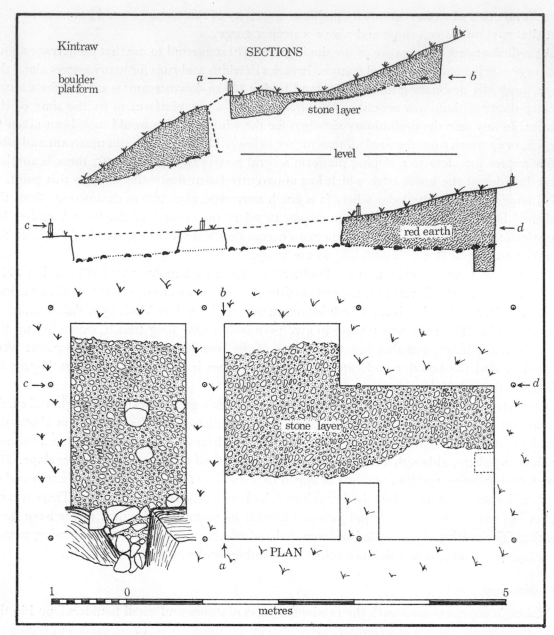

FIGURE 5. Plan and sections of the trenches cut at the boulder platform at Kintraw in 1970 and 1971. The long trench running up the hillside, seen in figure 6, plate 25, is not shown: it was 0.5 m to the left of the others.

the two forming the notch and in similar positions at the downhill edge of the terrace, poised above the steep slope down to the stream. The first rock was about 20 m downstream, at a point where the terrace narrowed to the width of the sheep path: it proved to have an accumulation of small stones in the small space between its vertical uphill side and the steeply sloping rock face of the hill immediately behind it. No firm conclusions can be drawn about this boulder because any small stones falling down and striking the rock would be certain to fall back into the narrow space behind it and form an accumulation there. The conditions are not the same as at the boulder notch where there is a relatively broad area of terrace behind the boulders and where the ground is almost flat immediately next to them.

Figure 6. View of the boulder notch, and the stone layer behind it, at an early stage in the excavations (1970). The stone layer can be seen coming to an end at the right of the boulders; it was subsequently traced for another 4.0 m to the left (upstream) and may go further.

The other rock is 46 m upstream, again at the downhill edge of the sloping terrace at a point where it has widened to about 13 m (figure 3). An excavation 1.5 m deep behind this boulder revealed the same deep deposit of red earth containing a random scatter of stones and small boulders throughout. Bedrock was not reached, not even by probing a further 0.5 m into the base of the trench. Nowhere in the deposit was there any stratum of stones of any kind, nothing comparable to the stone layer found behind the boulder notch. Two probes out of three went into the ground from the surface for the full length of the instrument (0.8 m) without striking any stones at all. Since small boulders and stones have become incorporated into the soil deposit here at all depths, it would appear that these were able to roll down the slope above as far as the large boulder. Hence if the stone layer behind the boulder notch is a deposit of scree there would seem to be no obvious reason why a similar layer did not accumulate behind the southeastern boulder; yet one did not.

That the stone layer behind the boulder notch is not in any case an accumulation of scree is implied by its angle of rest down the slope – about 10° (figure 5) – and demonstrated by the petrofabric analysis (appendix). Observation of any pile of scree at the foot of a slope shows that its surface comes to rest at an angle of about 30° to the horizontal. There appear in fact to be only two plausible explanations for the stone layer: it could either be a man-made deposit or it could be a natural platform formed under periglacial conditions. Such natural platforms are known and have been studied in Scotland (appendix). The diagnosis of the layer as man-made would, for reasons given earlier, obviously have profound significance for the astronomical interpretation of the site advanced by Thom and, by implication, for his astronomical theories in general. However, such a diagnosis is handicapped by the complete lack of what one might call the normal archaeological signs of human activity – no potsherds or artefacts of any kind were found during the excavations, nor any fragments of charcoal such as might have come from fires. Neither were there any signs of post-holes, kerb-stones or any other obvious structural remains. Only the stone layer itself, and the two adjacent boulders up against which it runs, are there to be considered.

The boulder-notch

As noted earlier, excavation showed that what originally appeared to be a single large stone, possibly fallen from the upright position, was in fact two boulders which had clearly remained in the position in which they were when the stone layer was formed. This layer ran up against them, and into the notch formed by their pointed ends, and a cutting through the layer showed that it rested on red earth which itself rested against the lower part of the boulders (figure 6, plate 25). Excavation down in front of the boulders showed that these were standing more or less upright in the soil, though at different depths, and with their front faces vertical and more or less parallel. Stones at the base of the smaller, upstream boulder could have been set there to wedge its base in position; a similar cluster of possible packing stones were found at the end of the right hand, larger boulder. If these boulders were a natural formation it must be assumed that they rolled down the hill and by chance came to rest together in the red soil – touching, more or less upright and in line. Their appearance suits the interpretation that they are a man-made formation but the evidence is perhaps not conclusive.

The petrofabric analysis

The absence of all but circumstantial evidence relevant to the problem of whether the boulders and stone layer at Kintraw were man made or natural led to the opinion of a soil scientist being sought. Mr J. S. Bibby, of the Macaulay Institute for Soil Research, came to the site and performed two analyses of the orientation and dip of the long axes of the stones of the platform layer. As he describes in his appendix (p. 191 below) the pattern taken up by stones varies according to the origin of the layer, the angle of the slope and so on. The orientations and dips of a hundred adjacent stones are plotted on a Schmidt net (figure 8a). The easiest way to visualize this net is by thinking that the circle represents a view of half the hollow interior of the earth, with the north pole at the centre. The stones are plotted so that their dips are represented by latitudes as they would thus appear in two dimensions – vertical at the 'pole' in the centre, horizontal at the rim or equator – and their orientations appear on the lines of longitude, due north being at the top of the circle. The diagrams are contoured to show the relative frequency of these orientations and dips.

The two contoured diagrams from the Kintraw platform are shown (figure 8b, c) together with one of a known solifluction platform on Broad Law in Peebles-shire (figure 8f), one from a scree on Broad Law (figure 8e) and with another taken from a rubble floor layer of known human origin in the Sheep Hill vitrified fort in Dunbartonshire (figure 8d) (MacKie 1974a, forthcoming). Comparison of the five diagrams will show how closely the Kintraw patterns resemble that obtained from the Sheep Hill fort and how different they are from the two Broad Law diagrams. The slope of the Kintraw platform, about 10°, and the much steeper slope of the hillside above mean that the analogy with the 17° slope at Broad Law is close. The similarities between Kintraw and Sheep Hill are rendered more striking, as Bibby points out, by the fact that the conditions are in many ways quite different at the two sites (appendix).

Conclusions

The relatively clear results from this analysis agree well with the circumstantial evidence obtained from the excavations and described earlier, and also with the implications of the coincidence of the declinations of three of the foresights seen from Kintraw and Ballochroy. The combined effects of all this evidence is surely to leave very little room for doubt that the boulder notch and platform at Kintraw together form an artificial construction. Since the existence of an artificial platform of some kind at this spot was in effect predicted by Thom's astronomical interpretation of Kintraw – which held that observations *must* have been taken from this particular point on the hillside – this interpretation would appear to have been veri-fied and also, by implication, that of Ballochroy. There can now be little doubt that these two sites were laid out as solstice observatories at some time in the past when the Earth's axis was tilted 23° 54′ away from the vertical.

The massiveness and permanence of this observation point seem to carry another implication. It looks like something more than a temporary flat surface from which the preliminary sight-ings were made in one year on a succession of evenings before midwinter simply to establish the correct line to Jura which was to be projected on to the field below (Thom 1971, p. 37). The structure could well be the permanent observing point from which the day of the midwinter solstice was regularly checked. If the site is interpreted in this way its various features seem to become more straightforward. The boulder notch is in fact easy to reach by the path along the side of the stream valley from where the road bridge now is. The view of Jura is unobstructed from the platform, whereas it is only just visible from beside the standing stone. If the cairn beside the stone was, as suggested, once higher and served as a viewing platform there would be little room for manoeuvre on top of it, in contrast to the hill platform. From the boulder notch the standing stone serves the useful function of directing the eye towards the foresight on Jura; the alinement would then be an indicated rather than an inferred one (figure 4). Altogether the Kintraw site would surely have been a more convenient and efficient observatory if the hill platform was the primary viewing point.

DATING THE STANDING STONE SITES

As far as the solstitial sites are concerned the assumption is that the date of the construction of these can be discovered by retrospective calculation. It is claimed that the formula of Newcomb and others can project the present variations in the obliquity of the ecliptic – detected with the most advanced instruments – backwards into the past (Newton, this volume p. 99). According to this formula the obliquity (equivalent to the angle between the Earth's axis and a line vertical to the plane of its orbit) was about 23° 54' in 1800 B.C., and this should be the date of the two solstitial sites at Ballochroy and Kintraw. Unfortunately no charcoal was found during the excavations at Kintraw and so it was not possible to date the construction of the boulder notch platform by radiocarbon. No isolated standing stones (as opposed to stone circles, discussed below) have yet been dated by radiocarbon except, tentatively, those at Duntreath in Stirlingshire (MacKie 1973). It seems reasonable to suppose that the standing stones and stone circles are of similar age and that the archaeological and radiocarbon dating for the latter apply also to the former.

Stone circles

Two well excavated sites – Cairnpapple Hill in West Lothian (Piggott 1948) and Stonehenge in Wiltshire (Atkinson 1956) – have produced information about the cultural context in which the stone circles there were erected. At both sites the building of the circle was not the first activity which was traceable: late Neolithic structures were also found. The stone circle at Cairnpapple and the first stone circle at Stonehenge (the bluestones) were associated with sherds of Beaker pottery, a characteristic ware almost certainly brought to Britain by immigrants at the beginning of the Bronze Age. The earliest radiocarbon dates for Beaker sites indicate a date of around 1900 to 1800 B.C. (Lavell 1970), which would be two or three centuries earlier in real years according to the tree-ring calibration (Libby 1970). Indeed the bluestone circle at Stonehenge received a carbon-14 date of 1620 ± 110 B.C. (I–2384) while charcoal below the bank surrounding the circle at Cefn Coch, Penmaenmawr, Caernarvonshire, was dated at 1520 ± 145 B.C. (NPL–11) and 1405 ± 155 B.C. (NPL–10) (Lavell 1970). The equivalent in real years of these three dates mentioned is probably two or three centuries older.

Thus, the limited amount of direct archaeological and radiocarbon evidence for the date of the stone circles suggests that they belong to the Early Bronze Age, early in the second millennium B.C. Yet there can be no guarantee that the period during which stone circles were built was not a long one, which extended back into the Neolithic period or forward into the Middle Bronze Age. For example, there is a stone circle surrounding the New Grange passage grave in Co. Meath which is unlikely to have been erected later than the massive cairn it encloses. The construction of the cairn has been well dated by two radiocarbon measurements; these

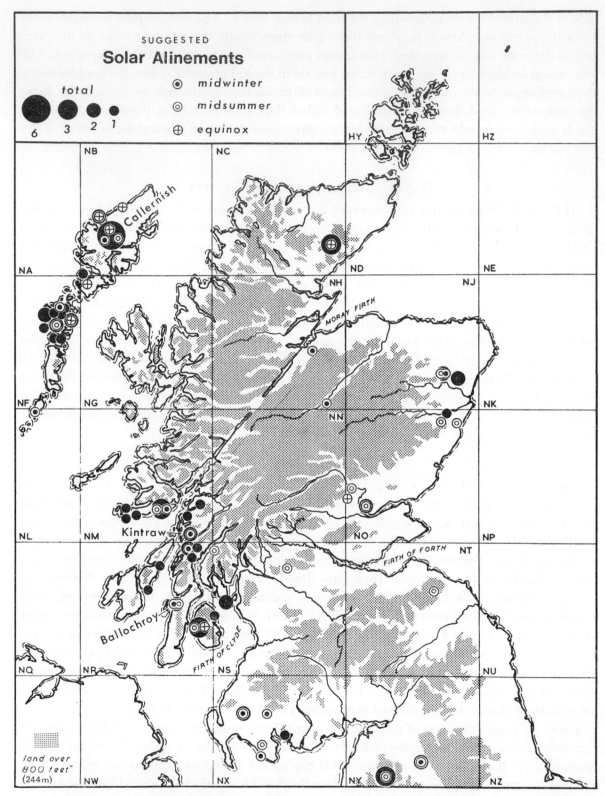

FIGURE 7. Map of Scotland showing standing stone and stone circle sites which are claimed by Thom to have solar alinements. Midwinter, midsummer and equinox lines are distinguished and the total of solar lines at each site is indicated by the size of the black disk. Information about the three line site on the isle of Arran was supplied by Dr A. E. Roy.

were 2465 ± 40 B.C. (GrN–5463) and 2550 ± 45 B.C. (GrN–5462), indicating a date in real years of perhaps nearer 3000 B.C. (Libby 1970; Lavell 1970).

Standing stones, either isolated or in groups, are notoriously difficult to date precisely. It seems reasonable to suppose that they belong to the same period of building activity as the stone circles, that is probably to the early second millennium B.C., and this is supported by a few examples of such stones with cup-and-ring marks on them. Such rock-carving are also found occasionally on graves of Early Bronze Age type (Childe 1935, pp. 116–117). Thus what little direct evidence there is seems to support a date for the standing stones which fits well with the date assigned to their alinements by the Newcomb formula for calculating the obliquity of the ecliptic in the past.

Duntreath

Yet here again the lack of evidence that standing stones were erected later than the Early Bronze Age does not mean that such evidence will not be found. Equally it is possible that some were much earlier, as the dates from New Grange imply. In the summer of 1972 excavations were carried out at the Duntreath standing stones in Stirlingshire (MacKie 1973). This is a group of six stones, five of which are now prone, and which may once have stood in a row. From the stones a notch in the nearby Strathblane hills to the east gives a declination of $+24° 0'$, and is claimed by Thom as a midsummer sunrise alinement (Thom 1967). The single upright stone has an extremely flat northern face which is precisely alined on a notch in the east. This notch (not measured by Thom) has a declination of $+1° 37'$: thus if the upper edge of the rising sun was just showing in this notch the centre of its disk would have a declination of about $+1° 21'$ which is close to, but certainly not at, the equinoctial position. (It marks the sunrise 2 days before the autumnal equinox and 2 days after the spring one). This is curious if the site is indeed an astronomical one, as the alinement to the notch is exactly indicated. The suggested summer solstice alinement – even though, being about $1\frac{1}{2}$ km long, it is rather short for accuracy – ought, if genuine, to be indicating a similar age for the stones as those at Ballochroy and Kintraw which have similar declinations.

The excavations revealed a sequence of four distinct layers in the soil around the stones and a study of the base of the standing stone suggested that the monoliths were inserted into layer 3 (layer 1 being the topsoil) and that the upper two strata accumulated later. Elsewhere on top of layer 3, a spread of whitish ash and charcoal fragments was found and the charcoal – presumably the remains of fires lit near the stones – gave a radiocarbon date of 2860 ± 270 B.C. (GX–2781). This would be equivalent to perhaps 3200 or 3300 B.C. according to the tree-ring calibration graph (Libby 1970). It is possible of course that the fires could have been lit on layer 3 (when it was the ground surface) at some time before the standing stones were put up, but the site is an exposed one which seems an unsuitable place for a camp or settlement. Although there are certain unsolved problems about the site, it seems reasonably probable at present that the fires were associated with the stones, and that this particular group was erected and used some 1500 years before the time inferred for the solstice sites at Ballochroy and Kintraw. No firm conclusions can be drawn on the basis of a single date, but Duntreath does illustrate the dangers of assuming that all problems are solved simply because none are apparent in the small quantity of hard evidence available.

ASTRONOMER PRIESTS IN IRON AGE BRITAIN

The concept that advanced geometrical and astronomical work – albeit of an empirical, practical kind – was undertaken in Neolithic and Bronze Age Britain inevitably raised the question of what the type of society was which could produce and support the specialists needed to do this work and how indeed the masses of knowledge which these activities must have accumulated was stored and passed on to new recruits to what was presumably a priestly caste. It must be more than a coincidence that we have plenty of historical evidence for the existence of such a priestly class, well versed in astronomy and other knowledge, in Iron Age Britain. Indeed, such intellectual classes are not uncommon in non-literate, barbarian societies as the studies of the Polynesian navigators have shown in a spectacular manner (Lewis, this volume, p. 133). Piggott (1968) and Burn (1969) have recently reviewed the evidence for this élite intellectual class consisting of jurists, poets and holy men – which was flourishing in Gaul and Britain in the first century B.C. Burn (1969, p. 5) concludes, for example, that 'Celtic Christian monasticism, with its frequent choice of islands for settlement, . . . was continuing a practice well known to religious men of pre-Christian times'. The British and Gaulish élites are of course known to us as the Druids, a term which has had so many fanciful connotations added to it in post-Roman times that it might be thought of dubious value to the present discussion. Yet this name, Burn suggests, probably simply means 'wise' – derived from the root *id* or *wid* – with the intensifying prefix *dri* or *tri* added to it to give 'thrice wise'. The *Drus* or 'oak tree' origin of the name is, he thinks, a Graeco-Roman guess with little to recommend it.

Piggott cites the evidence of classical authors such as Strabo who said that the Druids were well versed in astronomy and calendrical computation; Strabo called this *Physiologia* or natural science. Julius Caesar wrote that the Druids have 'much knowledge of the stars and their motion, of the size of the world and the Earth, of natural philosophy'. Hippolytus attributed the status of the Druids as prophets to the fact that 'they can foretell certain events by the Pythagorean reckoning and calculations' (Piggott 1968, p. 122). There seems little reason to doubt from these contemporary descriptions that the Druids included competent practical astronomers in their orders. There may also be evidence of the survival of an astronomical temple in Britain as late as the fourth century B.C. in the well known and oft-quoted (Hawkes 1967, pp. 129–130; Atkinson 1960, pp. 183–184; Piggott 1968) passage from the lost *History of the Hyperboreans* of Hecataeus of Abdera (*ca.* 300 B.C.), preserved in the writings of Diodorus Siculus. The passage describes a spherical temple and a sacred enclosure dedicated to Apollo (the Sun) in the island of the Hyperboreans. It is probable, though by no means certain, that this island is Britain in which case the temples mentioned could either be Stonehenge or Avebury, or both. Diodorus also describes the legend that Apollo visited the island every 19 years, a circumstance which is usually taken to be a reference to the Metonic cycle of the same length of time, at the end of which the solar and lunar calendars coincide (235 lunar months total one day less than 19 solar years). If this story really refers to Britain, it would seem to imply the practice here of a fairly advanced form of observational astronomy of the Sun and the Moon at about 300 B.C. Druidical knowledge of the Metonic cycle seems also to be implied by the evidence of the Coligny bronze calendar (Piggott 1968, p. 123).

A considerable span of time, perhaps as much as 1500 years, lies between these Iron Age priest-astronomers and wise men and the comparable class of people which, Thom's researches suggest, existed in Britain in the late Neolithic and early Bronze periods. Yet it would be wrong to assume that there could be no continuity of traditional learning between even such chronologically widely separated and culturally distinct epochs: indeed quite independent evidence is now emerging that the geometry and metrology of the stone circles re-appeared in the Iron Age brochs of Scotland (MacKie 1974 b). The Druids' expertise in *physiologia* is usually assumed to have been acquired from Greek science by way of the Greek trading colony at Massilia (Marseilles) from about 600 B.C. onwards (Piggott 1968, p. 125) but the existence of an advanced and presumably indigenous astronomy and metrology in Britain and Brittany at least as early as the Early Bronze Age would strongly suggest that the Druids were the inheritors of this ancient tradition, far older than Pythagoras.

In this context Caesar's descriptions of the Druids in *De Bello Gallico* are interesting. The reliability of his information has often been questioned because it was supposed to have been derived from the work of Poseidonios of Rhodes. However, Burn argues that this is unlikely, first, because the general theme of Caesar's writings is that he was doing what no Roman had done before and he is unlikely to have detracted from this impression with an obvious crib from a well-known book when on the subject of the Druids, and secondly, because he had a first class source of information to hand in the person of the great chief and Druid Divitiācus of the Aedui whom he had taken prisoner (Burn 1969, pp. 4–5). Caesar says of the Druidical learning, or *disciplina*, that it 'is believed to have been developed first in Britain, and thence introduced to Gaul: and to this day those who wish to pursue their studies of it more deeply usually go to Britain for the purpose'. The origin and more advanced development of the *disciplina* in Britain is harder to understand if much of the calendrical and cosmological knowledge is assumed to have diffused among the learned hierarchies of the northwest European Iron Age cultures outwards from Greece by way of Massilia, but the description exactly fits the state of affairs suggested by Thom's researches. The henge monuments, stone circles and standing stones of the period down to about 1500 B.C. (real years) are a phenomenon almost entirely restricted to the British Isles: Brittany alone elsewhere has a comparable concentration of standing stones and claimed astronomical alinements (Thom 1970, 1971, 1972) (though similar activities have been claimed for Germany – Müller 1970). The spread of the Neolithic or Early Bronze Age *disciplina* from Britain to Brittany now seems more than likely.

Lastly, the study of the Druids and allied non-literate intellectual classes helps to explain how a relatively sophisticated astronomical (and geometrical) knowledge could be acquired and transmitted without the use of writing. It has been pointed out that 'elaborate computations do not necessarily involve apparatus, or even the writing of figures. Among the Tamil calendar makers of South India in the last century the calculation of eclipses was done by arranging shells or pebbles on the ground in such a way as to recall to the mind of the operator the necessary algorithm, or steps in the process. One man, "who did not understand a word of the theories of Hindu mathematics, but was endowed with a retentive memory which enabled him to arrange very distinctly his operations in his mind and on the ground", predicted by such methods a lunar eclipse in 1825 within four minutes of its true time' (Piggott 1968, 124–125; Neugebauer 1952). It is easy to see how the presence of an intellectual class including men with such abilities in prehistoric Britain – abilities doubtless encouraged genetically by selective breeding and culturally by the appropriately favourable social environment – could have led to most of the achievements that Thom has inferred.

In a non-literate society specialized intellectual knowledge is likely to be transmitted by means of ballads and verse. Caesar also said of the Druids that 'Many are sent to join it (the learned order) by their parents and families. In it they are said to learn by heart huge quantities of verse. Some spend twenty years in this *disciplina*. They consider it impious to commit this matter to writing.' (Burn 1969, p. 4). No doubt this verse contained all the Druids accumulated knowledge and beliefs, including data on astronomy and cosmology. The oral transmissions of quantities of practical astronomical data for navigating canoes during long voyages in the Pacific by a Polynesian class of navigators has been described by Lewis (this volume, p. 133), and again seems to shed light on the practical problems of the storage and transmission of knowledge raised by Thom's interpretations.

I am most grateful to Mr John S. Bibby, of the Macaulay Institute for Soil Research in Aberdeen, for undertaking the petrofabric analyses at Kintraw and Sheep Hill, for analysing the data and preparing the appendix below. I thank Mr David Tait and Mr Richard Davis of the Department of Geography in the University of Glasgow for drawing the plan in figure 3 by photogrammetry from air photographs of the site. I had the benefit of many discussions about the site, and about the general problems of prehistoric astronomical practices, with Professor A. Thom himself and with Dr A. E. Roy of the Department of Astronomy in the University of Glasgow. I also thank Dr Roy for calculating the declinations of the alinement to Beinn à Chaolais at Ballochroy and of the notch indicated by the stone at Duntreath, both from data supplied by me. Dr James Dickson, of the Department of Botany in the University of Glasgow, kindly came to the site and took soil samples for pollen analysis, but unfortunately the pollen grains were too poorly preserved to be useful. I am grateful to Mr Hugh Mackay of Kintraw farm for ready permission to excavate on his land in 1970 and 1971. The following assisted in the excavations at various times, willingly though not always convinced of the validity of the project: Dr G. I. Crawford, Mr H. N. Hawley, Mr H. E. Kelly, Miss Sylvia Jackson, Miss Dorothy Milne and Dr J. C. Orkney.

REFERENCES (MacKie)

Atkinson, R. J. C. 1956 *Stonehenge*. London: Hamish Hamilton.
Baity, E. C. 1973 Archaeoastronomy and ethnoastronomy so far. *Curr. Anthrop.* **14**, 389–449.
Bergamini, D. 1971 *The Universe*. Time Life International (Nederland) N.V.
Burn, A. R. 1969 Holy Men on Islands in pre-Christian Britain. *Glasgow Arch. J.* **1**, 2–6.
Childe, V. G. 1935 *The prehistory of Scotland*. London: Kegan Paul.
Childe, V. G. 1955 'Comments' on Thom (1955), *Jl R. statist. Soc.* A **118**, 293–294.
Hawkes, J. 1967 God in the machine. *Antiquity* **41**, 174–180.
Hawkins, G. S. 1966 *Stonehenge decoded*. London: Souvenir Press.
Hogg, A. H. A. 1967 Review of A. Thom (1967) in *Arch. Cambrensis* **117**, 207–210.
Hoyle, F. 1966 Speculations on Stonehenge *Antiquity* **40**, 262–276.
Lavell, C., ed. 1971 *Archaeological site index to radiocarbon dates for Great Britain and Ireland*: yearly supplements. London: Council for British Archaeology.
Libby, W. F. 1970 Radiocarbon dating. *Phil. Trans. R. Soc. Lond.* A **269**, 1–10.
Lockyer, N. 1906 *Stonehenge and other British stone monuments astronomically considered*. London: Macmillan.
MacKie, E. W. 1968 Stone circles: for savages or savants? *Curr. Arch.* **2**, no. 5, pp. 279–283.
MacKie, E. W. 1973 The standing stones at Duntreath. *Curr. Arch.* **4**, 6–7.
MacKie, E. W. 1974a The vitrified forts of Scotland, in D. Harding, ed., *Hillforts*. London: Seminar Press.
MacKie, E. W. 1974b The brochs of Scotland, in P. Fowler, ed., *Recent work in rural archaeology*. Bath: Adams and Dart.
Morley, S. C. & G. W. Brainerd 1956 *The ancient Maya*. 3rd edn. Stanford, California: University Press.
Müller, R. 1970 *Der Himmel über die Menschen der Steinzeit*. Hamburg: Springer Verlag.
Neugebauer, O. 1952 Tamil astronomy. *Osiris* **10**, 252–276.
Piggott, S. 1948 The excavations at Cairnpapple Hill, West Lothian 1947–8. *P.S.A.S.* **82** (1947–8), 68–123.

Piggott, S. 1968 *The Druids*. London: Thames and Hudson.

Roy, A. E. 1963 New survey of the Tormore circles (Arran). *Trans. Glasgow Arch. Soc.* **15**, 59–67.

R.C.A.H.M.S. 1972 The Royal Commission on the Ancient and Historical Monuments of Scotland; *Argyll* **1**, *Kintyre*. Edinburgh: H.M.S.O.

Simpson, D. D. A. 1967 Excavations at Kintraw, Argyll, *P.S.A.S.* **94**, (1966–67), 54–59.

Somerville, B. 1912 Astronomical indications in the megalithic monument at Callanish. *J. B. astr. Ass.* **23**, 83.

Somerville, B. 1923 Instances of orientation in prehistoric monuments of the British Isles. *Archaeologia* **73**, 193–224.

Thom. A. 1954 The solar observatories of megalithic man. *J. Br. astr. Ass.* **64**, 396–404.

Thom. A. 1955 A statistical examination of the megalithic sites in Britain. *J. R. statist. Soc.* A **118**, 275–291.

Thom, A. 1961a The geometry of megalithic man. *Math. Gaz.* **45**, 83–93.

Thom, A. 1961b The egg-shaped standing stone rings of Britain. *Archs. Int. Hist. Sci.* **14**, 291.

Thom, A. 1962 The megalithic unit of length. *J. R. statist. Soc.* A **125**, 243–251.

Thom, A. 1964 The larger units of length of megalithic man. *J. R. statist. Soc.* A **127**, 527–533.

Thom, A. 1966a Megalithic astronomy: indications in standing stones. *Vistas Astron.* **7**, 1–57.

Thom, A. 1966b Megaliths and mathematics. *Antiquity* **40**, 121–128.

Thom, A. 1967 *Megalithic sites in Britain*. Oxford: University Press.

Thom, A. 1968 The metrology and geometry of cup and ring marks. *Systematics* **6**, 173–189.

Thom, A. 1969 The lunar observatories of megalithic man. *Vistas Astron.* **11**, 1–29.

Thom, A. 1971 *Megalithic lunar observatories*. Oxford: University Press.

Thom, A. & Thom, A. S. 1971 The astronomical significance of the large Carnac menhirs. *J. Hist. Astron.* **2**, 147–160.

Thom, A. & Thom, A. S. 1972a The Carnac alignments. *J. Hist. Astron.* **3**, 11–26.

Thom, A. & Thom, A. S. 1972b The uses of the alignments at Le Menec, Carnac. *J. Hist. Astron.* **3**, 151–164.

Thom, A. & Thom, A. S. 1973 A megalithic lunar observatory in Orkney. *J. Hist. Astron.* **4**, pp. 111–123.

Appendix:

Petrofabric analysis

By J. S. Bibby

The Macaulay Institute for Soil Research, Craigiebuckler, Aberdeen, Scotland

Petrofabric analysis is the study of the spatial relations of the units that comprise a rock, including a study of the movements that produced these elements (Glossary of geology 1966). The fact that particles in a moving medium tend to orientate themselves with their longest (or *a*) axes parallel to the direction of flow and their shortest (or *c*) axes transverse to it has long been utilized in studies concerning the depositional history of particular rocks. In the last 40 years the technique has been used increasingly to deal with problems concerning superficial deposits such as glacial till, outwash sands and gravels and solifluction deposits. In Scotland, Kirby (1968), in a study of the stratigraphy of drift deposits in the Esk basin, Midlothian, utilized the technique to elucidate the direction of flow of ice depositing different boulder clays: a distinction was drawn between raised beach deposits and glacial outwash materials in the western Highlands (McCann 1961); and the orientation of stones in solifluction deposits with increasing slope has been described from the Southern Uplands (Ragg & Bibby 1966).

During archaeological investigations on a site at Kintraw, Argyll, a stone pavement was uncovered. It was important to determine whether the pavement had been formed by geomorphological agencies or by the activities of man. Natural stone pavements are known from the hill areas of the British Isles, and during the Pleistocene period they were of far greater extent. Their remnants are encountered frequently during the investigations of the Soil Survey

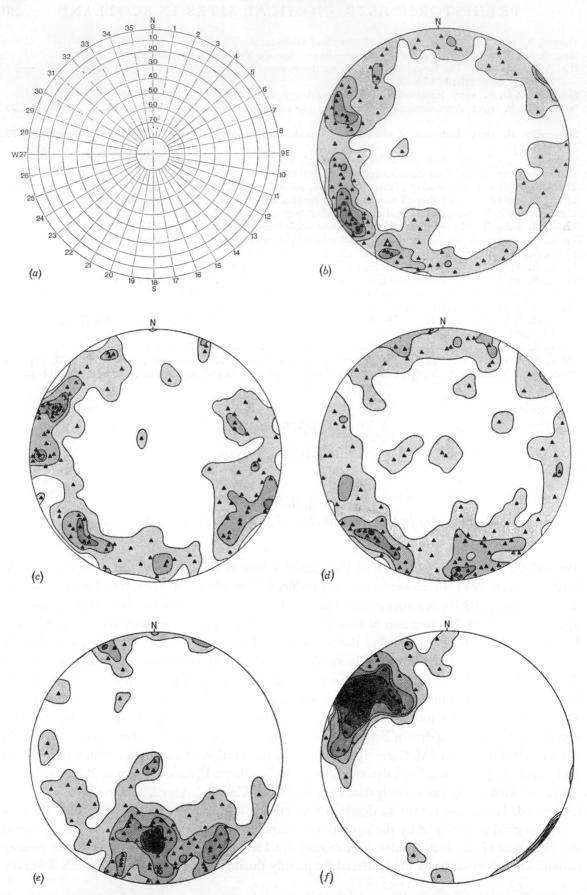

FIGURE 8. (a) The Lambert polar equal-area net. (b)–(f) Contoured petrofabric diagrams (contour interval 2%) (b) Kintraw no. 1 (Argyll): direction of slope, 240° (mag.); angle of slope, 21°. (c) Kintraw no. 2 (Argyll): direction of slope, 240° (mag.); angle of slope, 21°. (d) Man-made pavement (Sheep Hill Fort, Dunbarton-shire): direction of slope, 163° (mag.); angle of slope, 5°. (e) Scree (Broad Law, Peebles-shire): direction of slope, 175° (mag.); angle of slope, 37°. (f) Solifluoted stone horizon (Broad Law, Peebles-shire): direction of slope, 305° (mag.); angle of slope, 16°.

of Scotland (see, for example, Mitchell & Jarvis 1956, p. 71). The accumulation of stones on the surface of a deposit can be caused by frost heave of coarse particles with concomitant downward movement of fine material (Corte 1962). In the situation at Kintraw it could also have been formed as a scree accumulation. Other considerations were ruled out by the particular site factors obtaining, and the nature of the underlying deposits.

It is well established that the orientation diagrams produced during petrofabric studies of deposits of markedly different origin show characteristic patterns that can be associated with the different modes of origin. An investigation was therefore undertaken to determine the pattern within the deposit at Kintraw, and whether this pattern had any characteristics which would enable statements to be made concerning its mode of origin.

METHODS

Two areas were selected on the exposed pavement, each measuring 50 cm × 50 cm. At each site the *a* axis of each of 100 stones was determined by inspection, while the inclination from the horizontal was measured by Abney level and its azimuth determined with the aid of a prismatic compass. Only stones measuring over 3 cm in length were selected (smaller stones proved difficult to measure and handle effectively). They were derived mainly from schist and were dominantly tabular or wedge-shaped, forms which would normally react strongly to the stress forces responsible for orientation. The results were plotted on a polar-equiarea net (Lambert projection, figure 8*a*), which has a central point representing a pole at 90° to the plane of projection, with concentric circles representing angular declinations from this pole at 10° intervals, and radial rays the azimuths. Stereographic methods of data-presentation are described by Philips (1960). The inclination and orientation of the *a* axis of any stone can therefore be represented as a point on the stereogram; a horizontal stone would lie on the periphery of the circle and could be represented as either of two points 180° apart. The method of contouring (described by Philips 1960, p. 59) using a circle of 1 cm radius takes this anomaly into account.

RESULTS

Two contoured petrofabric diagrams produced from the Kintraw data are reproduced in figure 8*b* and *c*. The salient features are the wide variation in the direction of dip of the *a* axis, and the relatively low angles of inclination. There is a concentration associated with the slope of the ground surface but it is not a strong one. By contrast, figure 8*e* is derived from an area of scree and figure 8*f* from a stone pavement produced by frost action and since modified by solifluction processes. In the diagram derived from scree there is a marked association both with direction of slope and with its inclination; this trend is even more apparent in the diagram representing the structure within the soliflucted stone pavement. It appears that the process of solifluction, involving mass creep lubricated by melt water, imposes a strong degree of orientation on the constituent particles. The strong degrees of orientation shown by figure 8*e* and *f* are in obvious contrast to the weak orientations shown in figure 8*b* and *c*.

No information was available concerning patterns produced on fabric diagrams by data drawn from man-made stone pavements. In order to obtain some check, however tentative, a visit was made to the Sheep Hill vitrified fort, Milton, Dunbartonshire, where a pavement exists that has been identified as man-made by independent evidence (MacKie, in the press). The resulting diagram is shown as figure 8*d* which, allowing for the different direction of ground slope, is closely similar to figure 8 *b* and *c*. This resemblance is all the more remarkable because of the contrast in parent materials and stone shape between Kintraw and Sheep Hill, the former being dominantly tabular schists and the latter wedge-shaped basalt.

The evidence from petrofabric analysis indicates that the stone horizon discovered at Kintraw bears little resemblance in structure to superficially similar horizons known to have been formed by the action of frost heave or by scree accumulation. Other forms of genesis are rendered unlikely by the particular combination of lithological and site conditions obtaining. The available evidence supports the hypothesis that the Kintraw pavement was man-made.

REFERENCES (Appendix) (Bibby)

Corte, A. E. 1962 The frost behaviour of soils, field and laboratory data for a new concept. 1. Vertical sorting. U.S. Army Cold Regions Research and Engineering Labs. Research Report 85 Corps. of Engineers.

Glossary of geology and related sciences 1966 American Geological Institute.

Kirby, R. P. 1968 Ground moraines of Midlothian and East Lothian. *Scot. J. Geol.* **4**, 209–220.

Mackie, E. W. 1974 The vitrified forts of Scotland. In *Hillforts* (ed. D. Harding). London: Seminar Press.

McCann, S. B. 1961 Some supposed 'raised beach' deposits at Corran, Loch Linnhe and Loch Etive. *Geol. Mag.* **98**, no. 2.

Mitchell, B. D. & Jarvis, R. A. 1956 The soils of the country round Kilmarnock. *Mem. Soil Surv. Gt Britain.* H.M.S.O.

Philips, F. C. 1960 *The use of stereographic projection in structural geology.* Edward Arnold.

Ragg, J. M. & Bibby, J. S. 1966 Frost weathering and solifluction products in southern Scotland. *Geogr. Annlr* **48**, Ser. A 1.

Phil. Trans. R. Soc. Lond. A. **276**, 195–230 (1974)   [195]

Printed in Great Britain

Climate, vegetation and forest limits in early civilized times

By H. H. Lamb

Climatic Research Unit, School of Environmental Sciences, University of East Anglia, Norwich

CONTENTS

After reviewing the basic conditions which govern climate and the distribution of climates over the globe, with particular attention to the large-scale circulation of the atmosphere and the variations which it undergoes, this contribution proceeds to consider the sites of the early centres of civilization (particularly those that flourished between about 3000 and 1000 B.C.) and routes of travel by land and sea, some of which stand in a surprising relationship to the natural environment as it exists today. The sequences of variations of prevailing temperature, of sea level, of forest limits and rainfall, cloudiness, etc., and of the levels of great inland waters are reconstructed and lead to a consistent picture of the broad sequence of climatic régimes: for those régimes which differed most from the present-day northern hemisphere maps can be given. Climatic fluctuations on time scales from a decade or two to a few centuries are then considered: fluctuations tending to repeat at 200 or 400-year intervals seem rather prominent.

Introduction

Weather and climate are forever changing. Sometimes the changes are sharp, sometimes gradual. The spectrum of time-scales involved runs from minutes to millions of years. Not only does one year's experience differ from another in a middle latitudes climate such as Britain's: there are differences also in the Sahara desert, in the islands of the equatorial Pacific and in the heart of the polar regions. Moreover, the climatic experience of each decade, each century, each millennium in any one part of the world differs.

There has been some confusion about this. Many kinds of physical as well as biological evidence (see, for example, Ahlmann 1949, Dansgaard, Johnsen, Clausen & Langway 1971; Imbrie & Kipp 1971; Turekian 1971) now contribute to our reconstructions of the past climatic record. Yet many people who are not well acquainted with the multifarious nature of the evidence still hold to the view that there have been no significant changes of climate since the end of the last ice age some 10000 years ago (see, for example, Raikes 1967). It is true that Tacitus's writings about the climates of Britain and Germany, and Herodotus's description of the Crimea, register the same impressions that would strike visitors from Italy and Greece

13-2

today; but this need mean no more than that the spatial differences were in the same sense then as now. Moreover, much evidence suggests that about 2000 years ago the climate in many countries was passing through a phase quite similar to today's. Some centuries earlier there had been a colder régime (as is also true today). It subsequently became milder, and the bridge of many stone piers with a wooden superstructure built for the Emperor Trajan in A.D. 106 across the Danube at the Iron Gates stood for about 170 years (before being destroyed by the barbarian tribes) in a position where today it might be expected to be carried away by ice within any period of 50 years or less.

Archaeologists, botanists, and some meteorologists, have been sometimes suspected of calling in entirely hypothetical climatic changes, as a sort of *deus ex machina*, to explain shifts of vegetation limits which might more properly be ascribed to the works of Man, clearing and burning forests, grazing with goats, building and later neglecting irrigation works, and so on. But, since Roman times, the occurrence of a period in the early Middle Ages that was moister than now in the Mediterranean is well attested by bridges (e.g. at Palermo, Sicily) built to span rivers that were bigger than now, and by widespread evidence of more active stream erosion and deposition. The warmth of the same medieval period in higher latitudes is registered in the absence of ice on the Viking sailing routes to Greenland, particularly in the years between about A.D. 980 and 1200, and by the increase of ice which followed there and in the glaciers in the Alps. There was a lowering of the upper tree line on the mountains in central Europe by 200 m between about A.D. 1300 and 1600 (Firbas & Losert 1949) and a parallel change in the upper limit of trees in the United States Rockies (Lamarche 1972). Within the last few years, the ascription of these changes to changes of prevailing temperature has been effectively verified by oxygen isotope measurements on the ice of the times concerned, still present in the Greenland ice cap (Johnsen *et al.* 1970; Dansgaard *et al.* 1971). Moreover, significant shifts of the northern (and, in some cases, also the southern) limits of various species of birds and fish, of forest trees and wild flowers, in response to the general climatic warming in the early decades of the present century and to the cooling since about 1950, have been widely observed and reported (e.g. in Perring & Walters 1962; Perring 1965).

The natural limits of different types of vegetation – and of the faunas whose habitats the vegetation provides – are thus seen to undergo shifts from century to century and sometimes over shorter periods. The species composition of the vegetation had undergone some changes in response to earlier, and in some cases harsher, changes of climate than those so far mentioned. Therewith, the possibilities of agriculture and husbandry must have changed, particularly in areas near the thermal or arid limits of particular crops and grasslands. Human society, especially primitive human society, and economies based on a narrow range of resources, must have had great difficulty in adapting to some of the changes, and in a number of cases migration may have been the only solution. This suggestion, first put forward by Huntington (1907) in relation to drying up of the pastures in the homelands of the nomads of central Asia around A.D. 300 and the *Völkerwanderungen* of the time of the demise of the Roman empire, looks sensible in the light of the rainfall changes in central Asia in the last 130 years (figure 1) (cf. also Chappell 1970). Sixty years between 1890 and 1950 of a climatic régime in Khazakstan moist enough for grain cultivation were followed in the 1950s and later 1960s by a lower rainfall, as in the mid-nineteenth century, only one-half to three-quarters as great.

The climatic connexion of some important environmental changes is obscured by lag in the response. When a warming up of climates opens up the possibility of forest spreading poleward,

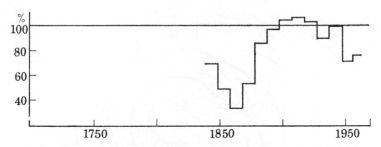

FIGURE 1. Rainfall averages for successive decades between 1840 and 1960 at Barnaul (53° N, 84° E), representing the grasslands of central Asia. Units: per cent of the 1900 to 1939 yearly average rainfall (482 mm).

much time may be required before all the new areas are colonized because of the limited spread of seed and the shelter needed for young trees to survive, needed also by the birds and insect life which help to spread the seed. And, after a turn to more arid climates in the fringes of the desert zone, the old forests may be able to survive for hundreds, and even thousands of years, thanks to the local preservation of soil moisture in the shade and moisture exchanges within the forest itself. But forests in this situation are 'sub-fossil', precarious relics of a past climatic régime: once they have been felled, or burnt, or encroached upon by grazing animals, the change is likely to be irreversible.

BASIC CONDITIONS OF CLIMATE

The climate of any place is produced by:

(i) *The radiation balance* there, the incoming radiation being graduated according to solar elevation and length of day, i.e. latitude and season.

(ii) *The heat and moisture brought and carried away by the winds and ocean currents.*

(iii) The *local conditions of aspect* towards the midday sun and the prevailing winds, *thermal characteristics* (specific heat and thermal conductivity) *of the soil and vegetation cover*, also the *albedo* (reflectivity) of the surface and hence the amount of radiation actually absorbed.

The local thermal characteristics are most of all affected by:

(*a*) *Wetness or dryness of the surface*, dry ground being subject to much the quicker temperature changes (owing to its low specific heat and poorer thermal conductivity below the surface).

(*b*) *Snow-cover* or the lack of it, since the high albedo of a snow surface means that most of the solar radiation falling on it is reflected and lost; the snow surface, moreover, radiates away heat to the sky and to space; and because of the air trapped within a snow layer its conductivity is poor and the temperature fall is concentrated near the surface.

The global climatic pattern is at all times generated by the radiation distribution and the general circulation of the atmosphere and oceans. This large-scale circulation of the winds, and the ocean currents which they drive, is set up by the unequal heating of different latitudes. Vertical expansion of the air columns over the heated zones of the Earth, and contraction of the air columns in high latitudes where most heat is lost, determines a rather simple global distribution of pressure through a great depth of the atmosphere between about 2 km and 15 to 20 km above the Earth's surface: high pressure generally over the warmest zones in low latitudes and low pressure near the poles. This pressure distribution, over the rotating Earth, maintains a rather simple pattern of wind flow, of prevailing upper westerly winds, throughout the same deep layer of the atmosphere: i.e. a circumpolar vortex over each hemisphere.

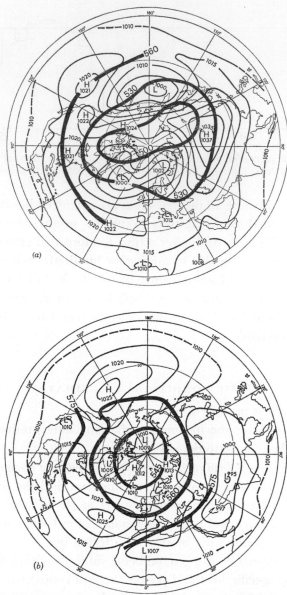

FIGURE 2. (*a*) Average atmospheric pressure over the northern hemisphere at sea level (in millibars) in January in the 1950s and (bold lines) thickness (in decametres) of the lower half of the atmosphere (1000 to 500 mbar layer), the latter to indicate the upper westerly wind flow. Winds at both levels blow (except where diverted by surface friction) nearly along the isopleths counterclockwise around the low-pressure regions (clockwise in the southern hemisphere). (*b*) Average atmospheric pressure and thickness in July, 1950s.

There are waves in the upper westerlies, as the flow meanders round the barriers of the Rocky Mountains, Greenland, and the mountains of Asia and is influenced by the warmth of the northeast Pacific and the Atlantic–Norwegian Sea as well as by the persistent cold surfaces of Arctic Canada and icy waters elsewhere, particularly in the tundra zone in northern Siberia and the Okhotsk Sea. The strongest thermal gradients and the strongest upper wind flow are in general in middle latitudes, though constrained here and there (and more at some times and seasons than others) by the geography, as mentioned. The wave train in the upper westerlies also develops its own dynamics, so that there is a tendency for a trough in the flow somewhere near the European sector as a resonance effect downstream from the trough maintained in the lee of the Rockies and over the cold surface of northern Canada. This downstream trough itself tends to induce a régime that is cold for its latitude wherever it lies at any given time over northwest Siberia, Europe or the eastern Atlantic. The longitude position of the resonance trough depends on the preferred wavelength in the upper westerlies, which increases with the latitude and strength of the mainstream of the upper wind flow.

The relationship between the circumpolar vortex of upper westerly winds and the more complex, but more familiar, distribution of high and low pressure on the surface weather map is intimate. It is this which makes it possible to derive much of the hemispheric climatic pattern from fragmentary surface data. It is illustrated here by the average situations in January and July in the 1950s seen in figure 2a and b. Convergences and divergences in the upper flow (at points where the pressure gradient changes) continually generate accumulations and losses of atmospheric mass over the areas affected, and the resulting surface-level pressure systems and wind circulations are steered and carried along by the general pattern of the upper westerlies. Surface low-pressure (cyclonic) systems are generated mostly at the forward (i.e. eastern) side of the troughs in the upper flow and are carried along, and towards, the cold flank of the mainstream of the upper westerlies. Surface high-pressure (anticyclonic) systems are generated mostly in the converse positions, at the rear of the upper troughs and are steered along, and towards, the warm flank of the upper flow.

These associations produce the following *prevailing distribution of atmospheric pressure at the Earth's surface:*

(a) A subtropical high-pressure (anticyclonic) belt along the warm side of the mainstream of the upper westerlies, developing the clear skies of the desert zone.

(b) A subpolar low-pressure belt, the locus of the centres of the main depressions, or travelling cyclones, associated with belts of cloud and rain (or snow) continually sweeping forward in the zone of prevailing westerly winds and near the cyclone centres.

The distribution also leaves room for:

(c) Higher pressure near the pole.

(d) An equatorial low-pressure trough where the surface Trade Winds from the two hemispheres converge.

The *prevailing surface wind zones* defined by this pressure distribution are

(i) Easterly (in the northern hemisphere mainly northeasterly) Trade Winds, between the subtropical high-pressure maximum and the equatorial low-pressure trough.

(ii) Westerly (in the northern hemisphere mainly southwesterly) surface winds in middle latitudes.

(iii) Polar easterly surface winds, prevailing between the latitude of the subpolar depression centres and the polar high-pressure maximum.

This outline of general principles (for further elucidation see Lamb (1972) and refs. therein) is essential to an understanding of variations in the pattern which occur from time to time, including those tendencies which seem to have prevailed in different eras in the past.

The types of variation to which the circumpolar vortex is liable may be listed as follows:

(1) Changes of strength of the upper westerly windstream.

(2) Changes of latitude of the main flow.

(3) Changes of wavelength (or spacing between the troughs) downstream from the nearly fixed disturbances at the main mountain barriers: hence also changes of the trough positions and of the total number of waves around the hemisphere.

(4) Changes of amplitude of the waves, i.e. of the north–south range, or meridionality, of the flow.

(5) Eccentricity: the main centre of the circumpolar vortex may move quite far from the geographical pole, occasionally to latitude 60–70°, though in such extreme cases the situation usually tends to become bipolar (or even tripolar).

The surface pressure and wind patterns become displaced and distorted in sympathy with these variations.

A common variant type of situation is known as *Blocking of the westerlies*, with an anticyclone developing (in association with a meridionally extended warm ridge aloft) in the usual zone of subpolar depressions, and low pressure in the subtropical zone, usually near the entrance to a confluent trough in the upper westerlies (which may become considerably distorted). This is the most abnormal-looking situation in that it produces easterly winds in middle latitudes and twin belts of westerlies near the Arctic circle and in subtropical latitudes, while the Trade Winds are liable to be disrupted. But blocking seldom lasts more than a few days to a few weeks without the pressure systems concerned undergoing a considerable change of longitude. Hence, periods of frequent blocking are represented in the climatic mean mainly by weaker than normal middle latitudes westerlies and weakness of the Trade winds.

The influences that affect the amount of rain falling at a given place in different epochs may be listed as:

(i) The prevailing temperatures of the water surfaces (chiefly the ocean) from which the air's moisture supply is drawn, the absolute humidity of the air increasing with the water temperature.

(ii) The positions of the mainstream of the upper westerlies: (*a*) latitude; (*b*) longitudes of troughs and ridges; (*c*) distorted, 'meridional' and 'blocking' patterns; and hence the locations of frequent cyclone formation and the paths along which these systems are steered, the rainfall increasing with the cyclonicity.

(iii) The most frequent surface wind directions and their aspect to the slopes of the terrain, the rainfall being greatest with upslope winds.

The climatic shift that took place in the early 1960s, portrayed in figure 3 by its effects on the world rainfall distribution, illustrates types of displacement which, in the light of the foregoing principles, may be looked for in other climatic régimes. The shaded areas on the map mark regions where the rainfall from mid-1961 to mid-1964 was above the previous 30-year (1931–60) average; elsewhere it was below the previous average – save that rainfall over the oceans beyond the range of the island observations used could not be surveyed. This map shows the years 1961–4 as characterized by

(i) Increased rainfall in most parts of the equatorial zone.

(ii) Reduced rainfall rather generally in zones on either side of the equator, especially between latitudes about 15 and 25° (presumably indicating less seasonal intrusion of the equatorial rains into these zones).

(iii) Increased rainfall in subtropical latitudes, especially about 40° (more winter rainfall, encroachment of the travelling disturbances of middle latitudes in both hemispheres and frequent blocking with low pressure in the subtropical zone).

(iv) In middle latitudes alternate north–south ('meridional') stripes of excess and deficient rainfall, associated with changed wavelength (and changes in the most frequent trough positions) in the upper westerlies.

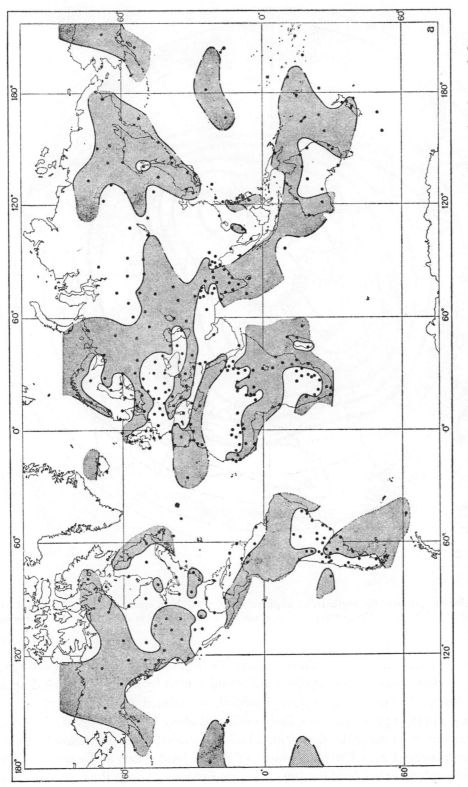

FIGURE 3. World rainfall July 1961 to June 1964: departures from the 1931 to 1960 averages. Dots show stations with complete data used. Ocean areas generally not included in the survey except where the indications from island observations were adequate. Stippled areas: rainfall exceeded 1931 to 1960 averages.

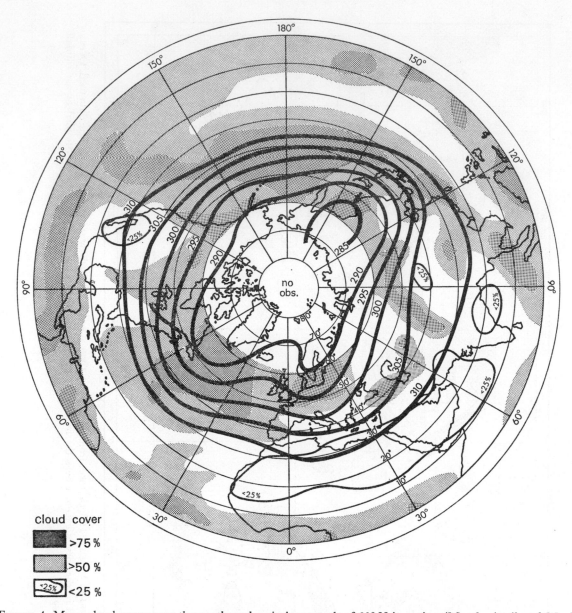

FIGURE 4. Mean cloud cover over the northern hemisphere south of 60° N in spring (March, April and May) 1962 from *Tiros* satellite data (Clapp 1964) and average 700 mbar heights in decametres during the same months.

Converse shifts are known to have occurred about the beginning of the present century.

Cloudiness (other than fogs over cold surfaces and cloud caused by uplift of the wind against the windward slopes of mountain ranges), and *rainfall*, are related to the forward (eastern) sides of the troughs in the upper westerlies, and are carried along in the strong windstream, in the same general way as cyclonic development. This is illustrated here in figure 4 by the first published 3-month survey by satellite photographs of global cloud cover, in the spring of 1962, together with the isopleths of height of the 700 mbar pressure level to show the flow of the upper westerlies at about 3 km height.

The sites of the early civilizations

Figure 5 is a world map of the beginnings of civilization. The areas of origin of the group of civilizations that flourished between 4000 B.C. and the time of Christ are, except in the special case of Egypt and the lower Nile, areas of rather intricate geography, within which the sites then occupied range today from the natural domain of woodland to steppe or desert. The changes in the environment that have taken place there could hardly be shown on a world map and have hardly been probed so closely that valid regional maps of the former vegetation could yet be drawn.

In at least three regions, however, there are unmistakable indications of formerly more extensive water supply and vegetation:

(i) The western and central Sahara, where rock drawings of a wide range of animals (Butzer 1958), and finds of skeletons, including elephants (Monod 1963), show that there was enough surface water for animals to pass across the terrain that is now desert and for at least some human occupation.

(ii) The Rajasthan (Thar) desert areas of India and Pakistan, where rivers have disappeared (Singh 1971) since the Indus valley civilization and the cities of Mohenjodaro and Harappa (31° N, 72½° E) were at their height before 2000 B.C.; also the other desert areas in southern Asia crossed by Alexander's army on expeditions in Iran and on the march to the Indus between 330 and 323 B.C. (On their return, the Indus was crossed in a flotilla built of timber from the local forests (Wadia 1960).) The fauna of the present Thar desert in Harappan times included rhinoceros, water-buffalo and elephant.

(iii) Sinkiang, particularly the Tarim basin in central Asia near 36–42° N, 80–90° E, was crossed by the ancient Silk Route which was used by trading caravans between China and the west in Roman times, when there was a chain of cities and settlements there and remnants of forest. Today it is largely a sandy desert (Wadia 1960; Chappell 1970, 1971).

Other places, where conquest of the area by desert conditions within the last 2000 years suggests a continued trend towards natural desiccation, include Palmyra (34° 40′ N, 38° 0′ E) and Petra (30° 20′ N, 35° 18′ E), on the fringe of Palestine and the Syrian desert, and parts of the Anatolian plateau in central Turkey (Carpenter 1966). It is unlikely that rainfall has been continuously decreasing for thousands of years, but probable that it has been fluctuating with a net downward trend; we shall see evidence of such a trend affecting wide areas of the sub-tropical zone over the last few thousand years. The vegetation may, therefore, long have been in a subfossil condition, in which disturbance by Man might easily bring on its final demise. Radiocarbon ages of 20 000 to 25 000 years since last contact with the atmosphere found in the water in a Saharan oasis, and in water-bearing strata under the Sinai and Negev deserts (see, for example, Valéry 1972), make it clear that the water-table in the arid zone in the Near East is still influenced by the supply of rain which fell in previous climatic régimes, particularly those of the last ice age. Certainly, the extent of the oases has been declining over the last 6000 years (Butzer 1958).

A curious feature of figure 5 is the spread of groups building stone circles as early as about 2000 B.C., and trading, evidently by a sea route through the western parts of the British Isles as far north as Orkney (59° N). This may suggest not only some skill in navigation, but an era of rather frequent quiet seas. In apparent support of this suggestion, there is now much evidence from radiocarbon dated pine stumps in the peat (H. H. Birks 1972, personal communication

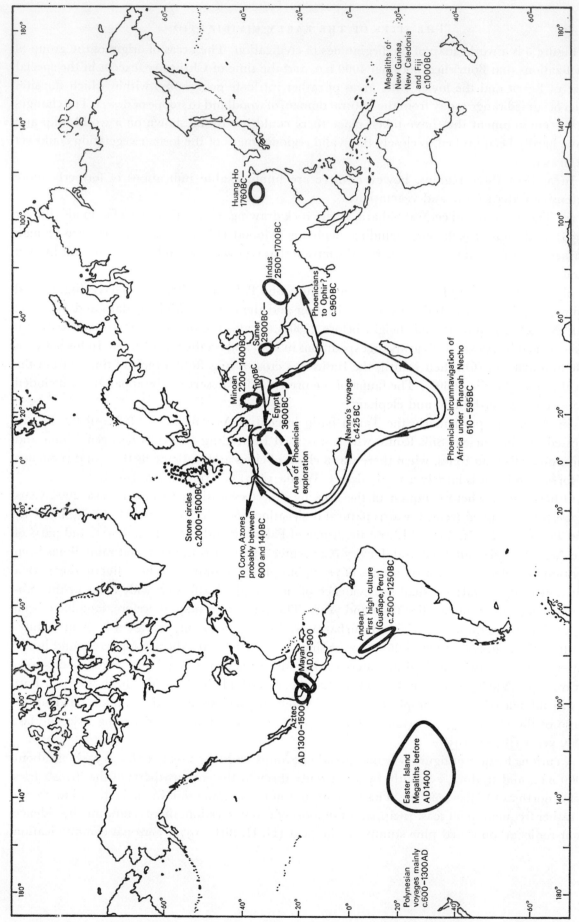

FIGURE 5. World map of the beginnings of civilization.

5 July 1972; Lamb 1964, 1965 a; Pennington, Haworth, Bonny & Lishman 1972) that between about 5000 and 2000 B.C., forest grew much nearer to the open Atlantic coast of northwest Scotland than at any time since and also in parts of the Hebrides and northern isles.

SETTING IN THE SEQUENCE OF CLIMATE AND VEGETATION HISTORY

The ancient civilizations grew up and decayed in times when a previous extensive glaciation of the northern hemisphere was much more recent than is now the case. The demise of the last major ice sheet in Scandinavia and of the smaller glaciers in northern Britain may be placed a

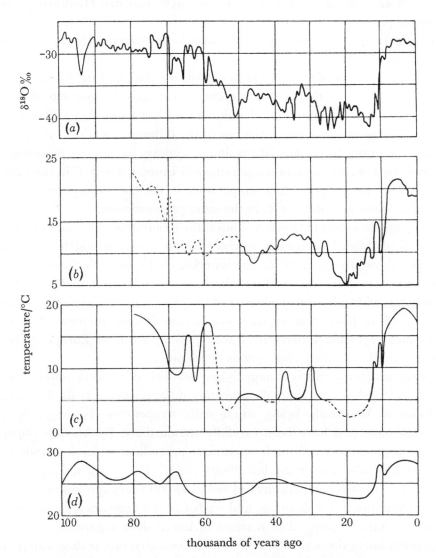

thousands of years ago

FIGURE 6. Variations of prevailing temperature during the last 100000 years: (*a*) N. Greenland (77° N, 56° W) – temperature changes indicated by differences in the $^{18}O/^{16}O$ oxygen isotope ratio ($\delta^{18}O$). A change of $\delta^{18}O$ by one part per thousand (‰) has been found to correspond to a change of mean annual temperature by 1.3 to 1.4 °C (Dansgaard 1964; Dansgaard *et al.* 1971). (*b*) Central Europe – summer temperatures deduced from pollen analysis (Gross 1958). (*c*) Netherlands – summer temperatures deduced from pollen analysis (van der Hammen *et al.* 1967). (*d*) Tropical Atlantic – annual temperatures of the surface waters derived by oxygen isotope measurements on the remains of surface-dwelling *foraminifera* identified in the bottom deposits (Emiliani 1955, 1961).

little before 8000 B.C., and that of the last remnants of the ice sheet on the North American mainland, east and west of Hudson's Bay, perhaps as late as 3000 B.C. (Bryson & Wendland 1967).

(i) *Temperature*

The sequence of prevailing temperatures over the last 100 000 years, through the last ice age and since, has been derived from various kinds of field data (see figure 6).

(*a*) For the surface waters of the tropical Atlantic – from studies of the remains of the minute biological organisms (*foraminifera*) in the deposits on the ocean bed (Emiliani 1961).

(*b*) For the summers in the Netherlands and in central Europe – by interpretation of the composition of the flora from pollen analyses (Gross 1958; van der Hammen, Maarleveld, Vogel & Zagwijn 1967).

(*c*) For the air over the ice-cap in northwest Greenland – from oxygen isotope measurements on the ice still present.

The results are remarkably consistent except as regards the timing of the fluctuations beyond the workable limit of radiocarbon dating 30 000 to 50 000 years ago. There is parallelism of most details of the changes from millennium to millennium in the last 15 000 years. Not surprisingly, there are differences in the range of the variations in the different locations: the temperature range from ice-age minimum to postglacial maximum appears to amount to nearly 20 °C in northern Greenland, 16 °C in the summer temperatures in Europe, 5 to 6 °C in the waters of the tropical Atlantic; the decline of prevailing temperature in the last 4000 years is indicated as 1 °C in the tropical Atlantic and about 2 °C in the other places named.

Judged by the changes in height of the snow-line on the mountains, this range of temperature changes in the tropical Atlantic is similar to those which occurred widely in tropical and subtropical latitudes. A peculiarly great ice-age lowering of the snow-line (*last* glaciation) by as much as 1200 m in the eastern Taurus Mountains in southern Turkey, and by 1800 m in the Zagros Mountains (Algurd Dagh) near the border of Iraq and Iran (Wright 1961), is attributed to the combined effects of lower temperatures and a considerable increase of precipitation accompanying frequent cyclonic activity in a zone passing south of Europe. (These areas are close to where a number of early civilizations developed.)

The course of postglacial temperature changes is best established for northwestern and central Europe from botanical, and in later times documentary data, particularly for England (Lamb 1965*b*, Lamb, Lewis & Woodroffe 1966), where daily temperature readings are available for almost 300 years past, and have been carefully standardized by Manley (1959, 1961). Figure 7 shows the sequence derived for temperatures prevailing in the lowlands of central England in summer and winter over the last 20 000 years. (There is a discrepancy, not yet resolved, regarding the summer temperatures 11 000 to 15 000 years ago indicated by the evidence of insects, broken line in the diagram, and studies of the vegetation by pollen analysis; but this need not concern us here.) The margins of uncertainty regarding the dating and temperature-interpretation of the botanical evidence are indicated on the diagram by the ovals, each of which represents the data relating to some régime that seems to have lasted for the duration indicated by the horizontal bars. The curves are drawn to indicate the probable course of the 1000-year average temperature. The dots within the last 1000 years indicate the individual century averages (derived by the methods reported in Lamb 1965*b*) and serve to indicate the range of the century-by-century variations; the decade values and the individual years, of course, vary more widely – the latter apparently, since 1680, show just 7 times the range of the

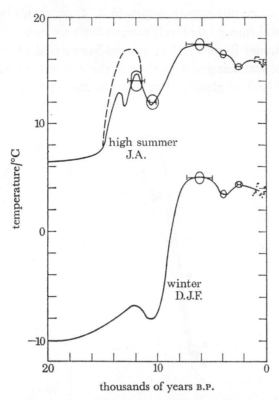

FIGURE 7. Variations of prevailing temperature in central England (lowland sites) over the past 20 000 years. (Estimates mainly from botanical evidence.) *Oval plots* indicate the range within which the mean value derived from botanical data must lie, within the uncertainties of carbon-14 dating and temperature implications of the botanical data. Each plot represents the data of a well-marked régime which appears to have lasted for a period indicated by the horizontal bars. *Dots* indicate temperature values derived by statistical analysis of documentary weather reports (for the method see Lamb 1965 *b*, 1972) and, in later times, directly from thermometer readings. *Broken line*: temperatures indicated by analysis of the beetle fauna (Coope, Morgan & Osborne 1971). *Whole lines*: estimated course of the thousand-year average temperatures.

century means of the last 1000 years. The variability in England, near the ice limit, in glacial times was probably three times as great, but is unlikely to have changed significantly in the last 6000 years, or rather more, since the geography of Europe became more or less as it now is. Moreover, England being exposed to prevailing winds from the Gulf-Stream–North Atlantic Drift water in the ocean in middle latitudes is likely to experience temperature changes which are fairly representative of a wide range of latitudes: within the past 100 to 150 years the temperature trends in England seem to have been always in the same sense as, and very close in magnitude to, the world average. This analysis confirms the diagnosis that the prevailing temperature level between 4000 and 1000 B.C. was mostly 1 to 2 °C above that in recent historical times.

(ii) *Sea level*

The melting of the great ice sheets that had covered the northern parts of Europe and North America caused a great ('eustatic') rise of world sea level. The most careful computations of the rise by Godwin, Suggate & Willis (1958), Shepard (1963) and Schofield & Thompson (1964) indicate that it amounted to 35 to 40 m between 7000 and 2000 B.C., an average rate of 70 to 80 cm a century: most of the rise took place between 7000 and 4000 B.C. During those times the North Sea was reconstituted much as we know it today, and many former coastal lowlands in other parts of the world became submerged. Schofield & Thompson's curves, seen in figure 8, indicate that the general level (mean tide level) around 2000 B.C. was probably 2 to 3 m higher than today. The building under Rameses II of a proto-Suez canal around 1230 B.C. may well be related to the level of the seas being at its highest at the end of the warmest postglacial times and the continued melting back of glaciers that went on during those times.

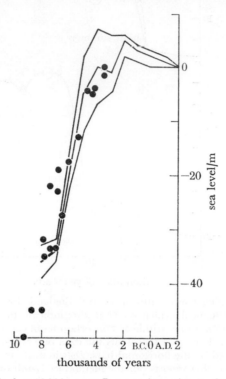

FIGURE 8. World sea-level changes in the last 12000 years. *Dots* – point estimates from widely scattered parts of the world, chosen for their tectonic stability; radiocarbon dated values (Godwin *et al.* 1958). *Curves* – calculated results from raised beach levels in Scandinavia, by identifying the local isostatic and general eustatic components of change at different places along the same former strand-line (Schofield & Thompson 1964). The middle curve is the most probable result. All the results came within the area bounded by the outer curves.

In those parts of the world that had been subjected to a great ice-load, postglacial times have seen a gradual 'isostatic' recovery of the Earth's crust resulting in local, or regional, rises of the land relative to the sea. In Scandinavia, and even in Scotland, this isostatic land-rise has over long periods overtaken the rise of world sea level, so that the land has continued to emerge despite the rising ocean. Only between about 6000 and 4000 B.C. in Scotland, and between 6000 and about 2000 B.C. in Norway, was the sea level rising relative to the land – in both these areas by about 10 to 15 m over the period named (see figure 7 in Donner 1970). From 4000 to 2000 B.C. sea level was approximately stationary. In the southern half of England and Wales and in the Netherlands a local sinking of the land relative to the sea is observed, and between East Anglia and Holland the sea now stands higher than it did about 4000 B.C., according to one careful analysis about 9 to 10 m higher (Churchill 1965).

In the Mediterranean (and perhaps elsewhere in the Near and Middle East) there have certainly been many erratic local changes of sea level, due to the tectonic instability of the region; but the region as a whole has experienced the world-wide changes to which figure 8 refers. Bloch (1970) has traced important effects of changes of sea level and of (evaporation) climate in the Mediterranean and the Near East on the salt industry, which was at certain times related to sea-shore installations (salt-pans) that were later submerged, and at other times related to sources inland, particularly the Dead Sea. He gives as phases of high general sea level 2200 to 1800 and 1500 to 1200 B.C., also around A.D. 400 and 1100 to 1500, and lowered sea-level 500 to 0 B.C. and around A.D. 700. Sea levels higher than those of today, derived from the data generally between about 3000 and 1000 B.C., presumably checked the maximum rates of discharge of which the rivers were capable and should therefore have tended to increase the river floods in the lower parts of the valleys in the Near and Middle East, and in India and China, as in all other parts of the world except those affected by isostatic uplift.

(iii) *Forest and grassland limits*

Pollen-analysis applied to sample deposits found in peat-bogs and past and present lake-beds has by now provided the knowledge from which a fairly reliable general survey of the history of forest limits and vegetation boundary movements over the great lowland expanses of Europe and northern Asia, as well as the eastern and central sectors of North America, can be given. In these regions the limits are generally thermally controlled. Figure 9 depicts the withdrawal southwards of the northern forest limit and of the limit of the broad-leafed trees (mixed oak forest) from the latter part of the warmest postglacial times in 3000 to 1000 B.C. to the present day. The readvance of the limit of permafrost (permanently frozen subsoil) in northern Asia is also shown. Recent studies in the relationship of the pollen-rain at the present day to tree populations (Andersen 1972) are making it possible to go further than before in reconstructing the actual forest composition from the pollen percentages of different species in past ages. From this, it appears that in the warmest postglacial times the forests in Denmark may have been dominated by lime (linden) trees and to some extent by elm: these species gave way from about 3000 B.C. onwards, beeches becoming very prominent in Denmark and oaks becoming the dominant element in most parts of the broad-leafed forest zone. The change seems to coincide with the first colder climatic phase which lasted perhaps a few centuries in Europe and also with the appearance of the first Neolithic farmers over wide areas; there is as yet no certainty as to how far the change in the forests should be directly attributed to climate or possibly to some activities of Man and his adaptation to a climatic crisis (see, especially, Tauber 1965, pp. 54–61; Frenzel 1966).

The greatest postglacial spread of forest to the Atlantic fringe of Europe seems to have been somewhat before the times with which we are mainly concerned and apparently included some woods or thickets of birch and willow, and even oak, elm, hazel, and perhaps pine, in the Shetland Islands between about 5000 and 3500 B.C., according to a radiocarbon-dated pollen stratigraphy in peat that is now below sea level (Hoppe & Fries 1965). In the Orkney Islands only birch, hazel and pine are known, from macro remains, certainly to have been present but were extensive, even at the northwest coast (Traill 1868,† Moar 1969a). There is evidence of similar woods or thickets in the Outer Hebrides, e.g. South Uist, so far undated by modern methods but thought to be from about the same period. And pines grew close to some of the

---

† Traill quotes a dramatic account by a local resident at Otterswick (59° 15′ N, 2° 30′ W), on Sanday in the northern part of the Orkney Islands, of how a prolonged NE gale in the winter of 1838 caused the sea to scour the sand from about 50 acres of the bay just above the low-water mark, laying bare a forest floor of 'black moss' and fallen trees (up to 2 feet in trunk-width) all lying in the same direction from SW to NE. The same forest remains were then found to extend for $6\frac{1}{2}$ km across the shallow bay to Tuftsness. The forest had evidently grown close to the Atlantic shore of the time concerned and in a locality completely exposed to the ocean to the NW.

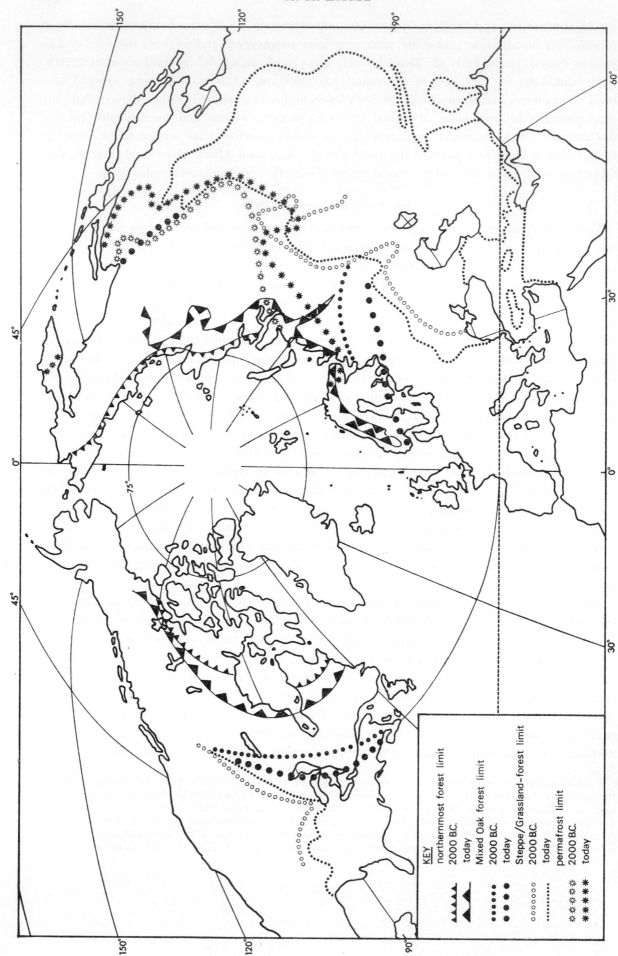

KEY

northernmost forest limit		
2000 B.C.		
today		
Mixed Oak forest limit		
2000 B.C.		
today		
Steppe/Grassland–forest limit		
2000 B.C.		
today		
permafrost limit		
2000 B.C.		
today		

FIGURE 9. Forest limits on the plains of N. America and Eurasia about 2000 B.C. and at the present day. Compiled from works by Bryson & Wendland (1967), Frenzel (1966, 1967), Nejstadt (1957) and Nichols (1967a).

most exposed parts of the Atlantic coast of the Scottish mainland, in Wester Ross and Suther-
land (Lamb 1964, 1965a). Though there was some decline of these woodlands from as early as
3500 to 3000 B.C. onwards (Moar 1965, 1969b), more rapid decline, which has been variously
attributed to increased wetness, soil acidification and the beginnings of bog growth, possibly
also to stronger winds, or to all these things, occurred between about 2600 and 1600 B.C.
(H. H. Birks personal communication, Moar 1965, 1969b). Godwin (1956, p. 338) records that,
as late as the Bronze Age, Cornwall, Wales and Ireland were forest-clad right to their western
shores and to higher altitudes on the hills than any present woods.

The farthest northern extensions of the forests in the continental sectors, in both North
America and Eurasia, are dated somewhat later, between about 3000 and 1500 B.C. This may
be because of the long aftermath of the ice age in these sectors, particularly the slow withdrawal
of the cold sea from the regions of land surface that had been depressed by the former ice load.
In latitudes near 80° N, radiocarbon dating of driftwood left on the now raised beaches of
Spitsbergen and the Canadian archipelago indicates that there may have been most open water
about 4000 B.C., though similarly open conditions continued until perhaps 1500 B.C. (Blake
1970). In the fiords of northernmost Greenland the period of most open water was not until
2000 to 1600 B.C. (Fredskild 1969). Nichols (1970) deduces from pollen diagrams for the North-
Western Territories that the onset of cooler summers in northern Canada may be dated about
1500 B.C., in good agreement with the time of change indicated in Greenland and elsewhere.
There was a warmer phase in both countries between about A.D. 900 and 1100 to 1200.
Throughout the times which we have been considering, the spruce (*Picea abies*) seems to have
been spreading westwards across Finland and Scandinavia at the expense of the earlier established trees (Tallantire 1972a, b). The advance of the spruce to dominance from the small
pockets which had existed from much earlier times in favourable sites throughout the region, is
probably to be attributed to this tree's competitive advantage in long cold winters, and therefore may be taken as indicating a fall in the general level of winter temperature. The spread of
spruce forest from the east which betokens this, as derived from pollen diagrams from sites all
over Fennoscandia, was not a smooth, continuous process but took place in distinct steps: (1)
around 3400 to 3000 B.C. to the Finnish–Russian border, (2) around 2300 to 2100 B.C. to much
of southern and central Finland, (3) between 1600 and 1300 B.C. a slight further advance, (4)
1000 to 400 B.C. across all central Sweden, (5) between A.D. 250 and 750 over southern Sweden
and (6) between about A.D. 1100 and 1400 in Norway across Trøndelag to Trondheimsfjord.
At each of these stages a renewed decline of the winter temperatures may be suspected.

The vegetation changes so far discussed probably owe little or nothing to human interference.
But in Britain and central and southern Europe from Neolithic times (say 4000 B.C.) onwards,
recession of the forest was increasingly due to Man (Godwin 1956, p. 332 ff; Turner 1965a;
Nichols 1967b; Pennington 1970).

In southern England, particularly in the inland districts near the chalk downs and the
Cotswolds and Mendip Hills, human activity has been considerable ever since Neolithic times.
A closely argued paper by Godwin & Tansley (1941) seems to have given a reliable picture of
the vegetation with implications about the climate despite all the human interference. Oak and
hazel were clearly present in some abundance near the chalk, and were the main species used
as firewood by the Neolithic inhabitants around 3000 to 2000 B.C.; yew constituted 8 % of the
charcoals analysed. But the evidence suggests that the trees were at the foot of the hills and on the
slopes. The Neolithic people settled the chalk uplands, including Salisbury Plain, grew crops

and kept sheep, and it seems likely, therefore, that the tops were already bare of trees. In the succeeding Bronze Age, the chalk plateau seems to have ceased to be used for settlement, though it was much used for grazing, for traffic and for burials: the implications of this archaeological evidence seem to be that the downs were already grassland and that the climate was drier than it had been earlier, in the Neolithic. The beautifully preserved form of the burial mounds could not have survived if their sites had been disturbed for long by the roots of woodland trees at any time since the barrows were made. There is more direct evidence of dryness of the Bronze Age climate in the digging at Wilsford, in the Salisbury Plain area, of a 33 m deep shaft, dated about the middle of the second millennium B.C.† to 1.5 to 2 m below the present water table‡ (Ashbee 1963, 1966). With the colder climates and increased wetness that set in from about 800 to 500 B.C. onwards, the early Iron Age peoples resettled the chalk uplands and tilled greater areas on the downs than ever before; it seems clear that the plateau remained largely bare of trees. Another type of archaeological evidence points to the same climatic sequence; wooden trackways were laid across the Somerset Levels (and other fens in the lowland areas of England), to keep open communications across the marshland when climatic conditions became wetter. The radiocarbon datings of these trackways are generally between about 3500 and 2500 B.C., and between 900 and 300 B.C. (see, for example, Godwin & Willis 1959).

　　Figure 9 also shows how the southern limit of forest, where broadleafed and mostly deciduous forest gives way to steppe, has shifted in the last 4000 to 5000 years on the plains of Eurasia and North America. In general, the steppe appears to have advanced, though the reasons for this are by no means clear: human action and increased windiness may have combined to clear the trees and reduce soil moisture at the margin of the forest zone. The great plains in the latitudes concerned east of the Rockies and in European Russia and Siberia had probably had a greater precipitation/evaporation ratio in glacial times, but the history of the Caspian Sea (see later, figure 10) suggests that any legacy of that moisture had vanished by about 5000 years ago. The water table may, however, have had a continued tendency to fall through most of postglacial time in the southern parts of these regions, as already noted (p. 203) in the Sahara and central and southern Asia, with the wastage of the subsoil and aquifer moisture left over from the climatic régimes of the last glaciation (or the moister stages of it).

　　To understand more about the vicissitudes of climate and vegetation, particularly in the more intricate terrains near, and in, the mountains of Asia, Africa and the Americas, it is necessary to use indirect methods to elucidate the general climatic régimes. The use of such arguments is fortified by the similarity of the main forest successions through postglacial time revealed by pollen-analysis of the stratigraphy of deposits in many parts of the world. This indicates, with broadly similar datings, a climax of warmth followed by a return to a cooler (or 'more boreal-type') environment – for example, in the Alpine foothills of northern Italy (see, for example, Beug 1964), Kashmir (see, for example, Singh 1963; Vishnu-Mittre 1966; Vishnu-Mittre & Sharma 1966), Japan (see for example, Tsukada 1967), and north and south America (Heusser 1966; Nichols 1967a, 1970) – like the successions observed in northwest and central Europe (see, for example, Averdieck & Döbling 1959; Firbas 1949; Godwin 1956;

---

† In the datings mentioned here, as elsewhere in this paper, corrections have been applied in accordance with the recent bristlecone-pine tree-ring calibration (Ralph & Michael 1967, Suess 1970a). Thus, the conventional carbon-14 date of the Wilsford shaft given as 3330 ± 90 years before present becomes about 1600 to 1700 B.C.

‡ 1.5 to 2 m below the present water table at its seasonal lowest, 9 m below the normal seasonal maximum at the present day.

Godwin, Walker & Willis 1957). Among the Anatolian mountains in northern Turkey Beug (1967) has found indications – a decline of beech and fir about 2000 B.C., while the oak, pine and juniper elements of the forests continued or gained ground – that the warmest postglacial millennia had been drier than the climatic régime after 2000 B.C. Farther north, at least in northwestern and central Europe, as also in the lower latitudes in Africa (the desert and the Nile), climates – though fluctuating – seem to have been becoming successively drier between 2500 and 2000 B.C.†

(iv) *Levels of inland waters – indications of moisture and cloudiness*

The changing levels of lakes and inland seas, which can be recognized by old strand-lines and dated either from historical records or radiocarbon tests on organic matter deposited, serve as a gauge of the changing balance between precipitation and inflow on the one hand, and evaporation and outflow on the other (always provided that the case has not been altered by erosion of the outlet, or by tectonic changes). The simplest cases are those like the Caspian Sea which has long had no outlet.

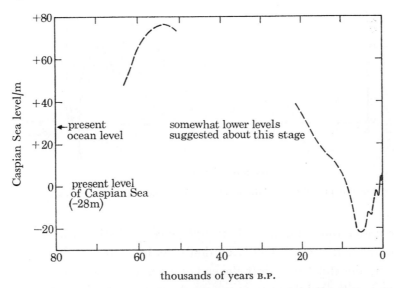

FIGURE 10. History of the level of the Caspian Sea. Scale in metres.

The history of the Caspian has been compiled by Berg (1934), Leontev & Federov (1953) and Nikolaeva & Han-Magomedov (1962) and is portrayed here in figure 10. The level was high in glacial times (when the area submerged was more extensive‡ than now), though it may have been somewhat lower in the long temperate period (interstadial) around 40 000 years ago. The level was much lower than now in the warmest postglacial times between 8000 and 5000 years ago, and has risen in several stages since. It continued to be lower than now throughout the period from 3000 B.C. to A.D. 1000, though, as with the forest evidence from nearby Turkey, the tendency then was always towards increasing moisture. The Caspian, however, is supplied by

---

† The opposite trend, towards increasing moistness, detected in the region of Turkey and the Caspian Sea was presumably related to positions commonly occupied by a trough in the upper westerlies at that time.

‡ The Caspian Sea at present reaches, at its extremes, from about 37° to 47° N and 47° to 55° E. In glacial times, at its greatest extent, it reached 50 °N and joined with the Aral Sea to form a water surface stretching from 45° to 62° E in the northern part, with an overflow channel to the Black Sea. Its area approached twice that of the present Caspian.

the flow of the River Volga, and its level is therefore partly representative of the régimes farther north, in European Russia.

The long history of Lake Chad (near 13° N, 13° E) on the south side of the Sahara, as illustrated here in figure 11, has been put together from information supplied by R. E. Moreau (personal communication December 1963) and that published, with radiocarbon dates, by Grove & Warren (1968). This lake also was much higher than now in glacial times, even after erosion had greatly changed the then existing outlet to the Atlantic, and its extent was such that it might better be described as an inland sea (Mega-Chad, in Moreau's nomenclature). Its

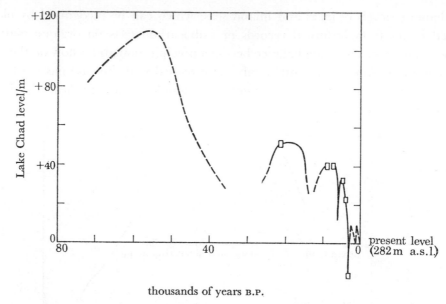

FIGURE 11. History of the level of Lake Chad. Scale in metres.

prevailing level has undergone a number of changes since, and still varies from year to year, though a net decline over the last 20000 years doubtless indicates the lowering of the water level in porous strata under the desert, since the end of the moistest régime at some stage during glacial times. (There is still doubtless some seepage through the underground strata: so Lake Chad is, strictly, not a lake without any outlet in the full sense that the Caspian is.) The history differs from that of the Caspian in that Lake Chad registered secondary high stands in the warmest postglacial times and, it is suggested, even in the more modest and short-lived warm periods of global climate about 1000 and 3000 years ago. The moisture maxima coinciding with these warm climatic phases are attributed to farther northward extension than now of the summer monsoon rains.

The comparative histories of the Caspian Sea and Lake Chad in recent millennia are made clearer in figure 12.

It is suggested that the Caspian may be taken as indicating moisture changes in latitudes 40 to 60 °N, at least in the Russian–west Siberian sector, and that Chad is representative of moisture changes in equatorial Africa. Butzer, Isaac, Richardson & Washbourn-Kamau (1972) have demonstrated that the dates of the high and low stands of Lake Chad and Lake Rudolf (3 to 5½° N near 36° E) and four other lakes between 0 and 1° S in eastern equatorial Africa over the last 20000 years or more are in general agreement (see figure 13). The gross

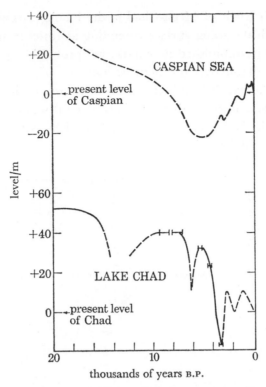

FIGURE 12. Variations of the Caspian Sea and Lake Chad over the last 20 000 years compared. Scale in metres.

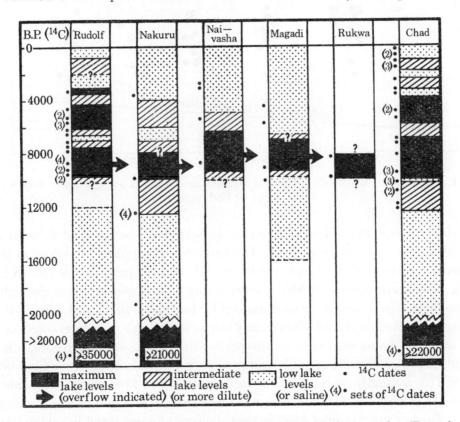

FIGURE 13. Fluctuations of the levels of lakes in tropical Africa that lack any outlet. (Reproduced from Butzer *et al.* (1972) by kind permission.)

differences between the high levels of Mega-Chad in the Ice Age, as well as around 8000 and even 4000 to 3000 B.C., with the water surface extending at times from 10° N 16° E to 18° N between 14 and 20° E, and its diminished state today are probably representative of the variations of general moisture available over northern Africa.

There may be a meaningful parallelism between the later rises of the level of Lake Chad and the spread of early settlements into the driest regions of southwestern Palestine around 6000 and 3000 B.C., reported by Blake (1969), each time followed by withdrawal in the succeeding centuries, in the latter case withdrawal for good.

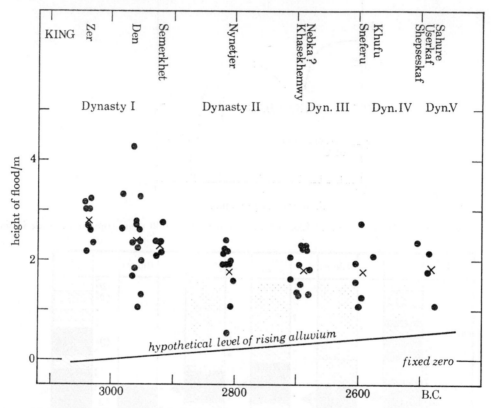

FIGURE 14. Height of the Nile floods in Egypt in sample runs of years in ancient times. (Reproduced from Bell (1970) by kind permission.)

The levels of the annual flood of the Nile river in Egypt have been recorded, either in inscriptions or documents, from before 3000 B.C. The lower Nile is supplied by two main branches of the river from farther south in Africa, the White Nile which emerges from Lake Victoria and drains much of eastern equatorial Africa, maintaining a fairly constant flow through the year, and the Blue Nile, which rises in the mountains of Ethiopia near 10° N and is fed by the summer monsoon rains. There is a record of the yearly low as well as the high level stages of the River Nile from A.D. 622 (Toussoun 1925). In interpreting this unique climatic record, it is necessary to allow for the progressive change of level due to the continued silting of the river bed and the flood-plain. This is usually taken, from the records of the last 1300 years, to amount to 10 cm per century. In reality, the rate of silting must have varied with the varying strength of flow of the river. Further uncertainties enter in over the type of gauge used in earlier times and its position. Nevertheless, Bell (1970, 1971) has shown that whatever assumptions are made about the datum, it is clear that the average flood level of the Nile in lower Egypt was always lower and less variable from year to year after about 2900 to 2800 B.C. than it had been in the centuries immediately before that (see figure 14). This decline of the flow of the Nile may reasonably be related to the end of the high-level stages of Lake Chad (figure 11) and the moist régime in subequatorial Africa that had obtained for some thousands of years previously.† It is also clear that there were several shorter periods marked by particularly low levels of the annual Nile flood, between 2180 and 2130 B.C., 2000 to 1990 B.C., soon after 1786 B.C. and again from about 1200 B.C. onwards. Several of these fluctuations are seen to be separated by an interval close to 200 years.

DERIVING THE DISTRIBUTIONS OF CLIMATE AND PREVAILING WINDS AT DIFFERENT EPOCHS

Any peculiarities of the global distribution of climates in any epoch are likely to be understood in terms of the peculiar features of the atmospheric (and ocean) circulation and anomalies of the local or regional condition of the surface (particularly ice-cover, waterlogging or parching), rather than by reference to subtle differences of the radiation available, which is always graded according to latitude and season. This is likely to be the case, even if the incoming radiation should differ from now, and be the ultimate reason for greater or less energy in the atmospheric circulation and a somewhat different pattern of heat transport from that now prevailing.

Attempts to reconstruct the atmospheric circulation patterns prevailing over the northern hemisphere at different stages of the last Ice Age and postglacial times have been made by Lamb *et al.* (1966, 1970). The method proceeded in a series of logical steps:

(*a*) Mapping the surface air temperatures prevailing in the warmest and coldest months, as indicated by the boundaries discovered for the occurrence of different species of flora (particularly the assemblages characteristic of different forest and vegetation zones) and of microorganisms (mainly *foraminifera*) living in the surface waters of the ocean, whose thermal limits today seem to have been reliably determined. (See particularly Iversen (1944), West (1968, chapters 7 and 10) and the use made by Lamb *et al.* (1966) of the forest maps of Nejstadt (1957); for the ocean data see Emiliani (1955, 1961) and Imbrie & Kipp (1971).)

(*b*) Next, mapping the upper air temperatures, based on those that accompany such surface temperatures in comparable geographical circumstances today and hence deriving (via the air density) the vertical thickness of the lower half of the atmosphere.

(*c*) From the distribution of pressure of the overlying atmosphere defined by (*b*), the corresponding flow of the upper westerlies around the hemisphere, and the pattern and intensity of the circumpolar vortex, are then arrived at.

(*d*) Next, the tendency for cyclonic or anticyclonic development at all points of the map is computed by a method related to that used on instantaneous synoptic weather maps in numerical daily weather forecasting, based on the work of Sutcliffe (1947) on the development of surface weather systems.

(*e*) Finally the probable pattern of prevailing atmospheric pressure at mean sea level is

---

† Bishop (1965) reports that the lowest of a series of old strand-lines of Lake Victoria near the Nile outlet, just above the highest levels of the lake observed in modern times, was dated 3720 ± 120 radiocarbon years before present (about 2200 B.C. when the bristlecone pine-calibration correction is applied); and it is thought that the lake never reached that level since.

sketched in from the analyst's experience by considering the derived distribution of areas of prevalent cyclone and anticyclone formation and the steering of these surface pressure systems by the mainstream of the upper westerly winds (the circumpolar vortex).

The January and July maps so derived for periods around 2000 B.C., 500 B.C. and in recent years (A.D. 1950s) for comparison are reproduced here as figures 15a, b; 16a, b and 2a, b respectively. The corresponding maps for 6500 and 4000 B.C. were published in Lamb et al. (1966). These maps may be read as maps of the prevailing winds, since under equilibrium conditions (i.e. the forces acting upon the moving air being in balance) the wind blows nearly along the lines of equal pressure, counterclockwise around the centres of low pressure in the northern hemisphere and clockwise around the high pressure centres.

The features indicated by these maps which seem relevant to our present discussions are:

(i) The weaker circulation and spread towards rather higher latitudes of the anticyclone belt around 2000 B.C. than in any period much after 1000 B.C. except probably for a few centuries in the early Middle Ages (approximately A.D. 950 to 1310, at longest). Around 4000 B.C. the anticyclonic influence already seemed to have spread to middle latitudes, almost as at 2000 B.C., but a more marked belt of westerly winds was indicated near 50 to 60° N. This difference probably means that around 2000 B.C. a greater frequency of blocked, meridional and rather stagnant circulation patterns was occurring: this suggestion would be in keeping with a high frequency of sunshine and quiet seas for sailing in latitudes as far north as 60° N. It would probably also account for greater variations (long- or short-term) of temperature and rainfall – attributable to fluctuations in the frequency and positions of blocking – in the co-called 'Sub-Boreal' period (3000 to 1000 B.C. approx.) than before about 3000 B.C.

(ii) The marked deterioration of climate which was affecting latitudes north of about 40° N by 500 B.C., according to many different types of evidence, is verified by withdrawal southwards of the anticyclonic influence, stronger and doubtless more often stormy winds, with the belt of cyclonic activity surrounding the polar cap having spread to rather lower latitudes than before. This would be accompanied by a considerable increase in cloudiness in the latitudes affected, particularly in the fifties, and in all those parts of Europe exposed to the increased westerly and northwesterly winds.

The features of this deterioration – i.e. lower temperatures and increased windiness – which was taken as defining the onset of the so-called 'Sub-Atlantic' climatic period in the older European literature have continued more or less to the present time, though with century to century variations which almost amounted to a short return of the Sub-Boreal warm climate in the early Middle Ages and a much colder climate associated with more blocking and a lower latitude of the cyclonic activity around the seventeenth century A.D.

When the average latitude of the subpolar low-pressure belt and that of the subtropical high-pressure belt in the European sector (longitudes 0 to 30° E) shown on the charts of this series are plotted against time, as in figure 17, the main trends are seen more clearly. This diagram brings to light a degree of parallelism between the latitude variations of these features and the changes of prevailing temperature (cf. figure 6), particularly in winter. Only the long-term changes, however, are reliably indicated in this way: analysis of the century-by-century (and shorter-term) variations of circulation régime within the last millennium shows that the varying incidence of blocking situations (with high pressure in the zone 50 to 70° N) may introduce wide variations of the latitudes of highest and lowest pressure averaged over periods as long as 20 to 30 years and perhaps longer. Indeed, a recently completed study by Bryson, Lamb & Donley

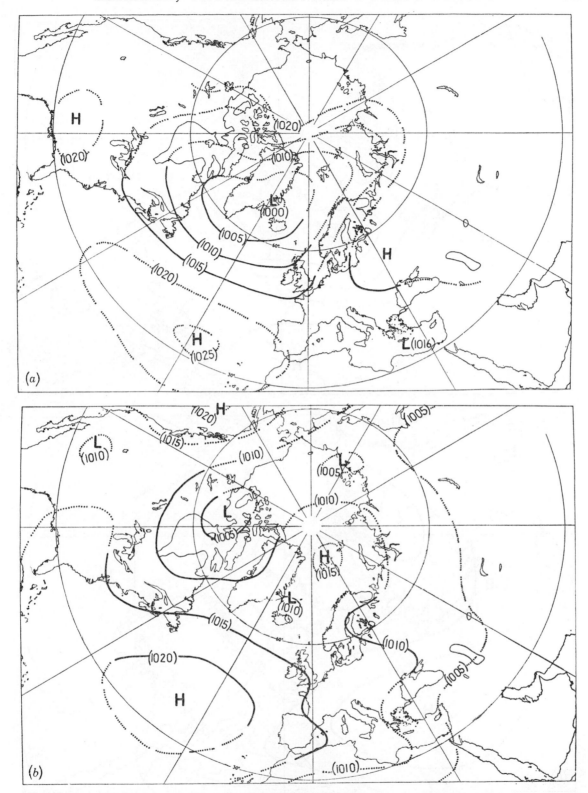

FIGURE 15. Average atmospheric pressure distribution at sea level derived by Lamb *et al.* (1966) for times around 2000 B.C.: (*a*) January, (*b*) July. Figures give suggested pressure values in millibars. Prevailing winds blow counterclockwise around the areas of low pressure, clockwise around areas of high pressure; the wind at the surface usually blows at an angle of 20 to 40° to the line of the isobar, i.e. blowing inwards towards the lower pressure.

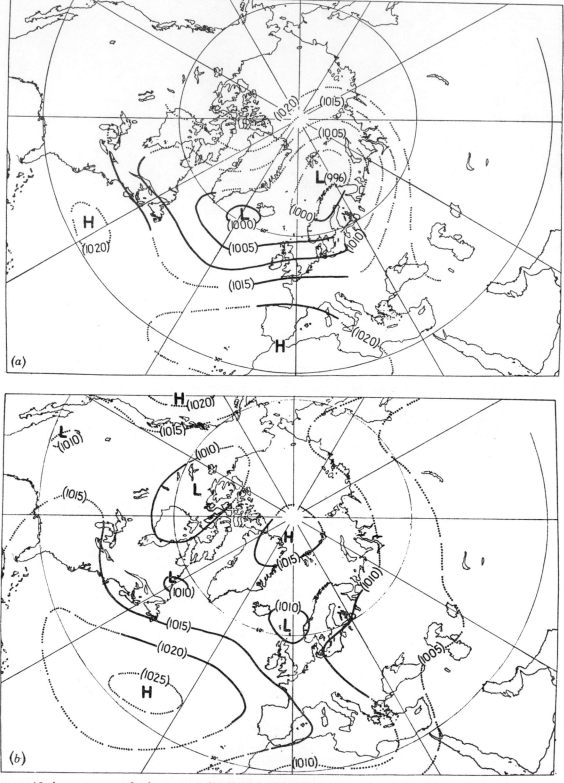

FIGURE 16. Average atmospheric pressure distribution at sea level derived by Lamb *et al.* (1966) for times around 500 B.C.: (*a*) January, (*b*) July.

(so far unpublished) of patterns of variation† of the northern hemisphere atmospheric circulation has built a case for believing that for some centuries around, and after, 1200 B.C. circulation patterns very like the particular, partly meridional pattern that characterized the 1954–5 winter were either abnormally frequent or, perhaps, the dominant situation.

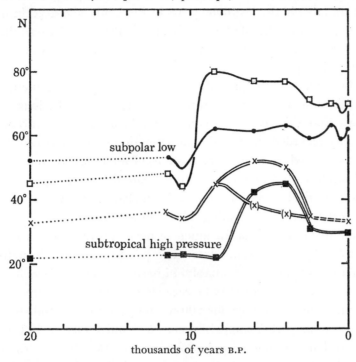

FIGURE 17. Average latitudes of the subpolar low-pressure belt and subtropical high-pressure belt in longitudes 0 to 30° E and their variations over the last 20 000 years. □, ×, winter (January); ●, ■, summer (July). (Derived from charts by Lamb *et al.* 1966, 1970.)

The high periods of the Mayan and Aztec civilizations in central America, respectively in the first nine centuries A.D. and 1300 to 1500, may be seen as periods of climate either close to the average of the last 3000 years or below that average (especially in the eighth and fifteenth centuries) in terms of temperature and general latitude of the subtropical high-pressure belt. They may therefore have been drier in the latitudes of Mexico and Yucatan than the warm centuries of the early Middle Ages which apparently coincided with the Dark Age in central America that separated the flowering of these two cultures. In the warm centuries the inter-tropical convergence, and its associated clouds and rain, probably ranged farther north than in the colder periods.

Some evidence of the succession of climate and circulation régimes over the southern hemisphere during the last glaciation and through postglacial times has been given by Auer (1960, see particularly the discussion pp. 533–538), Derbyshire (1971) and Hastenrath (1971).

† Empirically found eigenvectors of the mean sea-level pressure distribution, from Kutzbach (1967), were used. The 1954–5 winter pattern was dominated by an eigenvector associated with more frequent than usual surface northerly and northwesterly winds from the Norwegian Sea to the western Mediterranean (and a sharp upper cold trough with its axis from the Baltic to Sicily – the other two winter-time cold troughs, over eastern Asia and Canada, were also sharper than usual) and abnormal development of high pressure over all northern Siberia, with a ridge extending towards the eastern Mediterranean and covering all southwest Asia. Under these conditions, excess warmth and drought developed over much of Greece, Asia Minor and neighbouring areas to the south and west.

The fluctuations of climate in early civilized times

As indicated on p. 207, the ranges of year-to-year variation of temperature (and, it is reasonable to suggest, those of rainfall) have probably not differed by more than a factor of 2 from their present values at all times within the last 6000 years in England, or indeed in most parts of the world except the Arctic, the northern Baltic and Hudson's Bay area (where the distribution of land and sea has changed), and in some mountain areas where the extent of glaciation has undergone considerable changes.

Reference has also been made in the preceding pages to longer-lasting fluctuations of climate, over a few decades, or in some cases a few centuries, superposed on the long-term trends. It is with these fluctuations that this section is concerned. Workers with various types of field evidence have commonly regarded the Sub-Boreal climatic period, between 3500 or 3000 B.C. and 1000 or 500 B.C. as marked by secular fluctuations of temperature and rainfall, at least in middle latitudes, of significantly greater amplitude than the earlier part of the postglacial warmest era. The most marked variations generally indicated are a cooler period of several centuries duration that defined the beginning of the Sub-Boreal (Frenzel 1966) and a period (or periods) of extreme dryness, at least in much of northern Europe (Brooks 1949, p. 298; Godwin 1956, pp. 227–228), towards its close. The period of great drought in Iran, Asia Minor, the eastern Mediterranean and, probably, in parts of Greece (Carpenter 1966) for some centuries from about 1200 B.C. onwards may be regarded as one of these fluctuations.

The evidence of drying out of the bogs for 200 to 300 years in late Sub-Boreal time, about 1000 B.C., and perhaps at other earlier stages between 2500 and 1200 B.C., in England, Scotland, also in some places in Ireland (Jessen 1949, p. 259; but see also Mitchell 1956, pp. 238–244), and other parts of northern Europe, includes stumps of birch and pine in the peat (Godwin 1954; Jessen 1949), trees which evidently spread onto the bog surface at that time. Close above the layer at which these tree-stumps are now buried is peat which is shown by sequences of radiocarbon dates to have grown rapidly, in west Wales with extreme rapidity (90 cm of peat formed in about 300 years from around 700 to 400 B.C. in Tregaron bog, according to Turner (1965 b)). This 'recurrence surface' marking renewed onset of bog growth, manifestly in a wetter climate with cooler summers, and dated around 500 B.C., is found over much of northern and central Europe, where it is widely known as the *Grenzhorizont*. A sequence of similar recurrence surfaces, dated about 2300 B.C., 1200 B.C., 500 B.C., A.D. 400 and A.D. 1200 was recognized in the Swedish bogs by Granlund as long ago as 1932, and most of these seem to be traceable in widely separated areas of Europe. The dates have stood up well to comparisons with later radiocarbon tests; though Tauber (1965) gives 2200 B.C. for the earliest one, and A.D. 1250 to 1300 may be a more widely representative timing for the last of the series.

Most peat bogs show a stratigraphy of recurrence surfaces overlying peat layers heavily decomposed by drying out and darkened. Additional recurrence surfaces are found in the bogs in many parts of Europe. In many areas local factors confuse the sequence. Differences of bedrock topography and drainage underneath the peat vary the sensitivity of different parts of the bog surface to changes of the evaporation/precipitation ratio and so introduce a spread of dates into the responses of different areas to any general change of climate. In Ireland, particularly, there are additional recurrence surfaces, some of them quite localized, perhaps because of the country's immediate exposure to moisture-bearing winds from the Atlantic; this is also the likeliest reason why the timing of the main groups of recurrence surfaces (neatly summarized

TABLE 1. CLIMATIC FLUCTUATIONS REGISTERED IN THE STRATIGRAPHY OF PEAT BOGS AND SOME PARALLEL DATA

Note. Dates derived from radiocarbon measurements are rounded approximations, after correction by comparison with Egyptian and Scandinavian varve chronologies and using the calibration curves published by Ralph & Michael (1967) and Suess (1970a). Those radiocarbon dates the most probable value of which required correction by more than a few decades are marked *

Ireland (data from Jessen 1949; Mitchell 1956; Smith *et al.* 1971)	Britain (data from Godwin 1956; Godwin & Willis 1959)	Scandinavia (data from Granlund 1932; Tauber 1965)	Germany and Central Europe (data from Frenzel 1966; Overbeck *et al.* 1957)	Greece (after Carpenter 1966)	Nile (data on annual floods in Egypt from Bell 1970, 1971)
			3400–3000 B.C. cold phase		
ca. 2800* B.C. wetter, peat growth setting in		2800 B.C. wet phase beginning			
		ca. 2500 B.C. probably dry			
ca. 2400 B.C. becoming wetter, pines decline, bog growth increasing					
ca. 2200 B.C. further spread of bog growth		2200 B.C. wet phase beginning			
					2180–2130 B.C. low floods
ca. 2000* B.C. further spread of bog growth					
					2000–1990 B.C. low floods 1786 B.C. ff low floods
ca. 1500 B.C. renewal of bog growth			1400 B.C. ff. adequate moisture		
	ca. 1200 B.C. onset of wetter conditions	*ca.* 1200* B.C. wet phase beginning	1250–1200* B.C. wet phase	1230–1100 B.C. cultural decline and depopulation, possibly indicating drought: worsening after 1100 B.C.	*ca.* 1200 B.C. low floods phase sets in again
ca. 800 B.C. renewal of bog growth			*ca.* 900* B.C. wet phase beginning		
	ca. 700–500 B.C. conditions becoming wetter	*ca.* 500 B.C. wet phase beginning	*ca.* 600 B.C. wet phase beginning		
			ca. 400 B.C. wet phase beginning		
	ca. 350–100 B.C. wet conditions		*ca.* 150–100 B.C. wet phase beginning		
B.C. —— A.D.	60 B.C.–A.D. 50 becoming drier				
		A.D. 400–500 wet phase beginning	several indications of further recurrence surfaces (regrowth of bogs) in N. Germany A.D. 565–595 and A.D. 690–770		
ca. A.D. 500 renewal of bog growth					
		A.D. 1200–1300 wet phase beginning			

in a diagram by Mitchell 1956, p. 238) differs from the rest of Europe. In Ireland, the main dates when bog growth set in again, after a drier period in which the peat had become humified and darkened, seem to have been about 1500 B.C., 800 B.C. and A.D. 500. Another type of local or regional peculiarity affects the sequence of recurrence surfaces near the Baltic in Russia and Finland, where the water table varied largely in accordance with the changes of sea level that depended on the balance between the world-wide post-glacial rise of sea level and the strong isostatic uplift of the land surface which had underlain the ice-sheet. Nevertheless, Khotinsky (1971) reports that similar decomposed peat layers with lighter-coloured peat and regrowth above a boundary horizon are found in peat bogs throughout the forest zone of the Russian plain, and in western Siberia, as well as in the Far East and Alaska; some of these recurrence events are dated as far back as the so-called 'Atlantic' climatic period, well before 3000 B.C.

Table 1 is an attempt to summarize the data on fluctuations of climate within these millennia from a limited number of reliable sources for Europe and the Near East. The resulting survey appears self-consistent, but must be regarded as tentative in detail until many more firm datings are available. Correction of the radiocarbon dates by applying the calibration adjustments suggested by the curves published by Ralph & Michael (1967) and Suess (1970a) has brought many of the dates into straightforward agreement with the dates given for events elsewhere in Europe which had not been arrived at by carbon-14 tests, and has thereby simplified the apparent message of the table; but the adjustments amount to several hundred years before 900 B.C. and further confirmation seems desirable.

It seems safe to conclude that there were well-marked wet, and presumably cloudy, phases in northern Europe during these millennia, those which set in about 2200, 1200 and 500 B.C., as well as in A.D. 1250 to 1300, bringing the sharpest deteriorations and being the most widely registered. It appears that there may have been a 'preferred' (i.e. most common) interval of 200 or 400 years between successive wet phases. There seem to have been several centuries of drier conditions at least between about 1800 and 1400 B.C., possibly from 2000 B.C. or soon after, till 1200 B.C., and, at a more modest level, from about 80 to 60 B.C. till A.D. 550. Since drought in Greece can occur with westerly winds over Europe, especially if these alternate with the occasional spread of anticyclonic conditions over southern and eastern Europe and the Mediterranean, Carpenter's (1966) suggestion that drought set in in Greece at the time of the decline of Mycenae, from 1230 B.C. onwards, may well concur with a long period of wetness in north-western and central Europe, accompanied by frequent westerly and northwesterly winds over Britain and Scandinavia. Evidently, the situation over Greece in the centuries between 1230 and 800 or 750 B.C., when the land remained to a considerable extent depopulated and in cultural decline (if the cause was drought) was more frequently anticyclonic than during the next wet phase in northwestern and central Europe around 500 to 100 B.C. This latter period probably more often saw the spread of cold air from the north to all parts of Europe: there were considerable glacier advances, though rarely matching those of A.D. 1550 to 1850. There is scattered evidence of a marked warming in Europe from about 100 B.C. onward.

Figure 18 reproduces the analysis, thought to represent rainfall, by Schostakowitsch (1934) of the remarkable series of annual layers in the mud deposit on the bottom of a small salty lagoon on the west coast of the Crimea. The layers are presumed mainly due to the run-off from the land caused by the heavier rainstorms; but most of the layers are only a few millimetres thick, and some may have been missed in the counting. The earliest layer was counted as 2294 B.C., but could have been slightly earlier. A rainfall scale suggested by Brooks (1949,

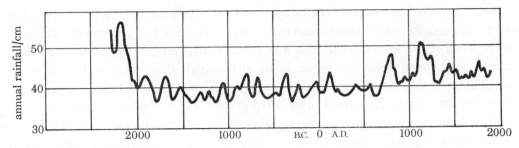

FIGURE 18. Rainfall variations in the Crimea since about 2300 B.C. as indicated by the thickness of the yearly mud layers in the bottom deposit of Lake Saki, 45° 07′ N, 33° 33′ E. (From Schostakowitsch (1934), with rainfall scale suggested by Brooks (1949).)

p. 299) has been added to the diagram. This record suggests (*a*) that variations occurring 4 to 6 times in a thousand years are important, (*b*) that a few centuries between A.D. 800 and 1250 represented a temporary return to moister conditions rather as they were before 2200 B.C. The fluctuations of rainfall in the Crimea appear often inverse to the changes in the prevalence of ground moisture found in northern, western and central Europe: this lake indicates dryness setting in about 2200 B.C. and no return to the former level except in the early Middle Ages. Particularly dry spells occur around 600 to 500 B.C. and a sharp return to drier conditions about A.D. 1250, both being times of marked wetness setting in (and renewal of the growth of the peat bogs) in western and northern Europe.

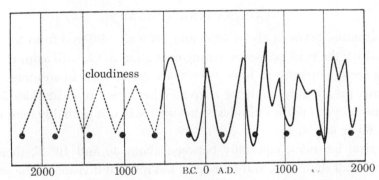

FIGURE 19. Variations of night cloudiness in latitudes about 30 to 50° N since 2300 B.C. indicated by Link's (1958) analysis of the frequency of discoveries of comets.

One other derived climatic series covering the whole time from 2300 B.C. to the present is illustrated in figure 19. Link (1958) has surveyed separately the available collections of reported discoveries of comets by observers in China and in western countries. The number of comet discoveries per century depends on:

(i) k, the number of comets actually within observable range of the Earth.

(ii) C, the frequency of cloudless nights, so that the number of comets visible in practice is Ck, where C is expressed as a fraction.

(iii) h, the human social or 'cultural' factor, describes the observation skill of the existing state of civilization; so that, with h expressed as a fraction, hCk is the number of comets actually observed and of which records were kept.

When the number (N) of comet discoveries per century ($N = hCk$) is plotted on a graph against the date, the resulting curve shows a sequence of waves superposed on a generally upward trend. The waves are in remarkable agreement in China and the West (for the period from 500 B.C. onwards for which continuous curves for both areas can be drawn). The level of the curves represents the human skill: this and its upward trend differs as long as the two communities were isolated from each other, being at first higher in China and after the ninth century A.D. higher in the West. Thus, the long-term mean level of the curve corresponds to the value of h, the human factor. Link therefore plotted graphs of N/h, thus eliminating the human factor, and obtained curves which showed only the waves. And since $N/h = Ck$, if k is (in the long-term) constant, these variations (similar in China and the West) must represent variations of night cloudiness in latitudes about 30 to 50° N. This is the curve seen in figure 19, plotted so that cloudiness increases as the curve rises. Groups of comet discoveries make it possible to place the minima of the curve as far back as 2300 B.C. The curve suggests a fairly regular oscillation, of period length about 400 years; the detail after A.D. 600 points to the possibility of a superposed oscillation of period 200 years. Such indirect and incomplete data cannot be regarded as establishing identity with a periodicity affecting weather data, and glimpsed in the Nile data and the apparently rather more irregular fluctuations of the lake in the Crimea which we have shown. But all these fragments suggest there may be a single underlying phenomenon. Link's suggestion was that the fluctuations were of solar origin, and there may be some support for this in the history of variations of radioactive carbon in the Earth's atmosphere which has since come to light (Stuiver 1961; Suess 1970b).

CONCLUDING SUMMARY

The climatic situation between about 3000 and 1000 B.C. differed from today's in at least a few ways about which the evidence seems strong enough to speak with assurance. In particular

(i) Conditions seem generally to have been warmer, at least in summer, and more anticyclonic than now, presumably implying a greater frequency of clear skies and less frequent stormy winds, especially in middle latitudes – the difference was probably most marked between latitudes 40 and 60 or 65° N.

(ii) In subtropical latitudes, especially between about 30 and 40° N, there was probably somewhat more rainfall than now, and perhaps a less marked division of the year into dry and rainy seasons, though longer-term rainfall fluctuations involving runs of drought years separated by long periods with more frequent rain and cloud were important. A 200-year oscillation seems to have underlain at least some of the most severe of these fluctuations. At other times a 400-year spacing between similar fluctuations is apparent.

(iii) The equatorial rains seem to have ranged rather farther north than now in their seasonal migration for some long time ending about 2900 to 2800 B.C.: while this régime lasted, it presumably constituted a regular monsoon rainfall in Africa and SW Asia reaching to about 20° N, and an erratic incidence of more rainfall than now even farther north.

(iv) There was more subsoil moisture than now, and presumably more numerous and more extensive oases, in the present deserts and arid lands generally, partly as legacy of the climates of earlier millennia.

In consequence of these differences, there was more vegetation than now in the present arid regions, which even included forest in some areas that are now desert, though this vegetation was becoming increasingly vulnerable to human interference as climates became drier in tropical and subtropical latitudes and the soil moisture level sank (particularly in open country).

Forest also extended farther north than now, and nearer the coasts of Europe that are most exposed to Atlantic gales: recession of the forest from the most exposed positions was beginning quite early in the millennia here discussed, particularly from the Atlantic coasts, presumably due to climatic fluctuations involving increased windiness. Elsewhere, the main forest recession was probably after 1000 B.C., when the prevailing temperatures also seem to have dropped rather sharply. For some centuries about 600 to 400 B.C., but perhaps with forerunners as early as about 1600, 1200 and 800 B.C., and some further incidents as late as 120 to 100 B.C., there was evidently a particularly inclement period, with great wetness and frequent westerly and northerly or northwesterly storms near the Atlantic coasts, and penetration of the western Mediterranean also by colder and occasionally disturbed conditions.

The chalk uplands of southern England seem to have been grassland essentially bare of trees from as early as 2000, or even 2500 B.C., perhaps basically due to Man's activities, though a sequence of moisture changes can be traced, with mostly drier climate in the second millennium B.C. than before or after.

Meteorological knowledge and modern understanding of the behaviour and dynamics of the atmosphere can now be applied at such different levels of elaboration to the reconstruction of past climatic conditions that they amount to a variety of techniques, which contribute in quite different ways to the interpretation of past climatic régimes. The methods employed in this paper, and advocated by the present writer elsewhere, make the minimum necessary use of theory to derive the simplest outline picture of the complete pattern of flow of the atmosphere over the northern hemisphere, from the scattered factual data on prevailing surface conditions supplied by other branches of science and learning. In this quest for knowledge of the past, there is no room for barriers between the arts and sciences, since contributions from the most diverse studies and fields of expertise are needed. There is indeed a danger with the most sophisticated theoretical constructions of atmospheric models (in which the abilities of computers of great capacity are employed to take account of action and reaction and second-order effects), that the elaborate computation may go too far beyond our real knowledge of the detailed impact of past environmental conditions upon the atmosphere. If our reconstructions of past climates are to be realistic, there is a continuing need for all those field studies, and for all those attempts to cull relevant data from ancient documents and inscriptions, which can contribute facts about the actual conditions at an ever greater number of points on the map and at ever closer, and more closely dated, intervals of time.

The author expresses his gratitude to Dr H. J. B. Birks of the Botany Department, Cambridge University, for introducing him to many of the botanical works used, to Mrs H. H. Birks and Mrs Tutin (W. Pennington) for various helpful communications relating to forest history in Scotland, to Professor G. F. Mitchell of Trinity College, Dublin, for the loan of papers relating to recurrence surfaces in the peat bogs of Ireland, Europe and the Soviet Union, and to his own colleague Professor K. M. Clayton of the University of East Anglia, Norwich, for comments and suggestions in the realms of geomorphology and hydrology.

REFERENCES (Lamb)

Ahlmann, H. W. 1949 The present climatic fluctuation. *Geogr. J.* **112**, 165–195.
Andersen, S. T. 1972 The differential pollen productivity of trees and its significance for the interpretation of pollen diagrams from forested regions. *British Ecological Society Symposium on Quaternary Plant Ecology*, Cambridge, 9–12 April 1972. (To be published 1973 in *Quaternary Plant Ecology*, ed. H. J. B. Birks & R. G. West. Blackwell Scientific Publications.)

Ashbee, P. 1963 The Wilsford shaft. *Antiquity* **37**, 116–120.

Ashbee, P. 1966 The dating of the Wilsford shaft. *Antiquity* **40**, 227–8.

Auer, V. 1960 The Quaternary history of Fuego-Patagonia. *Proc. R. Soc. Lond.* B **152**, 507–516 (and discussion) 533–538.

Averdieck, F.-R. & Döbling, H. 1959 Das Spätglazial am Niederrhein. *Fortschr. Geol. Rheinld. Westf.* **4**, 341–362.

Bell, B. 1970 The oldest records of the Nile floods. *Geogr. J.* **136**, 569–573.

Bell, B. 1971 The Dark Ages in ancient history I. The first Dark Age in Egypt. *Am. J. Archaeol.* **75**, 1–26.

Berg, L. S. 1934 The level of the Caspian Sea in historical times. *Problemy fizic. geograf.* **1**, (1). (In Russian.)

Beug, H.-J. 1964 Untersuchungen zur spät- und postglazialen Vegetationsgeschichte im Gardaseegebiet unter besonderer Berücksichtigung der mediterranen Arten. *Flora* **154** (Neue Folge), 401–444.

Beug, H.-J. 1967 Contributions to the postglacial vegetational history of northern Turkey. In *Quaternary palaeoecology* (ed. E. J. Cushing & H. E. Wright), pp. 349–356. New Haven and London: Yale University Press.

Birks, H. H. 1972 Studies in the vegetational history of Scotland. III. A radiocarbon dated pollen diagram from Loch Maree, Ross and Cromarty. *New Phytol.* **71**, 731–754.

Bishop, W. 1965 Quaternary geology and geomorphology in the Albertine Rift Valley, Uganda. *Geol. Soc. of Am. Spec. Pap.* 84 (Inqua 1965 – ed. H. E. Wright and D. G. Frey).

Blake, I. 1969 Climate, survival and the second-class societies in Palestine before 3000 BC. *Adv. Sci.* **25**, 409–421.

Blake, J. W. 1970 Studies of glacial history in Arctic Canada. I. Pumice, radiocarbon dates, and differential postglacial uplift in the eastern Queen Elizabeth Islands. *Can. J. Earth Sci.* **7**, 634–664.

Bloch, M. R. 1970 Zur Entwicklung der von Salz abhängigen Technologien. *Saeculum* **21**, 1–33.

Brooks, C. E. P. 1949 *Climate through the ages*, 2nd edn. London: Benn.

Bryson, R. A., Lamb, H. H. & Donley, D. L. 1973 Drought and the decline of Mycenae. (Unpublished manuscript, completed October 1972: to be published in *Antiquity*.)

Bryson, R. A. & Wendland, W. M. 1967 Radiocarbon isochrones of the retreat of the Laurentide ice sheet. *Tech. Rep.* **35**. Nonr. 1202 (07). Dep. Meteor. University Wisconsin, Madison.

Butzer, K. W. 1958 Studien zum vor- und frühgeschichtlichen Landschaftswandel im Sahara. *Akad. der Wiss. Lit. in Mainz, Math.-naturw. Klasse*, no. 1.

Butzer, K. W., Isaac, G. L., Richardson, J. L. & Washbourn-Kamau, C. 1972 Radiocarbon dating of East African lake levels, *Science, N.Y.* **175**, 1069–1076.

Carpenter, R. 1966 *Discontinuity in Greek civilization*. Cambridge University Press.

Chappell, J. E. 1970 Climatic change reconsidered: another look at 'The pulse of Asia'. *Geogr. Rev.* **60**, 347–373.

Chappell, J. E. 1971 Climatic pulsations in inner Asia and correlations between sunspots and weather. *Palaeogeogr. Palaeoclim. Palaeoecol.* **10**, 177–197.

Churchill, D. M. 1965 The displacement of deposits formed at sea level 6500 years ago in southern Britain. *Quaternaria* **7**, 239–249.

Clapp, P. F. 1964 Global cloud cover for seasons using TIROS nephanalyses. *Monthly Weather Rev.* **92**, 495–507.

Coope, G. R., Morgan, A. & Osborne, P. J. 1971 Fossil Coleoptera as indicators of climatic fluctuations during the last glaciation in Britain. *Palaeogeogr. Palaeoclim. Palaeoecol.* **10**, 87–101.

Dansgaard, W. 1964 Stable isotopes in precipitation. *Tellus* **16**, 436–468.

Dansgaard, W., Johnsen, S. J., Clausen, H. B. & Langway, C. C. 1971 Climatic record revealed by the Camp Century ice core. *The Late Cenozoic ice ages* (ed. K. K. Turekian), pp. 37–56. New Haven, Conn.: Yale University Press.

Derbyshire, E. 1971 A synoptic approach to the atmospheric circulation of the last glacial maximum in southeastern Australia. *Palaeogeogr. Palaeoclim. Palaeoecol.* **10**, 103–124.

Donner, J. J. 1970 Land/sea level changes in Scotland. *Studies in the vegetational history of the British Isles (Essays in honour of Harry Godwin)* (ed. D. Walker and R. G. West), pp. 23–39. Cambridge University Press.

Emiliani, C. 1955 Pleistocene temperatures. *J. Geol.* **63**, 538–578.

Emiliani, C. 1961 Cenozoic climatic changes as indicated by the stratigraphy and chronology of deep-sea cores of *Globerigina*-ooze facies. *Ann. N.Y. Acad. Sci.* **95**, 521–536.

Firbas, F. 1949 *Spät- und nacheiszeitliche Waldgeschichte Mitteleuropas nördlich der Alpen: I. Allgemeine Waldgeschichte*. Jena: Fischer.

Firbas, F. & Losert, H. 1949 Untersuchungen über die Entstehung der heutigen Waldstufen in den Sudeten. *Planta* **36**, 478–506.

Fredskild, B. 1969 A postglacial standard pollen diagram from Peary Land, north Greenland. *Pollen et spores* **11**, 573–583.

Frenzel, B. 1966 Climatic change in the Atlantic/sub-Boreal transition on the northern hemisphere: botanical evidence. *Proc. Int. Symp. World Climate 8000–0 B.C.* (ed. J. S. Sawyer), pp. 99–123. London: Royal Meteorological Society.

Frenzel, B. 1967 *Die Klimaschwankungen des Eiszeitalters*. Braunschweig (Vieweg – *Die Wissenschaft* Band **129**).

Godwin, H. 1954 Recurrence surfaces. *Danmarks Geol. Undersøgelse*. II. Raekke, no. 80.

Godwin, H. 1956 *The history of the British flora*. Cambridge University Press.

Godwin, H., Suggate, R. P. & Willis, E. H. 1958 Radiocarbon dating of the eustatic rise in ocean level. *Nature, Lond.* **181**, 1518–1519.

Godwin, H. & Tansley, E. G. 1941 Prehistoric charcoals as evidence of former vegetation, soil and climate. *J. Ecol.* **29**, 117–126.

Godwin, H., Walker, D. & Willis, E. H. 1957 Radiocarbon dating and post-glacial vegetational history: Scaleby Moss. *Proc. R. Soc. Lond.* B **1947**, 352–366.

Godwin, H. & Willis, E. H. 1959 Radiocarbon dating of prehistoric wooden trackways. *Nature, Lond.* **184**, 490–491.

Granlund, E. 1932 De svenska högmossernas geologi. *Sveriges Geol Undersökning Afhandl.* Ser. C **26**, 373.

Gross, H. 1958 Die bisherigen Ergebnisse von C 14 Messungen. *Eiszeitalter und Gegenwart*, **9**, 155–187.

Grove, A. T. & Warren, A. 1968 Quaternary landforms and forest on the south side of the Sahara. *Geogr. J.* **134**, 194–208.

van der Hammen, T., Maarleveld, G. C., Vogel, J. C. & Zagwijn, W. H. 1967 Stratigraphy, climatic succession and radiocarbon dating of the last glacial in the Netherlands. *Geol. Mijnbouw* **46**, 79–95.

Hastenrath, S. 1971 On snow-line depression and atmospheric circulation in the tropical Americas during the Pleistocene. *S. Afr. Geogr. J.* **53**, 53–68.

Heusser, C. J. 1966 Late Pleistocene pollen diagrams from the province of Llanquihue, southern Chile. *Proc. Am. Phil. Soc.* **110**, 269–305.

Hoppe, G. & Fries, M. 1965 Submarine peat in the Shetland Islands. *Geogr. Annaler* **47**A, 195–203.

Huntington, E. 1907 *The pulse of Asia*. Boston and New York: Houghton Mifflin

Imbrie, J. & Kipp, N. G. 1971 A new palaeontological method for quantitative palaeoclimatology: application to a Late Pleistocene Carribean core. *The Late Cenozoic ice ages* (ed. K. K. Turekian), pp. 71–181. New Haven, Connecticut: Yale University Press.

Iversen, J. 1944 *Viscum, Hedera* and *Ilex* as climate indicators. *Danmarks Geol. Undersøgelse*. II. Raekke, **3**, no. 6.

Jessen, K. 1949 Studies in Late Quaternary deposits and flora-history of Ireland. *Proc. R. Irish Acad.* **52** B, 85–290.

Johnsen, S. J., Dansgaard, W., Clausen, H. B. & Langway, C. C. 1970 Climatic oscillations A.D. 1200–2000. *Nature, Lond.* **227**, 482–483.

Khotinsky, N. A. 1971 The problem of the boundary horizon with special reference to the Shuvaloff peat bog. *III Internat. Palynological Conference, Appendix to Guide for Field Route No.* 1 B. Novosibirsk.

Kutzbach, J. E. 1967 Empirical eigenvectors of sea-level pressure, surface temperature and precipitation complexes over North America. *J. appl. Met.* **6**, 791–802.

LaMarche, V. C. 1972 Climatic history since 5100 B.C. from treeline fluctuations, White Mountains, east-central California. *Geol. Soc. Am. Abstracts with Programs* **4**, (no. 3), 189.

Lamb, H. H. 1964 Trees and climatic history in Scotland. *Q. Jl R. Met. Soc.* **90**, 382–394.

Lamb, H. H. 1965a *Q. Jl R. Met. Soc.* (Discussion) **91**, 542–550.

Lamb, H. H. 1965b The early medieval warm epoch and its sequel. *Palaeogeogr. Palaeoclim. Palaeoecol.* **1**, 13–37.

Lamb, H. H. 1972 *Climate: present, past and future*. Vol. 1. *Fundamentals and climate now*. London: Methuen.

Lamb, H. H., Lewis, R. P. W. & Woodroffe, A. 1966 Atmospheric circulation and the main climatic variables. *Proc. Int. Symp. World Climate 8000–0 B.C.* (ed. J. S. Sawyer), pp. 174–217. London: Royal Meteorological Society.

Lamb, H. H. & Woodroffe, A. 1970 Atmospheric circulation during the last ice age. *Quaternary Res.* **1**, 29–58.

Leontev, O. K. & Federov, P. V. 1953 History of the Caspian Sea in Late and Post-Hvalynsk time. *Izvestia, ser. geograf.* 1953, no. 4, pp. 64–74. Moscow (Akad. Nauk). (In Russian.)

Link, F. 1958 Kometen, Sonnentätigkeit und Klimaschwankungen. *Die Sterne*, **34**, 129–140. Leipzig: Ambrosius Barth.

Manley, G. 1959 Temperature trends in England, 1698–1957. *Archiv Meteor. Geophys. Biokl.* B **9**, 413–433.

Manley, G. 1961 A preliminary note on early meteorological observations in the London region, 1680–1717, with estimates of monthly mean temperatures, 1680–1706. *Met. Mag.* **90**, 303–310.

Mitchell, G. F. 1956 Post-Boreal pollen-diagrams from Irish raised bogs. *Proc. R. Irish Acad.* **57** B, 185–251.

Moar, N. T. 1965 Contribution to the discussion in Lamb 1965a, p. 543. (See also N. T. Moar 1969b.)

Moar, N. T. 1969a Two pollen diagrams from the mainland, Orkney Islands. *New Phytol.* **68**, 201–208.

Moar, N. T. 1969b A radiocarbon-dated pollen diagram from northwest Scotland. *New Phytol.* **68**, 209–214.

Monod, T. 1963 The late Tertiary and Pleistocene in the Sahara and adjacent southerly regions. *African ecology and human evolution* (eds. F. C. Howell and F. Bourlière), pp. 117–229. Chicago: Aldine.

Nejstadt, M. I. 1957 *History of the forests and palaeogeography of the Soviet Union in the Holocene*. (In Russian.) Moscow: Izdat. Akad. Nauk.

Nichols, H. 1967a Central Canadian palynology and its relevance to northwestern Europe in the late Quaternary period. *Rev. Palaeobotan. Palynol.* **2**, 231–243.

Nichols, H. 1967b Vegetational change, shoreline displacement and the human factor in the late Quaternary history of SW. Scotland. *Trans. R. Soc. Edinb.* **67**, 145–187.

Nichols, H. 1970 Late Quaternary pollen diagrams from the Canadian Arctic barren grounds at Pelly Lake, northern Keewatin, N.W.T. *Arctic and alpine Res.* **2**, 43–61.

Nikolaeva, R. V. & Han-Magomedov, S. O. 1962 New data on the level of the Caspian Sea during the historical period. *Trudy* **60**, 178–188. Moscow (Akad. Nauk, Inst. Okeanologii). (In Russian.)

Overbeck, F., Münnich, K. O., Aletsee, L. & Averdieck, F. R. 1957 Das Alter des Grenzhorizonts norddeutscher Hochmoore nach Radio-Carbon Datierungen. *Flora* **145**, 37–71.

Pennington, W. 1970 Vegetation history in the northwest of England: a regional synthesis. *Studies in the vegetational history of the British Isles (Essays in honour of Harry Godwin)* (ed. D. Walker & R. G. West), pp. 41–79. Cambridge University Press.

Pennington, W., Haworth, E. Y., Bonny, A. P. & Lishman, J. P. 1972 Lake sediments in northern Scotland. *Phil. Trans. R. Soc. Lond.* B **264**, 191–294.

Perring, F. H. 1965 The advance and retreat of the British Flora. *The biological significance of climatic changes in Britain* (ed. C. G. Johnson & L. P. Smith), pp. 51–62. London: Institute Biology and Academic Press.

Perring, F. H. & Walters, S. M. 1962 *Atlas of the British flora.* London: Nelson.

Raikes, R. 1967 *Water, weather and prehistory.* London: John Baker.

Ralph, E. K. & Michael, H. N. 1967 Problems of the radiocarbon calendar. *Archaeometry* **10**, 3–11.

Schofield, J. C. & Thompson, H. R. 1964 Post-glacial sea-levels and isostatic uplift. *N.Z. J. Geol. Geophys.* **7**, 359–370.

Schostakowitsch, W. B. 1934 Bodenablagerungen der Seen und periodische Schwankungen der Naturerscheinungen. Leningrad (Mem. Hydr. Inst.) (In Russian, with German summary: English language presentation given by C. E. P. Brooks 1935 in *Met. Mag.* **61**, 134–139).

Shepard, F. P. 1963 Thirty-five thousand years of sea level. In *Essays in marine geology.* Los Angeles: University of California Press.

Singh, G. 1963 A preliminary survey of the postglacial vegetational history of the Kashmir valley. *Palaeobotanist* **12**, 73–108.

Singh, G. 1971 The Indus valley culture. *Archaeol. Phys. Anthropol. in Oceania* **6**, 177–189.

Smith, A. G., Pearson, G. W. & Pilcher, J. R. 1971 Belfast radiocarbon dates III. *Radiocarbon* **13**, 103–125.

Stuiver, M. 1961 Variations in radiocarbon concentration and sunspot activity. *J. geophys. Res.* **66**, 273–276.

Suess, H. E. 1970a Bristlecone pine calibration of the radiocarbon time-scale 5200 B.C. to the present. *Radiocarbon variations and absolute chronology (Proc. Twelfth Nobel Symposium, Uppsala* 11–15 *August* 1969) (ed. I. U. Olsson), pp. 303–31. New York: Wiley; Stockholm: Almqvist and Wiksell.

Suess, H. E. 1970b The three causes of the secular C 14 fluctuations and their time constants. *Radiocarbon variations and absolute chronology (Proc. Twelfth Nobel Symposium* 11–15 *August* 1969, *Uppsala),* (ed. I. U. Olsson), pp. 595–605. New York: Wiley; Stockholm: Almqvist and Wiksell.

Sutcliffe, R. C. 1947 A contribution to the problem of development. *Q. Jl R. met. Soc.* **73**, 519–524.

Tallantire, P. A. 1972a The regional spread of spruce (*Picea abies* (L.) Karst.) within Fennoscandia: a reassessment. *Norw. J. Bot.* **19**, 1–16.

Tallantire, P. A. 1972b Spread of spruce (*Picea abies* (L.) Karst.) in Fennoscandia and possible climatic implications. *Nature, Lond.* **236**, 64–65.

Tauber, H. 1965 Differential pollen dispersal and the interpretation of pollen diagrams. *Danmarks geol. Undersøgelse, III Raekke,* no. **89**.

Toussoun, Prince O. 1925 Mémoire sur l'histoire du Nil. *Mém. Inst. Egypt* **9**. Caire.

Traill, W. 1868 On submarine forests and other remains of indigenous woods in Orkney. *Trans. bot. Soc. Edinb.* **9**, 146–154.

Tsukada, M. 1967 Pollen succession, absolute pollen frequency and recurrence surfaces in central Japan. *Am. J. Bot.* **54**, 821–831.

Turekian, K. K. 1971 *The Late Cenozoic ice ages.* New Haven, Conn.: Yale University Press.

Turner, J. 1965a A contribution to the history of forest clearance. *Proc. R. Soc. Lond.* B **161**, 343–353.

Turner, J. 1965b A recent study of Tregaron bog, Cardiganshire. *Symposia in Agric. Meteorol.* Mem. no. 8, 33–40. Aberystwyth: University of Wales.

Valéry, N. 1972 Water mining to make the desert bloom. *New Scient.* **56** (819), 322–324. (9 Nov. 1972): reporting an isotopic age-survey of Israel's underground water systems by Professor J. Gat, Weizmann Institute of Science at Rehovot.

Vishnu-Mittre 1966 Some aspects concerning pollen-analytical investigations in the Kashmir valley. *Palaeobotanist* **15**, 157–175.

Vishnu-Mittre & Sharma, B. D. 1966 Studies of postglacial vegetational history from the Kashmir valley – 1. Haigan lake. *Palaeobotanist* **15**, 185–212.

Wadia, D. N. 1960 The post-glacial desiccation of central Asia: evolution of the arid zone of Asia. *Nat. Inst. Sci. India Monograph.* Delhi.

West, R. G. 1968 *Pleistocene geology and biology.* London: Longmans.

Wright, H. E. 1961 Late Pleistocene soil development, glaciation, and cultural change in the eastern Mediterranean region. *Ann. N.Y. Acad. Sci.* **95**, 718–728.

Phil. Trans. R. Soc. Lond. A. **276**, 231–266 (1974) [231]
Printed in Great Britain

Hunting quanta

By D. G. Kendall, F.R.S.
Statistical Laboratory, University of Cambridge

In loving memory of Noel Bryan Slater (1912–1973)

Thom considers that many of the dimensions of megalithic sites can be expressed in terms of a quantum of 5.44 ft (1.66 m). The problem of detecting a quantum when its size is not known in advance has been studied in pioneer papers by Broadbent (1955, 1956). With the greatly increased volume of evidence now available, a renewed attack on this problem seems called for, and a new approach, based on a Fourier analysis, will be outlined here. With samples of sufficient size, testing for the presence of a quantum is equivalent to testing a section of a realization of a Gaussian stochastic process, stationary under time shifts, for an unduly large supremum. If the test is not to be prejudiced by prior decisions about the fittings of 'eggs', 'fans', etc., then only data from circle-diameters can be used. This reduces the sample size, and so the asymptotic theory has to be supplemented by 'Monte Carlo' runs with simulated data (with and without a built-in quantum effect). It already seems clear that agreement on the data-set to be used in the analysis is the vital prerequisite for a decisive test of the quantum hypothesis.

1. Introduction

Although this is an archaeological lecture, spades will not appear in it, but if they were to do so I would make every effort to call them by their everyday name. There are two reasons why technical jargon is out of place on an occasion such as this. In the first place, those to whom it is unfamiliar may lose the thread of the argument, while secondly (and this is much more serious) others to whom it is only marginally familiar may be tempted to give to the argument greater weight than it deserves, for no better reason than that it is seductively incomprehensible.

The work of Professor Alexander Thom covers an enormous range, and raises a large number of statistical questions to most of which we can at present give no adequate answer. This is because of the imponderable influence of subjective selections which have to be made when choosing 'notches' to be associated with orientations, 'eggs' to be fitted to arrays of stones, or dimensions to be measured when dealing with monuments of complex shape. In order to take such selections fully into account one has to determine the universe of possibilities from which they have been drawn, perhaps with a due weighting for each item, so that one can assess the factor by which the significance-test probabilities must be diluted. If this is to be avoided, one is left with two alternatives. The first is, to review the thesis as an integral whole, and then to accept it as 'overwhelmingly convincing', or reject it as 'utterly preposterous'; both points of view are, perhaps, represented here today. Without committing myself to one or the other, I want to see what can be done by following the alternative path of picking out from the wealth of data collected by Professor Thom a particular group of observations which seems to be maximally free from the possibilities of conscious or unconscious prejudice, and then asking a straightforward 'yes' or 'no' question of this. We must of course be prepared to find that the data-set is too small to permit an unqualified answer – a constantly recurring situation in a branch of learning in which the data are in general limited by factors quite beyond our control.

2. The data

The data which I have chosen for this study are Thom's measurements of the *diameters* of *circles*. 'A circle is a circle is a circle' (if it *is* a circle). We may measure its diameter, or its radius, but if a unit of length underlies one set of measures then one half of that unit will underlie the others, and to the same extent. Thom has also been interested in the perimeters of the circles, and at first sight we might think that these are as rigidly linked to the diameters as the radii are, but this I consider to be a dangerous assumption. When *we* speak of the perimeter of a circle we mean its circumference in the specific sense of arc-length, but there is no guarantee that the constructors (as I shall call them) did so; they may have been interested rather in the perimeter of the polygon linking the stones which (for us) define the circle. It is plain that the ratio of the circular to the polygonal perimeter depends on the number of stones (i.e. on the number of vertices of the polygon), and as some stones may in individual cases have been lost, the danger in using perimeters is evident.

I am assuming without further inquiry that the circles listed as such in Tables 5.1 and 5.2 of Thom's book (Thom 1967) are 'true' circles, and that the diameters have been measured objectively, without reference to any prior opinions about the 'quantum'. Those who are unhappy about this assumption may feel, if the outcome of the present investigation should prove to be a positive one, that the next step ought to be a re-survey of the supposed circles, possibly by yet more sophisticated (and expensive) techniques (e.g. by aerial survey), and under the direction of an unprejudiced committee (that is, a committee with an unprejudiced distribution of prejudice).†

As those who are familiar with Thom's book will know, Tables 5.1 and 5.2 deal respectively with diameters 'known to ± 1 foot or better', and 'diameters known with less accuracy'. If we take the *circles* only, then there are 112 in the first set and 57 in the second, making a total of 169 in all. These Tables also include 42 'eggs' (I use this term here to include ellipses, flattened ovals, and so on), but I shall make only occasional reference to these; if they are included then the total number of measurements rises to 211. The data are separated into two groups: Scottish (S) and English & Welsh (EW). When the topographical distinction is ignored, the two sets being pooled, I shall indicate this by the symbol SEW. Data-set 1 will be the accurate circle-diameters, data-set 2 will be all circle-diameters, and data-set 3 will be all circle-diameters together with those measured from the 42 'eggs'. The complete breakdown of the number 211 of all observations is shown in table 1 which follows.

Table 1. The data-sets

	'good' circles (5.1)		all circles (5.1 and 5.2)		all data	
Scotland	S_1	(66)	S_2	(109)	S_3	(127)
England and Wales	EW_1	(46)	EW_2	(60)	EW_3	(84)
U.K.	SEW_1	(112)	SEW_2	(169)	SEW_3	(211)

† One contributor to the discussion asked if my analysis of the quantal effect could be conflated with the circle-survey procedure in one monster 'least squares' analysis, starting from Thom's individual unadjusted measures and sightings. I think he underestimated the technical difficulties of this (for my technique is not of 'least squares' type), and also failed to see how much less convincing (or perhaps how dangerously overconvincing) such an immensely complicated analysis would be. I feel myself that it is better for the argument to be broken down into small digestible morsels which can be contemplated one at a time.

Such a symbol as SEW_{2-1} will denote the less accurately measured circles only, for Scotland and England and Wales combined, and so on. *Our primary data will be that in the set SEW_2.* It must be emphasized that 169 is, statistically speaking, a small number. We may indeed learn something from this sample, but if we ask too many questions of it, we must expect to get some peculiar answers reflecting the peculiarity of the unique sample. The most straightforward procedure is to agree on a small number of simple and natural questions, and then to see what the data-set has to offer in respect of these. I shall be asking, in essence, the following question: is it reasonable to suppose that the circle-diameters could have arisen from a smooth (but not necessarily uniform) distribution over the range which they cover (from 10 ft, up to several hundreds of feet), *or* is it more reasonable to suppose that (apart from a small residual error) they are whole-number multiples of a basic unit (to be called, if it exists, the *quantum*)?

I shall work with the statute foot (≈ 0.305 m) as current unit of length throughout, because all of Thom's measurements are expressed in terms of this. By 'small' (with reference to the error component) we mean (*a*) small relative to the value of the quantum (for otherwise the second interpretation would be stupid), and (*b*) small in the sense that it might well have arisen from (i) errors made by the constructors; (ii) errors made by the surveyors; (iii) errors associated with the finite sizes of the stones themselves; (iv) errors associated with the re-erection of fallen stones etc.

As has very properly been pointed out by Professor Huber, a positive answer in favour of the second (the quantum) alternative would really amount merely to a rejection of the 'smooth distribution' hypothesis, and we can only accept this as a 'proof' that a quantum actually exists *if no other natural alternative hypothesis is available*. (For an analysis of what is 'natural', see the remarks of J. M. Hammersley in the printed discussion following Thom's 1955 paper.) [Professor Huber, in the present discussion, advanced one such further hypothesis, and although it seemed to me artificial rather than natural, I have subsequently examined it and will report briefly on it in the Appendix at the end of this paper.]

Perhaps it should be mentioned, even though it is obvious, that the hypothesis of the existence of a unit of length does not necessarily imply the existence of an invar rod, or a length of whalebone. The unit could indeed have been the height of a man (for example), and need not as such necessarily be excessively variable. If we can recruit guardsmen to within a fraction of an inch, I suppose that the equivalent task might have been achieved in neolithic times, if the matter were thought to be important enough. If we were to find evidence for a quantum approximating to some human dimension, that would not of necessity imply that the constructional work was in fact done 'anatomically', although it might be taken to indicate that at a much earlier stage this *may* have been so. After all, many of our English units (foot, span, cubit, etc.) have an anatomical reference with an interesting history behind it, but we do not therefore assume that there was at one time a 'standard human foot' kept in the Tower of London. If a unit were employed in the construction of these monuments, it could have been 'stored' and 'transmitted' from one generation to another, and transferred from one location to another, in a variety of ways. The primary question is not *how* measurements were made, but *whether* they were made. Of course the technique used (a whalebone rod being one extreme example, and human 'pacing' another) does have implications for the error terms, and to this we shall return later [see the Appendix].

3. The basic principles of quantal analysis

We now turn to the analysis of the circle-diameters. Here I will permit myself one or two mathematical formulae; as in this meeting we are apparently expected to be well read in the Chinese language, a little trigonometry may perhaps also be presumed to be common ground.

In its simplest form, Thom's quantum hypothesis is that an observed circle diameter X can be considered to be composed of two additive components:

$$X = Mq + \epsilon. \tag{3.1}$$

Here q is the quantum, in statute feet (or twice the quantum, if we ought to have been reckoning with radii), M is a whole number, and ϵ denotes the error component as at (i) to (iv) above in § 2. In a more refined analysis one might wish to break ϵ down into two independent components ϵ_1 and ϵ_2, where ϵ_1 is the cumulative effect of the errors to be attributed to each individual version of the M quanta making up X, so in average magnitude proportional to the square-root of M, whence to the square-root of X (this arises particularly in connexion with the 'pacing hypothesis'), and where ϵ_2 represents the variation in average quantum size from one circle to another (and in particular from one locality to another), and is thus in average magnitude proportional to M itself, and so to X itself. We shall, however, largely ignore the possibility of M (i.e. X) dependence in the error term, and leave that more technical matter to some other occasion.

Thom also worked for a time with a hypothesis slightly more complicated than (3.1) in that a constant $c = q\beta/(2\pi)$ was added to the right-hand side. I have examined this possibility also, as my analysis conveniently allows one to test the hypothesis $\beta = 0$ against $\beta > 0$. Like Thom, I found that the data did not support $\beta > 0$ (or rather, did not object to the simpler hypothesis $\beta = 0$), so I shall not go into such questions in any further detail today. (See, however, the Appendix.)

We have to test the hypothesis (3.1) against the alternative that the observed diameters X are better described by a smooth continuous distribution over the full range. This is a well-defined problem, and it is somewhat older than is generally believed. In 1815 Prout advanced such a proposal in respect of the atomic masses of the chemical elements, and a century later a statistician (von Mises, in 1918) tackled the matter with the aid of the statistical techniques available at that time. However, the problem was killed stone dead by Aston one year later, when he was able to announce that the atomic masses of all *isotopes* (save hydrogen) measured up to the date just mentioned were exactly quantal, the quantum being $\frac{1}{16}$th of the atomic mass of oxygen. (The anomalous position of hydrogen does not concern us here, but it is sobering to reflect that neither von Mises nor Aston can have realized all the consequences implicit in it.)

When Thom's interpretation of the metrology of the ancient British circles came to the notice of statisticians in the mid-1950s (Thom 1955), what happened was that Broadbent in two notable papers (1955, 1956) attacked the whole question *de novo*. Whether he was familiar with von Mises's earlier work I do not know, but I suspect that he was not; at all events Broadbent's analysis was entirely different in style and vastly more penetrating, and his work has dominated that of all others for the last 20 years. My own approach, worked out specifically for the present occasion, owes something to von Mises and something to Broadbent, and I should like to express my admiration for the pioneer work of these two writers. In its final form, however, my technique is quite different from either of its predecessors, although not all the points of difference will be apparent today, when emphasis on mere matters of technology would be quite out of place. Just because Broadbent's work has been used so extensively by archaeologists, however, it seems necessary to point out that devices introduced by him cannot necessarily be facilely transferred to my analysis (and vice versa).

The spirit of the present analysis can be very simply conveyed by the following 'thought experiment'. Suppose for the moment that we are interested in comparing the data with a *particular* suggested value q for the quantum. As Broadbent was the first to point out, and as I shall stress over and over again today, this is *not* the situation in which we actually find ourselves, but we adopt it temporarily for convenience of exposition.

This being so, we construct the following piece of apparatus, in structure simple enough to have been devised and built by many of the ancient civilizations which have been described to us by earlier lecturers. It consists of (1) a wheel and (2) a long transparent tape. The wheel is to have a circular perimeter (*not* diameter!) of exactly q feet. One point of the perimeter of the wheel (to be called the zero-point) is to be clearly marked, and one end of the tape is to be attached to this point. The tape is to carry a scale (e.g. of feet), with origin at the zero-point.

Suppose that we have a set of N measured diameters of ancient circles, or chemical atomic masses, or any other data which it is desired to test for quanticity *with quantum q*. Take each measurement X_j ($j = 1, 2, \ldots, N$) in turn, and lay out on the tape a length X_j (starting always at the zero-point, that is, at the fixed end of the tape), marking the tape boldly at the other end of the measured length X_j. Notice that the segments of length $X_1, X_2, \ldots$ are not laid 'end to end' along the tape, but partially overlap because they all start at the same point (the zero-point).

Now wind the tape round and round the wheel until one can proceed no further. (Which way round is irrelevant.) If the data were *exactly* quantal, *with quantum q* (so that all the ϵ's in formula (3.1) would be equal to zero), then it will I hope be obvious that on looking through the layers of transparent 'wound-up' tape we should see all the 'marks' lying on top of one another and coinciding with the zero-point. (If there is a 'β effect', then the marks would still coincide, but would no longer lie on top of the zero-point. However, we are not going into that matter today.)

Next suppose that the data are not quantal, but that instead they are smoothly distributed over the whole range of possibilities. Then of course the marks seen on the wound-up tape will be more or less uniformly but randomly distributed 'higgledy-piggledy' all around the wheel. Furthermore, much the same thing will happen even if the data *are* quantal, provided that the true quantum is neither q nor a rational multiple of q.

Finally return to the situation where the data are quantal, with quantum q, but suppose that the ϵ effect is no longer negligible. Then the marks now seen on the tape will lie, not exactly together, but still near to one another and near to the zero-point, and the amount of scatter which they show about the zero-point will increase with an increase in the statistical order of magnitude in the error-component ϵ, until, with a really large error-component, we shall once again have a uniform random distribution of marks all around the wheel.

Clearly in testing for quanticity with quantum q, we must assess how far the marked points on the wound-up tape are clustered round the zero-point.

Now most of the audience today, perhaps, are familiar with only one law of statistical distribution, namely that called Gaussian. But the Gaussian law has to do with errors scattered over the whole line from $-\infty$ to $+\infty$, and here we clearly have to do with an error-distribution confined to the perimeter of a circle. Statisticians know quite well how to modify the Gaussian distribution in order to produce its 'circular equivalent'. This time suppose that we have an *endless* (doubly infinite) transparent tape with a 'central point' marked on it and fixed to the wheel at the zero-point. Suppose further that this new tape is finely covered with narrow marks that are heavily clustered *on the tape* near to its central point, and less and less clustered as one moves away from it in either direction, in such a way that the density of marks at a particular place *on the tape* is proportional to the density of a particular Gaussian law (with given standard deviation σ) at that distance from the centre point.

This being arranged, we wind our new tape around the wheel (divested now of its old single-ended tape, which has served its purpose for the moment). We are in a new situation, however, because our new tape has two infinite 'arms', and what we are required to do is to wind one 'arm' infinitely often round the wheel in a clockwise direction, and the other arm infinitely often round the wheel in a counterclockwise direction. (For this to be physically possible the two arms of the tape have to be capable of passing through one another, but we are not to worry about that. Just imagine that you are wrapping the longest possible scarf round your neck on the coldest possible night.)

It will be clear, I hope, that after this new experiment has been carried out we shall see, on looking through the infinitely many layers of transparent tape, a distribution of marks *on the perimeter of the wheel* which are densest near the zero-point, and which fall off in density steadily and symmetrically as we move in either direction away from the zero-point towards its opposite or 'antipodal' point. How close the clustering is to the zero-point will of course depend on the magnitude of σ. If σ is very small, the concentration will be immense. If σ is very large, then we shall have a very nearly uniform distribution of marks round the whole perimeter of the wheel.

For given σ, *the density distribution of marks on the wheel follows what is called the circular Gaussian distribution* (or sometimes, as here, the *lumped Gaussian*).

The advantage of this construction is, of course, that it provides us with an adjustable standard, against which to assess the degree of concentration of marks about the zero-point in our first 'thought experiment' (that one using the real data, and the single-ended tape). If the concentration near the zero-point is high, then we shall be able to describe it quantitatively by quoting the σ value which best yields a matching distribution of marks in the associated lumped Gaussian distribution.

The lumped Gaussian distribution appears in the work of Broadbent, and in a sense it is the generating element therein. However, there is another frequently used error-distribution on the circle, first introduced by von Mises in the paper cited above (1918), but sometimes erroneously attributed to others. I will try to give an intuitive idea of this distribution by an appeal to the science of archery.

Suppose we have a skilled archer equipped with a good bow, and best quality arrows, and that he is trying to hit the conventional circular target embellished with the usual coloured rings concentric with a 'bull' occupying the centre of the target. If the range is short, he will of course hit the bull every time, but let us suppose that it is so large that he will only rarely hit the bull, and will sometimes miss the target altogether. As we are supposing that he is a first-class shot, with the best possible weapons, we may suppose that his strikes will be symmetrically disposed around the target, with highest density at the bull, and a density falling off steadily as one moves out from the bull to the successively larger surrounding circles. It will be a plausible assumption, and not in fact far from the truth, to take the law of distribution of arrow-strikes on the target to be the two-dimensional circular Gaussian law, centred on the bull.

If we fix our attention on one of the more extreme surrounding coloured rings, then we will expect the arrows which strike it to lie roughly uniformly around that ring.

But now suppose that a steady left-to-right wind is blowing, and that our archer, skilled as he is in all other respects, has not yet learned that one must allow for this. Then of course the pattern of his arrow-strikes will be shifted to the right of the bull, and the strikes on the particular coloured ring which interests us will no longer be uniformly disposed around that ring, but will have a higher density on its right-hand side, and a lower density on its left-hand side. *The angular distribution of strikes on the ring*, if one idealizes the situation by making the ring very thin in a radial sense, *will be a von Mises distribution*.

Those who wish to see a formula explicitly giving the von Mises density can easily be satisfied, for it is, when we work in radian units of angular measure on the ring,

$$K(k)\ \mathrm{e}^{k\cos\theta}\quad(-\pi < \theta < \pi). \tag{3.2}$$

Here k is positive, and it is large when the cross-wind is high and the asymmetry of the distribution greatest (maximum concentration at the right-hand extreme point, labelled here as $\theta = 0$). When there is little or no cross-wind then k is small and close to zero, and the distribution is nearly uniform on the ring. The constant $K(k)$ need not detain us; we know quite well what it is, and it is in any case uniquely determined by the requirement that the density at (3.2) must sum up to unity when we integrate round the circle.

It is a very curious and not quite adequately explained matter of *fact* (Stephens 1963) that if we take any one lumped Gaussian distribution, characterized by the appropriate value of σ, then we can find a von Mises distribution with appropriately corresponding k such that the two distributions are almost identical. Some care is necessary in giving numerical effect to this, because we have described the von Mises distribution in angular terms, and the lumped Gaussian distribution in linear terms. The relationship has to take into account the fact that the perimeter of our ring/wheel is 2π in the one case, and q in the other case; there is no difficulty in dealing with this, but it should not be forgotten.

Figure 1 shows a collection of lumped Gaussian distributions, and figure 2 a collection of von Mises distributions. Finally figure 3 shows a matched pair of distributions (one lumped Gaussian, one von Mises); they are not quite identical, but it will be evident that a very large sample of observations would be required to discriminate between them.

For analytical, computational, and statistical purposes it is sometimes the lumped Gaussian distribution which is the more convenient, and sometimes the von Mises. As they are practically indistinguishable, we shall have (and will accept) the option of using sometimes one, and sometimes the other, and we shall change horses more than once in crossing the broad stream that lies before us.

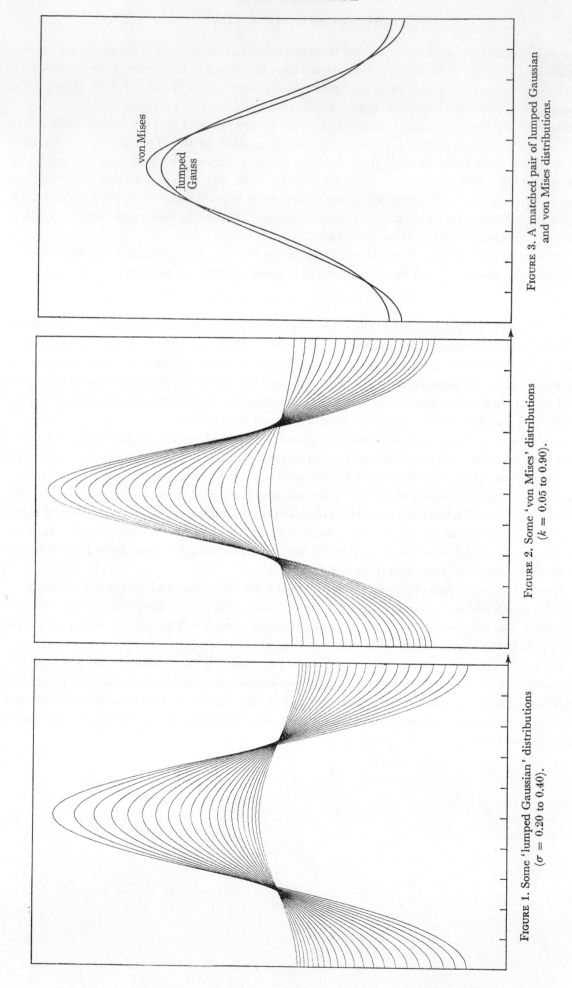

FIGURE 3. A matched pair of lumped Gaussian
and von Mises distributions.

FIGURE 2. Some 'von Mises' distributions
(k = 0.05 to 0.90).

FIGURE 1. Some 'lumped Gaussian' distributions
(σ = 0.20 to 0.40).

4. THE COSINE QUANTOGRAM

We have remarked that position on the wheel/ring carrying our markers can in a natural way be transformed into angular terms, using angular displacement θ in radians (measured from the zero-point) as a position-indicator on the circumference. An equally good position-indicator for our purposes is $\cos\theta$, which ranges from -1 to $+1$; it takes the value $+1$ at the zero-point, and falls off steadily to -1 as one proceeds away from the zero-point *in either direction*. Thus a natural (and in relation to the von Mises distribution, *the* natural) measure of clustering of marks around the zero-point will be the sum $\Sigma\cos\theta$, normed in some appropriate way so as to take into account the size N of the sample of measured values. If we follow this idea to its logical conclusion, we are led to the view that the appropriate 'score' for high clustering, in relation to the prescribed quantum q, is

$$\phi(\tau) = \sqrt{\left(\frac{2}{N}\right)} \sum_{j=1}^{N} \cos(2\pi X_j \tau), \qquad (4.1)$$

where $\tau = 1/q$ is the reciprocal of the quantum, measured in ft^{-1} (one can think of τ as a frequency measure). To understand this formula, notice that if as in (3.1) we have $X_j = M_j q + \epsilon_j$, then

$$\cos(2\pi X_j \tau) = \cos(2\pi\epsilon_j \tau),$$

and as ϵ_j is the error in linear arc-length on a wheel of perimeter q, so $2\pi\epsilon_j\tau$ is the corresponding error in angular terms (measured in radians). If all ϵ values are zero, then we shall have the large value $\sqrt{(2N)}$ for $\phi(\tau)$. If, however, the errors ϵ are very large, or if the quantum has been wrongly chosen, or if there is no quantal effect at all in the data, then the contributions of the cosines in (4.1) will largely cancel out through interference, and $\phi(\tau)$ will be small, and may be either positive or negative.

The purpose of the factor $\sqrt{(2/N)}$ is to allow for the dependence of the statistical 'size' of $\phi(\tau)$, when either there is no quantum, or when the quantum has been chosen incorrectly, on the size N of the sample. In fact, for not too small N, and for not too large q-values (i.e. for not too small τ-values), it can be shown that $\phi(\tau)$ *for fixed* τ is very nearly an ordinary Gaussian random variable with zero mean and unit standard deviation, so that a value in excess of $+4$ would only occur with a probability of about 0.000032.

We now have to make two exceedingly important points, both due to Broadbent, although he made them with reference to his own rather different method of analysis.

First, we do *not* know *a priori* what the quantum q is supposed to be (equivalently, we do not know what τ is supposed to be). Thus we have to consider *not* the single number $\phi(\tau)$, computed from the observations, but rather *the range of values of $\phi(\tau)$ as τ is varied*. We do not, however, have to consider the whole range $0 < \tau < \infty$, for this would correspond to $\infty > q > 0$, and clearly microscopically small and cosmologically large values of q are irrelevant. Thus the range must be truncated in some way, at each end. Indeed, if we do not do this, disaster will ensue, as is easily seen. For suppose we take q to be smaller and smaller; ultimately we will arrive at $q = 0.1$ ($\tau = 10$), and we will then certainly discover a quantum, because none of Thom's observations were quoted to more than one place of decimals! On the other hand, if we take q to be larger and larger, ultimately we will reach a situation in which (3.1) will be satisfied quite adequately by taking $M_j = 0$, and $\epsilon_j = X_j$, and once again we will have found a (ridiculous) quantum.

So much is easy to appreciate, once it has been pointed out. It is not quite so easy to decide how to choose the appropriate range, $\tau_0 < \tau < \tau_1$, or equivalently, $q_0 < q < q_1$, where $\tau_0 = 1/q_1$, $\tau_1 = 1/q_0$, although it is not hard to fix some bounds; the difficulty is in agreeing on precise ones. For the sake of argument we might choose $q_0 = 2$ ft (the dimensions of a single stone) and $q_1 = 10$ ft (the diameter of the smallest circle). This gives $\tau_0 = 0.1$ and $\tau_1 = 0.5$. For technical reasons it turns out to be convenient to use instead

$$\tau_0 = 0.09 \quad \text{and} \quad \tau_1 = 0.59,$$

thus giving $\tau_1 - \tau_0 = 0.50$. In what follows, this has been the basic choice of τ_0 and τ_1, although many computations have been performed with both wider and narrower ranges than this. The computer program implementing my procedure will of course accept any (τ_0, τ_1) values one chooses, perhaps at the cost of a greater expenditure of computer time; with a wide range of τ values it is necessary to compute and plot a very large number of points $(\tau, \phi(\tau))$ in order to be sure of picking up all the 'peaks' mentioned below. For this reason it is preferable, if a wide τ range has to be searched, to break this up into several subranges and then to search each one separately.

We are now in a position to appreciate Broadbent's second point, which is (in the present context) that we must look at the plot of $\phi(\tau)$ (plotted vertically) against τ (plotted horizontally) over the range $\tau_0 < \tau < \tau_1$, and pick out the principal 'peak'. But this plot is the plot of a sum of cosines, and will be wildly fluctuating; peaks will in fact abound. Which are we to choose? There is a latent difficulty of choice here, which Broadbent solved in one way, and I in another (see the appendix), but supposing (to make matters as simple as they could possibly be) that there is one overwhelmingly dominant peak, how are we to decide whether or not we have just got this big peak 'by chance'? On the face of it, if we find a peak greater than $+4$, we might say that the anti-quantum hypothesis has been rejected at the 0.0032 % level, but as Broadbent pointed out this is totally inadmissible because we *chose* the biggest peak in the range, so made it too easy for ourselves to get such a big one. (The position would of course be logically utterly different if the quantum q, and so the value of τ, were specified by our hypothesis *in advance*.) Thus the fact that we are choosing the highest peak in the whole of the τ interval requires us to water down the naïve significance level by a very considerable factor, and it is not altogether easy to say, by how much.

There are various ways of tackling this difficulty. For example (and here, in an aside for statisticians, jargon will creep in) we can take note of the fact that for not too small N and not too small τ_0, the process $\{\phi(\tau): \tau_0 < \tau < \tau_1\}$ is asymptotically a section of a stationary normalized Gaussian stochastic process for which the autocorrelation function can be estimated near $u = 0$ by setting $\rho(u) = \phi(|u|)/\sqrt{(2N)}$. Unfortunately, while $\rho(u)$ for the relevant small u can be (and has been) estimated in this way, we do not have a good usable theory for the distribution of the supremum of such a process. We could, of course, see how rapidly $\rho(u)$ approaches zero as u increases, and so estimate crudely how many 'independent pieces of information' are represented by the section (τ_0, τ_1) of the stochastic process. Alternatively we could invoke a limit theorem of Cramér concerning the supremum functional for stationary Gaussian processes. But we can only estimate $\rho(u)$ near to $u = 0$; also we do not know what are the circumstances in which Cramér's limit theorem can safely be used for a given finite N, and finite values of τ. (Here the aside to specialists ends.)

We are therefore driven, as was also Broadbent driven in his earlier but rather different investigation, to what are called *Monte Carlo techniques*. Despite the proximity of this building to a great palace devoted to the 'doctrine of chances', it will not be reasonable to suppose that everyone present is equally familiar with Monte Carlo procedures, so I will give a brief sketch of these in so far as they relate to our problem.

5. MONTE CARLO, AND 'MOUNT THOM'

Let us first take a look at a typical example of the plot of $\phi(\tau)$ against τ; I shall call such a plot a *cosine quantogram* (c.q.g.). That shown in figure 4 is taken over the very wide range

$$\tau_0 = 0.05, \quad \tau_1 = 2.55,$$

and here for once we are using the large data-set involving 211 measurements ('eggs and all'). As remarked above, such 'wide' plots are not very convenient and it is better to scan a wide τ range piece-by-piece, but it will shortly appear why we have deliberately taken a single wide range for our introductory example. Several large peaks will be observed. Of two on the right-hand side, the larger has

$$\tau = 0.05 + (7.8/10) \ (2.55 - 0.05) = 2.00, \quad \text{so} \quad q = 0.5,$$

and the large peak near the middle of the range has

$$\tau = 0.05 + (3.8/10) \ (2.55 - 0.05) = 1.00, \quad \text{so} \quad q = 1.0,$$

while similar measurements and calculations show that for the very large peak on the left-hand side of the range (indicated by vertical and horizontal markers),

$$q = 5.442.$$

Now the first two of these peaks obviously arise from the circumstance that when the less accurately measured circles are included, a number of the measurements will have been rounded by Thom to the nearest foot, or the nearest half-foot. It is exceedingly satisfactory that the c.q.g. has picked up these 'factitious' quantal effects, and it is also a very striking fact that the major peak, that at Thom's 'megalithic fathom' value $q = 5.44$, is actually more prominent than the two factitious peaks of whose reality we can be in no doubt.

We can eliminate the factitious peaks in either of two ways; by reducing τ_1 to a smaller and more appropriate value (and so just leaving the factitious peaks 'off the screen'), or we can *unround the data*, adding a random fraction of ± 0.5 to all measurements quoted to the nearest foot, and a random fraction of ± 0.25 to all measurements quoted to the nearest half-foot. The second alternative is to be preferred, and in order to be quite systematic about this I have also added a random fraction of ± 0.05 ft to all the 'accurate' measurements (i.e. those quoted to the nearest 0.1 ft); thus for an 'accurate' circle X is replaced by $X + 0.05U$, where U is chosen randomly (separately for each such X) from a uniform distribution over the range $(-1, 1)$, and similarly in the other cases. The program I have written allows one to choose the values of U once and for all, or separately and independently each time the data are re-analysed, and so 'the effect of unrounding' can be studied by varying the U values.

The reader is now invited to look at figure 5, which shows a cosine quantogram exactly similar to that in figure 4, save that the set of *unrounded* data has been used. The peak at $\frac{1}{2}$ ft has now collapsed, likewise that at 1 ft, but the remaining peak at 5.44 ft is as prominent as ever. The 'ladder' running up the middle of the cosine quantogram is in effect a scale of units with

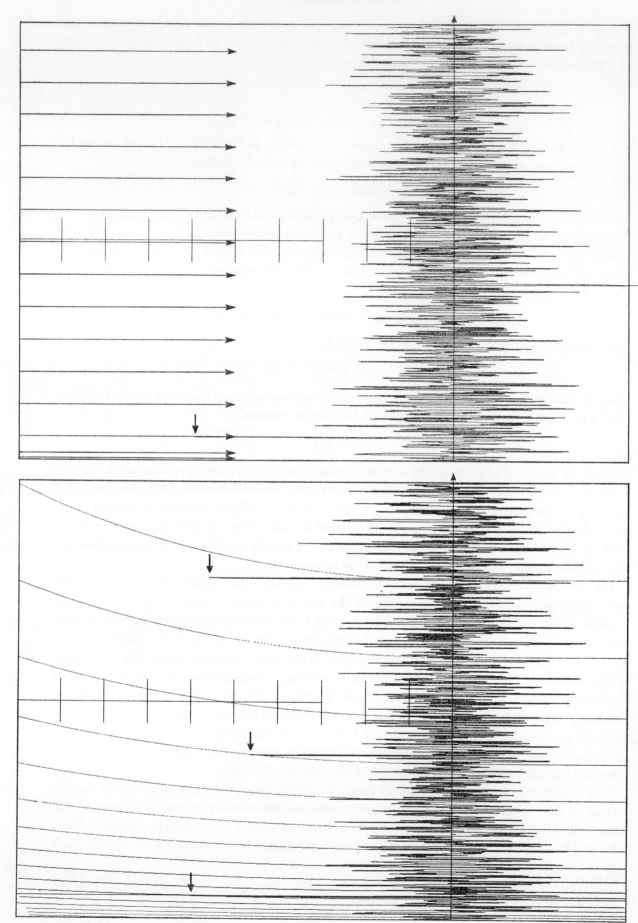

FIGURE 5. Cosine quantogram for SEW_3 (unrounded data) (τ_0, τ_1 as before).

FIGURE 4. Cosine quantogram for SEW_3 (raw data) ($\tau_0 = 0.05$, $\tau_1 = 2.55$).

respect to which the standard deviation of a single value of $\phi(\tau)$, that is, a single ordinate of the cosine quantogram, is 1 unit. Thus the peak at 5.44 ft is immensely significant, judged as a peak *by itself*, but the question is rather, is it significant when due allowance is made for the enormous number of peaks present, any one of which might have been chosen?

The various *vertical lines* on the new c.q.g. in figure 5 indicate the integer multiples and sub-multiples of 5.442 ft, and they are drawn automatically to enable the Thom peak and its under- and over-tones to be identified visually without having to go through a laborious procedure of measuring τ and converting to q. The *curves* (very steep, and on the left almost vertical straight lines) on the c.q.g. in figure 4 are *isopithons* (which I name after Πειθώ: the goddess of Persuasion); they represent loci of constant relative support in the sense of the theory of likelihood (Edwards 1972). The method of construction and use of these will be presented elsewhere (see the appendix). What is relevant here is that the two 'factitious' peaks at 1 ft and $\frac{1}{2}$ ft both lay to the right of and below the isopithon through the peak at 5.44, and this fact *of itself* would have suggested the priority of the 5.44 ft peak.

From analytical considerations it can be shown that large values of $\phi(\tau)$ (arbitrarily close to $\sqrt{(2N)}$) will occur for arbitrarily large τ. The isopithons show that these are to be regarded as dominated by 'moderate' peaks at 'moderate' τ values.

A glance at either of these two figures is sufficient to indicate that we here have to do with a 'signal:noise ratio' problem; we need to have some way of picking up genuine signals of quanticity from the general background of 'grass' in which they grow. As already remarked, the cosine quantograms in figures 4 and 5 cover much too wide a span of τ values, and were chosen to do so simply to illustrate the power of our technique in picking up factitious peaks like the ones at 1 ft and $\frac{1}{2}$ ft. But now we shall revert to the 'official' τ range,

$$0.09 < \tau < 0.59,$$

and we shall also revert to the 'official' data set, SEW_2, which, it will be remembered, contains only 169 observations.

Before leaving figure 5, it will be convenient to deal briefly with a question which may have formulated itself in the minds of some of those here: why do we work with cosines only, in (4.1) above; do not the *sines* also contain information of value? The answer to this question is readily given. If we define $\psi(\tau)$ exactly as at (4.1), but with the cosines everywhere replaced by sines, and if we then define

$$U(\tau) = \tfrac{1}{2}\{\phi(\tau)^2 + \psi(\tau)^2\} \quad (\tau_0 < \tau < \tau_1), \tag{5.1}$$

then we obtain in $U(\tau)$ a quantity which asymptotically has a negative-exponential distribution with unit parameter, and which 'peaks' at a value $\tau = 1/q$ when the observations follow a modified law, $$X_j = \beta + M_j q + \epsilon_j \quad (0 \leqslant \beta < q). \tag{5.2}$$

Another version of my computer program plots $U(\tau)$ against τ, complete with appropriate isopithons etc., (the *modular quantogram*) and also prints out an estimate of β, and an indication of whether it is significantly different from zero. The estimate of β not being significant for the Thom data, this aspect of the problem is not entered into any further here, but in the mathematical account of this investigation more details will be given (see appendix).

How are we to decide whether the peak at 5.44 ft in figure 5 is 'significant'; more exactly, how likely is it that we should find a peak as high or higher than this, *with its τ value within the agreed range* (0.09, 0.59), by the mere operations of chance?

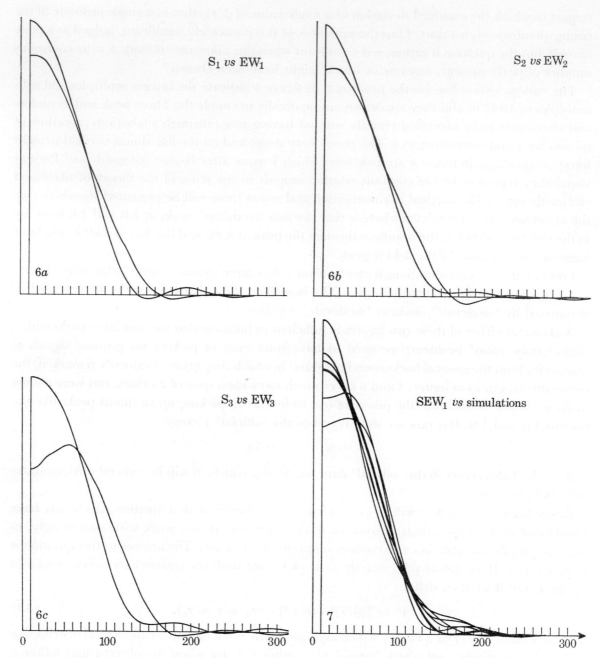

FIGURE 6. Spline transforms for the distribution of circle diameters (in feet): Scotland versus England and Wales, sets 1–3.

FIGURE 7. Spline transforms for the distribution of circle diameters (in feet): SEW_1 versus 5 simulations.

To make this question meaningful we must try to imagine what might happen if we took a 'random' set of data, similar in all statistical respects to the actual data save only in definitely *not* having any underlying quantal effect, and subjected it to the same analysis; i.c. computed its cosine quantogram over the agreed range (0.09, 0.59). Of course we should learn nothing worth knowing by doing this just once; we should have to do it at least, say, 100 times, the successive 'imitation' samples being all independent of one another.

We now find ourselves thrown back on a deeper-lying question: what do we mean by a sample, artificial and random in character, and definitely not involving a quantal effect, and yet 'similar in all statistical respects to the actual data'? Well, in the first place, it should be of the same size (i.e. have the same value for N), though this is not in fact an essential requirement because we have very largely eliminated the effect of sample-size by including the factor $\sqrt{(2/N)}$ in the definition of $\phi(\tau)$. (Statisticians please note; the value of N has little or no effect on the autocorrelation function nor the approximating stationary normalized Gaussian process.) So we need not be too particular about the N value, and this fact will serve us well at a later stage.

I submit that by 'similar in all statistical respects to the actual data' we should mean that in the imitating samples, there will be roughly the same proportions of small, medium-sized, and large diameters. That is, we must try to arrange that both the actual data and the imitating data will have the same *coarse-grained* distribution of diameter-values. To achieve this we must have a look at the coarse-grained distribution of diameter-values for the actual data-sets; figures 6 *a, b, c* show these for the S and EW data separately, for each of the three progressively enlarged data-sets 1, 2 and 3. It will be seen that there is indeed a sense in which one can speak of the coarse-grained distribution observed for X; it ranges from 10 ft (at which there is a natural cut-off) to several hundreds of feet, and the density when appropriately smoothed falls off in a fairly regular manner. An anomaly of SEW_3 need not trouble us because we shall not in fact make any serious use of that data-set.

It is a lucky accident that we can very well imitate this observed coarse-grained distribution of diameter values by one half of a Gaussian curve, with the peak shifted to the 'cut-off' at 10 ft. This is very fortunate because half-Gaussian random variables are extremely easy to simulate on a computer, so that we can produce as many replicates of the data as we like, and as they will all be derived from a half-Gaussian distribution there will be no question whatsoever of their being quantal with $0.09 < \tau < 0.59$.

I almost forgot to mention that the distributions in figure 6 are displayed not in *histogram* form, as would be more usual, but in *spline transform* form, as is far more convenient and appropriate. For the details of the spline transform, invented 3 years ago by Liliana Boneva and Ivan Stefanov (both of Sofia) and myself, see our paper (1971) which stemmed from a visit to Bulgaria under the Agreement between the Royal Society and the Bulgarian Academy of Sciences. All the reader needs to know is that the spline transform program converts a sample of observations in a unique invertible manner into a smooth curve, in such a way that the ordinate of the curve is large and positive where the observations lie thickly. Where they lie only thinly, or do not lie at all, the curve oscillates through small positive *and negative* values (part of the cost of this particular type of smoothness).

In figure 7 I show (i) the spline transform for the diameter distribution of the 112 circles in SEW_1, together with (ii) the spline transforms for 5 independent 'imitations' of SEW_1 (using the half-Gaussian simulation device). It will be observed that the curve for SEW_1 differs no more from the five imitations than they differ among themselves.

A word about the 'cut-off' at 10 ft, mentioned above: Thom's smallest circle has a diameter of 10.8 ft, and it is clear that there is some (perhaps not determinate) cut-off at the low-end of the diameter scale; one does not find, or even look for, neolithic circles of diameter 2 or 3 ft. I therefore treated 10 as an absolute cut-off for the data at the lower end, both for the observed and artificial samples, and the spline transforms have been computed taking this into account.

The formula generating the simulated circle diameters was

$$X = 10 + 57.176 |Z|, \tag{5.3}$$

where Z is an 'artificial' standard Gaussian variable. The numerical coefficient was chosen to make the average simulated X-value equal to the average X-value for SEW_1. It would in retrospect have been better to use the SEW_2-value, but figure 6 shows that this would have made little difference. For SEW_2, the average X value is 58.8, while for SEW_1 it is 55.6. (From the appendix it will appear that it would have been better to 'match' r.m.s. (X) rather than $E(X)$.)

In my first attempt at handling the Monte Carlo aspects of this problem I took what now seems to me a rather stupid path, but I will report on it here because it does, in fact, have something to teach us. I ran 100 simulations of the SEW_2 data, and computed the cosine quantograms for each one. Let us pause for a moment and think out exactly what this means. It is *as if* we said to the computer: go back 4000 years, observe the method and design then used for the construction of these monuments, build yourself 169 of them, with the appropriate proportions of small, medium-sized and large diameters, *but* totally disregard any quantal practices which may at that time have been customary. Then return to the present, simulate Professor Thom, have him survey the monuments, and report the diameters on the print-out device; having carried out these instructions to the full, go back and do the same thing again 99 times, *independently*. Then construct the 100 cosine quantograms for the data thus simulated, and plot these for examination.

What I then did, in this first analysis, was to require the computer to plot out max $\phi(\tau)$ against τ, and min $\phi(\tau)$ against τ, where the 'max' and 'min' operations were carried out over the whole set of 100 simulations of SEW_2. Finally the cosine quantogram of SEW_2 was drawn on top of the picture thus produced, and figure 8 was the result. It will be noticed that the 'max' and 'min' curves are 'almost' horizontal, thus confirming the essentially stationary character of the stochastic process, and that the 'min' curve is 'almost' a reflexion (in the line $\phi = 0$, i.e. the τ-axis) of the 'max' curve. This again is as one would expect, in view of what I know about the statistical structure of this particular stochastic process. (There are no large negative values in the estimate which we have of its autocorrelation function ρ.) We can think of the 'max' curve as a sort of flood-level, showing the greatest height to which the $\phi(\tau)$-waters have risen at each separate τ, and there, proudly showing its summit above them (but only just), is what we might well call Mount Thom, at a τ value corresponding to $q = 5.44$ ft.

This result was gratifying and persuasive, but one can in fact do better. First, in view of the near-horizontal nature of the 'max' and 'min' curves, which subsequent computations confirmed, it is clear that a secular trend with τ is not to be feared, and therefore that the relevant statistic from each separate simulation is just

$$S = \sup\{\phi(\tau): 0.09 < \tau < 0.59\}.$$

However,

$$I = \inf\{\phi(\tau): 0.09 < \tau < 0.59\}$$

is in fact just as interesting, and indeed just as useful, for as a study of the autocorrelation function of the stochastic process revealed only negligible negative ordinates, we can deduce that deep troughs will not (or hardly at all) be correlated with high peaks. Therefore, to a useful degree of approximation, we can double the amount of information at our disposal by treating each $(-I)$ as if it were the S derived from *a different independent simulation*, and so on collecting

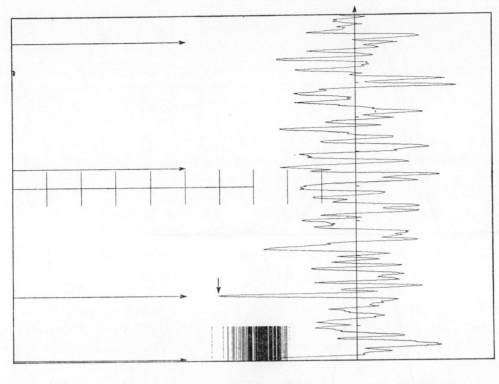

FIGURE 9. Cosine quantogram for SEW_2 ($\tau_0 = 0.09$; $\tau_1 = 0.59$), and the 200 simulated 'flood levels'. (The horizontal arrow indicates 'Mount Thom', and the vertical arrows indicate $q = 5.442$ ft and its multiples and submultiples. The scale running down the centre of the diagram is one of standard deviations.)

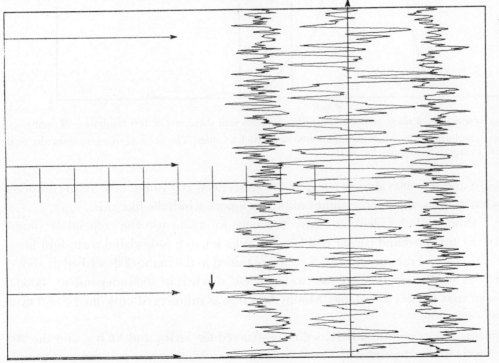

FIGURE 8. Cosine quantogram for SEW_2 ($\tau_0 = 0.09$, $\tau_1 = 0.59$), and the supremum and infimum curves for 100 simulations. The horizontal arrow indicates 'Mount Thom'.

(10)

(11)

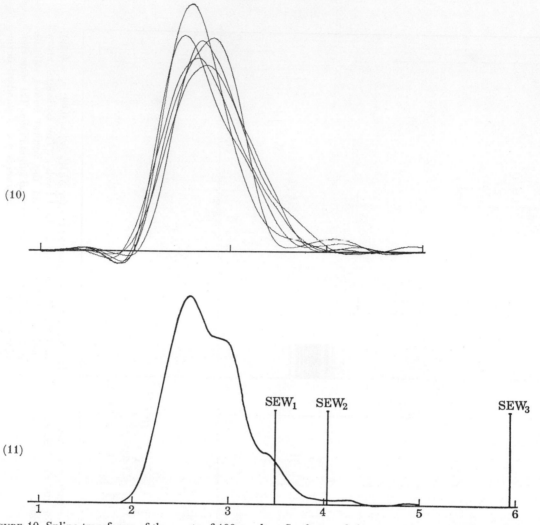

FIGURE 10. Spline transforms of three sets of 100 random S values and three sets of 100 random $(-I)$ values.

FIGURE 11. Spline transform of the whole sample of 600 simulated S values. The vertical markers show the peak heights in the cosine quantograms for $SEW_{1,2,3}$.

the 100 values of S and the 100 values of $(-I)$ we have, in effect, 200 values of S. With computer time costing what it does today, one cannot afford to ignore windfalls like this.

The reader is now invited to look at figure 9, where we again use the 'official' τ range, 0.09 to 0.59. In the top left-hand corner will be seen a thick batch of horizontal straight lines. There are 200 of these, and they are the 200 S-values obtained in the manner described as above, with $0.09 < \tau < 0.59$. That is, they are the 'water-levels' reached in 200 independent 'floods' and it will be seen that Mount Thom, like Mount Ararat, was submerged only once – and even then, only just.

Similar diagrams, not reproduced here, were constructed for SEW_1 and SEW_3, and the 200 S values were freshly and independently obtained each time. When it was realized that sample size had virtually no effect on the S distribution, it was then seen that we had in fact $200 + 200 + 200 = 600$ independent S values, and so a rather substantial amount of information about the S distribution over the τ range (0.09, 0.59) for the stationary normalized Gaussian stochastic process having this particular autocorrelation function.

Figure 10 shows a spline transform display for the distributions of the three batches of S values and the three batches of $-I$ values, in order to check that there was no appreciable difference between them. Figure 11 shows the spline transform for the S distribution using all 600 S values, and also on this picture are indicated, as vertical lines locating positions on the S scale, the height of Mount Thom for SEW_1, SEW_2 and SEW_3, respectively.

The significance of the evidence provided by SEW_3 would be quite overwhelming were it not for the fact that it is this datum which is compromised by the inclusion of 'eggs', etc.

The data set SEW_2, which I regard as the primary one, gives a pretty clear result. Inspection of the figures shows that the height of Mount Thom (4.033) was exceeded or equalled on just 5 occasions (out of 600 possible occasions), these five values being

$$4.887, 4.254, 4.293, 4.152, 4.208.$$

The hypothesis of a smooth, non-quantal distribution of circle diameters for SEW_2 is thus rejected at the 1 % level.

The data set SEW_1 (112 accurately measured circles) gave a peak at essentially the same position, and of height 3.427. This was exceeded on the five occasions noted above and also on 36 further occasions, out of the same possible total of 600. Thus the hypothesis of a smooth, non-quantal distribution of circle-diameters for SEW_1 cannot be rejected: the significance level attained is only about 7 %.

Whether one accepts the quantal hypothesis or not thus depends very much on which data-set one decides to use. However (see the appendix) the smaller N value for SEW_1 militates against the chance of detecting a real effect, if present, for (other things being equal) *the altitude of Mount Thom varies as the square-root of N*, and we cannot hope to pick it up, in relation to the general 'noise', if N is too small.

6. CONCLUSIONS

If 'a conclusion' is required, it will depend to some extent upon one's point of view. If the audience were to put a pistol to my head and demand an answer to the question: 'a quantum – yes or no?', then I should plead for more observations. A significance level of 1 % is normally regarded as a strong recommendation that the experiment be repeated on a larger scale, this being preceded perhaps by a cautious letter to *Nature*.

But if more observations are not to be had, then we must make the most of those with which we are provided. If the audience were to put the question somewhat differently, and say 'does the analysis of the available data justify the expenditure of public monies on a costly and sophisticated aerial re-survey of the circular sites?', then I think few would disagree with my affirmative reply.

APPENDIX

The purpose of this appendix is twofold; first, I wish to record one or two details of this new technique of quantal analysis for the benefit of those who may wish to employ it in similar problems (or to study in greater detail its application to the present problem); and secondly, I should like to offer at least a few comments on some of the very interesting points which were raised at the discussion on 8 December 1972.

There were many of these queries, some pressed very insistently; Oliver Twist himself could hardly have asked for more. I shall only attempt to deal with a few of them here, and must begin by saying straight away that the more searching the question, the more complicated the answer; it will no longer be possible, as it was in the main lecture, to avoid mathematical complexities entirely. However, I will start with some queries of a rather general character, and work gradually towards those whose answers involve technicalities.

(i) I have been asked whether, if it be accepted for the moment that 'there must be something in the quantum', does it follow *from this judgement alone* that Thom's claims with regard to astronomical alinements must be accepted in their entirety, without further examination? The answer is, no.

(ii) Professor Huber pointed out that in the data set SEW_2, consisting of 169 observations, there are no fewer than 12 in the range 20.5–22.0 ft inclusive; these are the diameters

$$20.5, \quad 20.6, \quad 20.9, \quad 21.0, \quad 21.0, \quad 21.3,$$
$$21.4, \quad 21.8, \quad 22.0, \quad 21.4, \quad 21.2, \quad 21.0.$$

I shall call this the data set H, and the remaining 157 observations the data set H^*. Professor Huber's remark is a very pertinent one, because this 'clumping' of the data is an observation *ex post facto*, just as was Thom's original observation of the 'quantal' effect, and so Huber is quite entitled to ask whether the 'clumping' is all that is really peculiar about the data, and whether we need to go to the extravagance of a quantal hypothesis. Mr Hogg made a rather similar point, when he spoke of 'a relatively few circles set out using some standard unit, mixed with a larger number where no standard was used'. It is largely a matter of taste whether we admit the Hogg or Huber 'hypotheses' as 'natural' alternatives to that of Thom, but let us do so, and see what happens.

To begin with, let us make some random (so *not* quantal) simulations of H^*, using the simulator (5.3) *but inhibiting this so that for X the interval of values* (20.5–22.0) *is 'taboo'*. Ten such simulations were created, and each separate set of 157 was then made up to 169 by adjoining the Huber set H of 12 observations to it. Figure A1 shows the c.q.g of the first five of these (those of the second five are quite similar). A rather high lateral 'magnification' has been used, so that here $\tau_0 = 0.155$ and $\tau_1 = 0.205$; the vertical scale is the same as before. It will be seen that there is no special clustering of peaks near the 'Thom' value $\tau = 0.1838$ (indicated by the arrow pointing down from the top of the diagram); also the peaks which do occur within this range (indicated by short horizontal arrows) do not reach significant heights – the largest of the set of ten has a height of 2.76, which would not look at all impressive when marked in on figure 11.

Next, let us see what happens to the c.q.g. of SEW_2 when the Huber observations H are *deleted* from it. This is shown by figure A2, where two c.q.gs are superimposed; both are viewed under the high lateral magnification ($\tau_0 = 0.155$, $\tau_1 = 0.205$) used in figure A1, and that with the highest peak is in fact the c.q.g. for SEW_2 itself (height 4.03). The other curve, with a peak of height 3.23 in almost exactly the same place ($q = 5.450$ instead of 5.441) is the c.q.g. of H^*; i.e. of the 157 original observations which remain when 'Huber's 12' have been 'cast out'. A height of 3.23 is not particularly impressive either, when judged against figure 11, but it is still large enough to excite interest, and the fact that it lies in almost exactly the same position as the original one says, to me, that 'Huber's 12' made very little contribution to the latter.

We have further to bear in mind that H^* contains only 157 observations and that a *real* quantal effect (as I shall demonstrate below) will in general yield a peak whose height is proportional to the square root of the sample size, N. Now $\sqrt{(169/157)} = 1.0375$, so that the

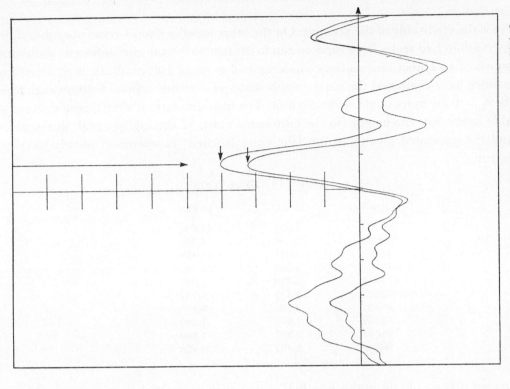

Figure A2. The cosine quantograms ($\tau_0 = 0.155$, $\tau_1 = 0.205$) for (i) SEW_2 and (ii) SEW_2 with the 12 'Huber' observations removed.

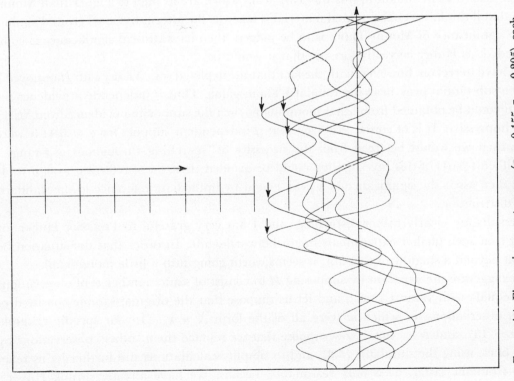

Figure A1. Five cosine quantograms ($\tau_0 = 0.155$, $\tau_1 = 0.205$), each being based on the 12 'Huber' diameters plus independent sets of 157 simulated observations.

comparison with figure 11 would be on a fairer basis if the peak height used were multiplied by this last factor, thus raising it to 3.35, when it begins to stand out rather more in the tail of the S distribution.

So much for the credit side of the account. On the other hand, we must remember that if the original observations had really been random (up to the first 157) from the 'tabooed' simulator, with the bunch of 12 Huber observations superimposed to make 169 in all, then we should in the first instance have examined the c.q.g. of this material over our official search range from $\tau_0 = 0.09$ to $\tau_1 = 0.59$, so we ought to do this now. Ten such simulations were therefore created, and the table below lists the height (in the customary units) of the highest peak in the given τ range, and the associated q value in feet (the entry labelled 'randomizer' merely serves to identify the run).

TABLE A1. $H+157$ SIMULATIONS

randomizer	height	q value
00293	4.934	11.055
05025	3.558	4.260
11126	3.042	10.499
12968	2.593	5.587
44972	2.720	1.977
53632	2.473	6.757
60561	2.730	3.118
72739	4.052	4.269
94054	3.327	1.950
96561	2.493	3.428

It will be seen that we obtain in this way two peaks which are as high or higher than Mount Thom, so that if this last set of simulated material is taken as the alternative background against which the significance of Mount Thom is to be judged, then its statistical significance is completely eroded, as Huber correctly foresaw that it would be.

We still have to reckon, however, with the fact that the depleted set, 'SEW_2 with H removed', produced a substantial peak near the standard Thom value. Thus *if* independent evidence for $q = 5.44$ ft could be obtained from other monuments, then the supporting evidence from SEW_2 would be impressive. It is of course also true that *if* independent support for $q = 5.44$ ft were available, then we would be freed from the necessity of 'searching throughout a τ-range', and the difficult parts of this investigation could be avoided. Instead of referring to figure 11, we would then assess the significance of a peak height by immediate reference to the standard Gaussian distribution.

These results are clearly rather disturbing, and I am very grateful to Professor Huber for suggesting that such further experiments might be worthwhile. In order that the situation be understood beyond a shadow of a doubt, it seems worth going into a little more detail.

Let us exaggerate the situation by supposing H to consist of some number m of observations all *exactly* equal (with value x_h, say), and let us suppose that the original sample consisted of $N = n+m$ observations, of which m were all of the form $X = x_h$. (In our specific example $m = 12$, $n = 157$, and $N = 169$.) Now suppose that we replace the n 'other' observations by simulated ones, using the simulator (5.3), and to simplify calculations still further let us relax the taboo from the latter, as is only reasonable, because we have already shrunk Huber's interval (20.5 to 22.0) down to a single point, x_h. What sort of c.q.g. would we expect from this material? Well, the m replicates of x_h will of course yield m identical cosine contributions to

(4.1), which will reinforce one another with zero interference, and at any τ value which is an integer multiple of $1/x_h$ they will make a contribution $+m$ to the summation occurring in that formula. It will be clear from our previous discussions that the remaining n terms will contribute to the same summation a term which we might write as

$$\sqrt{(\tfrac{1}{2}n)}\,S,$$

where S is a random variable drawn from the distribution pictured in figure 11. Thus we shall have

$$\phi(\tau) = \sqrt{(1-m/N)}\,S + \sqrt{(2/N)}\,m,$$

and therefore we shall obtain a peak of height 4 (equal to that of Mount Thom) if and only if

$$S = \frac{4}{\sqrt{(1-m/N)}} - \sqrt{\left(\frac{2}{N-m}\right)}\,m.$$

Let us now insert the values $m = 12$, $N = 169$. The formula then gives $S = 2.795$. Judged by figure 11, which is now appropriate to the problem, this is a typical rather than an extreme value for S, *but in order to obtain the value 4 for the peak-height in the c.q.g.* it is necessary for this S value to be attained at one of the multiples of $1/x_h$, and further at a multiple of $1/x_h$ which lies in the interval $0.09 < \tau < 0.59$.

Now in our problem x_h is about 21.25, and its submultiples x_h/k yield τ values in the required range for $k = 2, 3, \ldots, 12$; that is, at 11 sites within the range. To achieve a value $\phi = 4$ it is necessary and sufficient that for the simulated data, S should be about 2.795, *and that* this supremum should occur at or sufficiently near one of these 11 sites. Looked at in this way, our record of two peaks of height four or more out of the ten simulated trials will be seen to be quite reasonable.

(iii) We now turn to some other matters relating to other ways of breaking down the material, noting, as of course we must, that if there is one thing worse than having only limited data, it is a necessity to dissect it. Let us first consider the interesting question, what happens when we look separately at the data-sets

$$SEW_1, \quad SEW_{2-1}, \quad \text{and} \quad SEW_{3-2};$$

here the first consists of 112 accurate circles (diameters known to within ± 1 ft), the second consists of 57 less accurate circles, and the third consists of 42 'eggs'. It will be interesting to see how far they agree in pointing to 5.44 ft as a quantum, and also how far the statistical size of the 'errors', described as ϵ in (3.1), increase as we proceed from the accurate towards the less accurate or less accurately defined data. We must first explain how the statistical size of ϵ is to be calculated.

If we insert (3.1) into (4.1) with $\tau = 1/q$, we clearly obtain for $\phi(\tau)$ a multiple $\sqrt{(2N)}$ of an averaged value for $\cos(2\pi\tau\epsilon)$. Let us then adopt the lumped Gauss rather than the von Mises model, and suppose that the ϵ values have a standard deviation equal to σ ft. Then the expectation value for $\cos(2\pi\tau\epsilon)$ will be

$$\exp(-2\pi^2\sigma^2/q^2),$$

so that if we write the equation,

$$\text{peak height at quantum} = \sqrt{(2N)}\exp(-2\pi^2\sigma^2/q^2),$$

then this provides us with an estimate for σ, and the estimate obtained in this way for SEW_2 is, in fact,

$$\sigma = 1.507 \text{ ft.}$$

It is highly significant that this figure is of the order of the dimensions of a single stone.

Now let us apply the same technique to the 'split' data. We obtain the following results.

TABLE A2. SPLITTING THE DATA

data	N	height	q value	σ
SEW_1	112	3.482	5.435	1.477
SEW_{2-1}	57	2.242	5.459	1.535
SEW_{3-2}	42	5.359	5.435	0.896

These results are of great interest. They show, first, that the accurate and the less accurate circles indicate the same quantum, *and the same error*; that is, the effect of the greater error in measuring the less accurately defined circles is 'swamped' by the major source of error (which I believe to be stone-size). Secondly, they show that while the 'eggs' indicate the same value for the quantum, they also indicate an appreciably smaller size for the error.

There are clearly two possible interpretations here, and I have no inclination to take sides on this issue. *Either* all layouts are eggs, and the higher errors for the circles arise from trying to squeeze an egg-shaped foot into a circular shoe, *or* the greater freedom of choice available when fitting an egg (e.g. when deciding of how many circular arcs it is to be composed, and so on) has resulted in getting an unnaturally good fit to the data by unconscious selection.

At least it is clear that I was wise to leave the 'eggs' out of my primary analysis. We can present the situation in an alternative, pictorial, fashion because my computer program among other things works out the 'remainders' when the best integer multiple of 5.442 ft is removed from each of the individual observations, and then plots a periodic spline transform (see Boneva, Kendall & Stefanov (1971) for this) showing the distribution of these residuals on the same scale as figures 1 and 2 of the main text. The results for SEW_1, SEW_{2-1} and SEW_{3-2} are shown together in figure A3, with the lumped Gauss curves as a background. It will be seen that the distributions are very similar for SEW_1 and SEW_{2-1}, and of lumped Gauss/von Mises shape, but that the distribution for SEW_{3-2} is *very* much more highly peaked near zero, so much so that its plot goes off the scale altogether. This picture, of course, just tells us the same story in a different way.

(iv) We now dissect SEW_2 along another dimension, this time a geographically meaningful one, into the Scottish data (S_2) and the English and Welsh data (EW_2), the respective sample sizes being 109 and 60. As the sample sizes are so unequal, it seemed a useful idea to split S_2 into two sets of sizes 54 and 55 respectively; these were called S_2A and S_2B. Various kinds of c.q.g. analysis are now possible.

TABLE A3. SCOTLAND VERSUS ENGLAND AND WALES

data	N	height	q value	σ
S_2	109	3.748	5.435	1.433
EW_2	60	2.067/ 1.697	4.53/ 5.46	*/1.677 (two peaks)
$S_2A.EW_2$	114	3.680	5.411	1.447
$S_2B.EW_2$	115	2.917	5.459	1.578

It will be seen that the English and Welsh data does give some support to a quantum at the Thom value, but that it also produces a slightly higher peak at about 4.5 ft. However the sample size is here very small; this suggested 'mixing' it with each of the two halves of S_2 in turn, with the results shown above. There is then consistent evidence pointing to $q = 5.4$, with the usual σ of about 1.5 ft. Notice that S_2 and each of the two 'mixed' data-sets involve only about 113

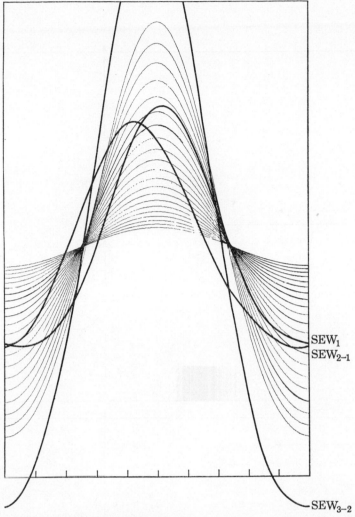

FIGURE A3. Periodic spline transforms for the 'residuals' in the cosine quantogram analysis of SEW_1, SEW_{2-1} and SEW_{3-2}, shown against a background of lumped Gauss distributions.

observations. Thus for comparison with figure 11 there would be a good argument for multiplying up the corresponding peak heights by the factor $\sqrt{(169/113)} = 1.223$, thus raising the least of them to 3.57 and the largest of them to 4.58.

As for EW_2 by itself, if we look only at the peak near the Thom quantum, the factor to allow for the small sample size (60) would be $\sqrt{(169/60)} = 1.678$, which would raise the peak height to only 2.85. My inclination is to suspect that the evidence for the quantum resides in the Scottish data *only*. If this be true, it is of some archaeological importance.

These results seem to indicate that the evidence for the quantum really resides in the Scottish data set S_2 of 109 observations. It is therefore desirable to repeat the c.q.g. analysis in full for this smaller sample, and to re-estimate the statistical significance of the result. Figure A4 shows a c.q.g. over the 'official' range $0.09 < \tau < 0.59$, from which it will be seen that the peak near 5.44 ft is still the dominant one. To permit a more accurate determination of peak height, the analysis was repeated with the narrower τ-range $0.13 < \tau < 0.23$, and the result is shown in the *upper* curve in figure A5. Here the peak occurs at $q = 5.435$ ft, and is of height 3.748 units; the estimated value of σ comes out to be 1.43 ft. On referring to our sample of 600 S values

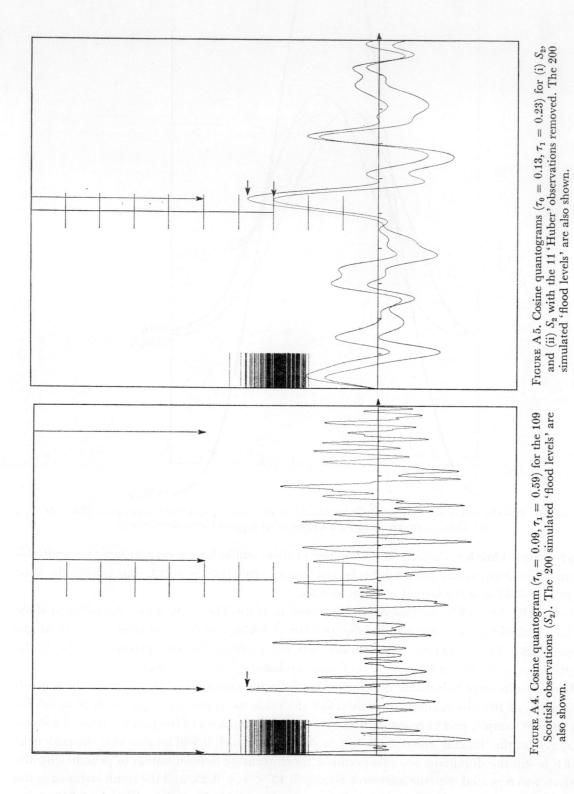

FIGURE A5. Cosine quantograms ($\tau_0 = 0.13$, $\tau_1 = 0.23$) for (i) S_2, and (ii) S_2 with the 11 'Huber' observations removed. The 200 simulated 'flood levels' are also shown.

FIGURE A4. Cosine quantogram ($\tau_0 = 0.09$, $\tau_1 = 0.59$) for the 109 Scottish observations (S_2). The 200 simulated 'flood levels' are also shown.

(whose distribution is, we know, very little affected by the value of N), we find that an S value of 3.748 is reached or exceeded just 13 times, so that the significance level is now 13/600 or about 2%. That is, *even if we throw away the English and Welsh data*, we still find that Mount Thom demands to be taken seriously. Unfortunately 11 of Huber's 'knot' of observations in the interval (20.5, 22.0) belong to S_2 so that as before we have to exercise considerable caution in interpreting these results. We therefore look at what remains of S_2 when the 11 diameters belonging to H are removed from it, leaving us now with only $109 - 11 = 98$ observations. The corresponding c.q.g. is to be seen in the *lower* curve in figure A5. Here the peak height is 3.004, the location of the peak is at $q = 5.441$ ft, and $\sigma = 1.52$ ft. Once again the familiar values of q and σ are found, but the loss of the 11 observations near $4 \times 5.442 = 21.77$ ft has eroded the peak to a much lower value, 'typical' rather than 'extreme'.

At the same time we must bear in mind that we have 'thrown away 11 of our best observations'; the successive paring away of observations from the initial sample size of 169, even if there *were* a genuine quantal effect, would by the 'square-root law' which we have explained above necessarily cut down the height of Mount Thom to a value comparable to the noise level. This is seen best by noting that

$$4.0 \times \sqrt{(98/169)} = 3.046.$$

Exactly the same point is made in another way by noting the stability of the σ value. Those who consider that the 'quantal effect' is merely an artefact caused by the fortuitous congregation of about a dozen circle diameters near to 21.77 ft have to find some convincing explanation for the invariability of σ.

(v) Professor Huber and several others mentioned the possibility of splitting the data by circle size, and this is obviously worth doing, although we must (to preserve sample numbers) work with the complete data set SEW_2, and avoid trying to split by *both* circle size and geography. To get three nearly equal groups I took three sets of SEW_2 circles:

> 58 small, diameters from the smallest up to 35.6 ft;
> 55 medium, diameters from 37.6 ft up to 68.4 ft;
> 56 large, diameters from 69.1 ft up to the largest.

These gave the following results.

TABLE A4. EFFECT OF CIRCLE SIZE

group	N	height	q value	σ
small	58	1.590	5.371	1.672
medium	55	2.586	5.429	1.446
large	56	2.937	5.447	1.388

The stability of the quantum is quite remarkable, and no less so is (*a*) the stability of the order of magnitude of σ, and (*b*) the fact that σ if anything *decreases* as the circle size increases. As has already been explained in the text, variation in the length of the quantum from one measuring out of a quantum's length to the next, along a radius, would produce a $\sqrt{X}$ dependence in σ, while a variation of the quantum from one site to another would produce an X dependence in σ, yet we find neither. This confirms what I have already suggested; that the 1.5 ft value of σ merely reflects the inherent uncertainty in the location of the starting or end point of the measurements (due to the non-negligible size of the stones).

(vi) This brings us to the vexed question of 'pacing'. Here I have received a great deal of help from Dr H. J. Case, who has consulted his contacts in the Brigade of Guards and reports as follows.

'My drill-sergeant friend tells me that trained Guardsmen should be able to maintain a pace of 30 inches $\pm \frac{1}{2}$ an inch on parade, and even when off duty in the vicinity of the parade ground. For Infantry of the Line and Royal Artillery about $\pm 1\frac{1}{2}$ to ± 2 inches would be acceptable, depending on the degree of training. These figures are heel to heel whereas ours would be toe to toe presumably, but I do not see how that would make a difference. However these military exercises are on asphalt, whereas ours would be on tussocky ground at best and on some degree of slope; but rough ground might not greatly affect the performance of fit and trained men. Guardsmen may be trained by lines painted on the parade ground 30 inches apart and their pace is checked against the drill-stick, but a competent drill-sergeant will detect consistent deviations of ± 1 inch instantly by eye. . . . I would be inclined to think that ± 2 inches was about right for trained men on rough grass.'

This fascinating information left one immediate impression upon me; if guardsmen 'pace' so accurately, but are trained by a drill-stick, then in their case there is very little difference between 'pacing' and 'quantum'; the drill-stick *is* their quantum, even if not physically present but only as a threat, as it were. However, it seemed that it would be interesting, having simulated Professor Thom, to proceed to simulate the Brigade of Guards, and accordingly I modified my simulation program so that it would act as follows.

First it constructs an X-value using the simulator (5.3), and then it computes M, the nearest whole number of Thom-quanta of 5.442 ft thereto. It then adds together M versions of the Thom quantum, *each one of these*, however, being perturbed by an error of mean value zero and standard deviation 0.2357 ft. This last figure was chosen because it is $\sqrt{2}$ multiplied by 2 in. The result obtained from a c.q.g. based on a sample of 169 such diameters was:

$$\text{height} = 13.589, \quad q \text{ value} = 5.441 \text{ ft}, \quad \sigma = 0.674 \text{ ft}.$$

Thus 'pacing' is *far too accurate*, or rather, 'pacing' may have been used, just as a whale-bone rod may have been used, but we shall never know, because the residual errors associated with the sizes of the stones completely swamp errors of the kind associated with good pacing.

Incidentally the reader may be curious to know why one obtains an estimated σ of 0.674 ft; the answer is that $0.674/0.2357 = 2.86$, and the square of 2.86 is about 8. Now 8 quanta make up some 43.5 ft, which is of the same order as the average circle-diameter. Thus there is consistency among the figures, after all.

(vii) If we assume that the errors ϵ are *not* associated in their average magnitude with circle size (and to this conclusion we were driven by the discussion at (v) above), then it is of course of some interest to simulate a set of 169 circles with the ϵ values deliberately controlled in this way. To be specific, we use the randomizer (5.3) to obtain 169 circle diameters X, we compute the nearest integer $\leqslant X/q$, and if this integer is M we replace X by

$$X' = Mq + \sigma\epsilon,$$

where σ is at our choice and ϵ is a random standardized Gaussian random variable. Here $q = 5.442$ and σ is fixed throughout the calculation, but the ϵ values vary randomly from one circle diameter to the next and represent the errors 'associated with the sizes of the stones'. Figure A6 shows the resulting c.q.g over the range $0.13 < \tau < 0.23$ (this narrow range was

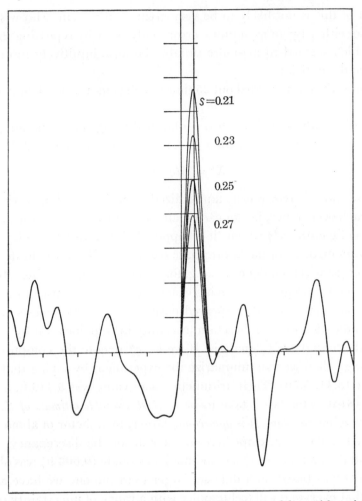

Figure A6. Cosine quantograms ($\tau_0 = 0.13, \tau_1 = 0.23$) for simulated *quantal* data with various values for $s = \sigma/q$.

chosen to show up the actual profile of the resulting peak); the four peaks shown have values of σ such that $s = \sigma/q = 0.21$, 0.23, 0.25 and 0.27. With $s = 0.27$ we will have $\sigma = 1.47$ ft, which is the 'observed' value, and it will be noticed that the simulated data then generates a peak of height 4 units, as of course it ought to do if our calculations and simulations are consistent with one another. It will be clear from figure A6 that the peak height is a very sensitive indicator of the true value of σ, so that we can have some considerable confidence in the value $1\frac{1}{2}$ ft which we have found for the latter. Incidentally it should be remarked that while the four peaks have been drawn separately, the c.q.g. throughout the rest of the range has here (for the sake of clarity) been drawn for one s value only. The differences elsewhere were very slight.

   (viii) We now turn to a question which has excited considerable controversy. If we accept that there *is* a quantum, of value about $q = 5.44$ ft, how *accurate* is this estimate? The matter is of great importance when we compare estimates of the quantum from different geographical regions (e.g. Scotland, and Brittany). We are not concerned with that question here (apart from our discovery that the evidence for any quantum at all in England and Wales is pretty shaky), but the standard error to be attached to the figure 5.44 remains an important matter to which we must give some attention. The analysis which follows was carried out before the primary importance of the Scottish data was recognized, and so uses an N value of 169 instead of 109. The effect of this is not likely to be very great, and anyone who wishes to examine it could readily do so, either by an asymptotic error analysis, or by repeating the type of calculation used here, which seemed to me more suitable because intuitively more readily comprehensible by non-mathematicians.

The following procedure was carried out 25 times, each computation being quite independent of the others.

First, a sample of X values of size 169 was obtained using the simulator (5.3). For each X value, the nearest integer M to X/q (with $q = 5.442$) was computed, and X was then replaced by

$$X' = Mq + \sigma\epsilon,$$

where $\sigma = 1.507$ ft and where ϵ was a standardized Gaussian random variable, drawn 'fresh from the well' at each occurrence, i.e. for each diameter. Thus each X in the original sample was replaced by a 'target length' Mq which (it is supposed) it was desired to lay out, and this was then perturbed by an error $\sigma\epsilon$ of the appropriate size (arising from the finite size of the stones, etc). On computing (as was then done) a c.q.g. for this sample of X'-values, 169 in number, we would *not* of course expect to get a peak lying exactly at 5.442, because of the perturbing effect of the ϵ's. The fluctuations among the observed positions of the 25 peaks so obtained provides one with a direct indication of the inherent uncertainty in peak location by this method, in the circumstances of the actual SEW_2 analysis. If we let $q*$ denote the computed location of the peak, using the c.q.g., then we can summarize the experiment by saying that the 25 values for $q*$ lay in the range (5.41, 5.50). Their arithmetic mean value was 5.446 ft, and their standard deviation was 0.0181 ft. This last I take to be *the best available estimate of the uncertainty in our knowledge of the value of the quantum*. It is *appreciably* larger, by a factor of about 3, than estimates offered previously. I make no attempt here to account for the discrepancy, because I do not claim to understand the method by which the smaller estimate (0.006 ft) was obtained, although it seems clear that it was based on a data-set larger than the one we have allowed ourselves. I hope, however, that everyone will feel happier with a range of uncertainty of about 1 in. The reader should note carefully that this is the degree of uncertainty *in our knowledge of the average value of the quantum*; it has *nothing to do* with how far one quantum may have varied from the next, as they were successively laid out along a radius or diameter, and it has *nothing to do* with how far one quantum-standard in the northwest of Scotland may have differed from another in, say, Carnac. I make these rather obvious remarks because the discussion at the meeting concerning 'the standard deviation of the quantum' was bedevilled by some lack of comprehension of these important (and really quite elementary) distinctions.

(ix) I conclude with a few very brief remarks intended for the mathematical reader. They are deliberately brief because this paper is already much longer than it strictly should have been, and also because my theory and algorithms are both so elementary and naïve that any mathematician could readily fill in the gaps and repeat these computations for himself, on the same or different data, if he were so minded and could spare the time.

The first point to make is that we can *either* accept (4.1) as a natural functional statistic to work with (and this, on the whole, is the point of view I personally favour), *or*, if we feel we must have a 'reason' for what we are doing, one can be found within the framework of likelihood theory (Edwards 1972). If X is an observed circle diameter, we write $X = (M+Y)q$ (where q is the quantum); here M is a random integer and Y is a random variable in the interval $(-\frac{1}{2} < Y < \frac{1}{2})$, and we take $2\pi Y$ to have the von Mises distribution with parameter k. As I have already explained, this is effectively equivalent to assuming a lumped Gaussian distribution for $\theta = 2\pi Y$, and that perhaps is how we should think of the matter in the first instance, but the switch to the von Mises distribution is analytically permissible (because of the famous empirical 'parrot' effect) and for the purposes of maximum likelihood it is a change for the better. We can have a β effect (in the sense of Thom) if we write $X = c + (M + Y)q$, where $0 \leqslant c < q$, and this will merely induce a phase shift of amount $\beta = 2\pi c/q$ radians into the von Mises distribution. For the sake of generality I will leave that effect in, for the moment.

The likelihood of the observations can now be written in the form

$$\prod_{j=1}^{N} p_{[x_j/q]} \, K(k)^N \exp\{kN(A\cos\beta + B\sin\beta)\},$$

where $[x]$ denotes 'the integer closest to x', $K(k)$ is the function in (3.2) (simply expressible in terms of modified Bessel functions), and

$$A = (1/N)\,\Sigma\cos\theta_j, \quad B = (1/N)\,\Sigma\sin\theta_j$$

(here θ_j is $2\pi Y_j \pmod{2\pi}$). The probabilities p_m which occur within the product represent the probability that a length mq will be 'required' by the 'architect'.

The maximum likelihood estimate of β can be obtained from

$$A = C\cos\hat{\beta}, \quad B = C\sin\hat{\beta},$$

where C is the non-negative square root of $A^2 + B^2$. If we carry out this preliminary maximization, we have next to maximize with respect to $k > 0$; it turns out that there *is* a unique maximum, given in fact by

$$I_1(\hat{k})/I_0(\hat{k}) = C,$$

or by

$$\hat{k} = 2C(1 + \tfrac{1}{2}C^2)$$

when, as is normal, C is small. Substitution of this value for k then gives the conditionally maximized likelihood

$$\prod_{j=1}^{N} p_{[x_j/q]} \exp\{NC^2\}.$$

We now have to compare this with the likelihood on the non-quantal hypothesis,

$$f(X_1)f(X_2)\ldots f(X_N), \text{ say.}$$

But for q not too small, we shall have, approximately,

$$\prod_{j=1}^{N} p_{[x_j/q]} = q^N \cdot f(X_1)f(X_2)\ldots f(X_N),$$

and so, constants apart, *the relative support for a value* $\tau = 1/q$ will be

$$\mathcal{S} = \tfrac{1}{2}\{\phi(\tau)^2 + \psi(\tau)^2\} - N\ln\tau.$$

We recall that we shall only be interested in the behaviour of this function over the range $\tau_0 < \tau < \tau_1$ agreed in advance.

We now observe that, when we are far enough away from $\tau = 0$, the stochastic processes $\{\phi(\tau)\colon \tau_0 < \tau < \tau_1\}$ and $\{\psi(\tau)\colon \tau_0 < \tau < \tau_1)\}$ are effectively stationary Gaussian stochastic processes with zero mean and unit variance, and what is more, their auto-correlation functions and cross-covariance functions can all be estimated (at least for small u) in terms of $A(u)$ and $B(u)$ (where here we recognize the dependence of A and B on $u = 1/q$). Figure A7 shows the

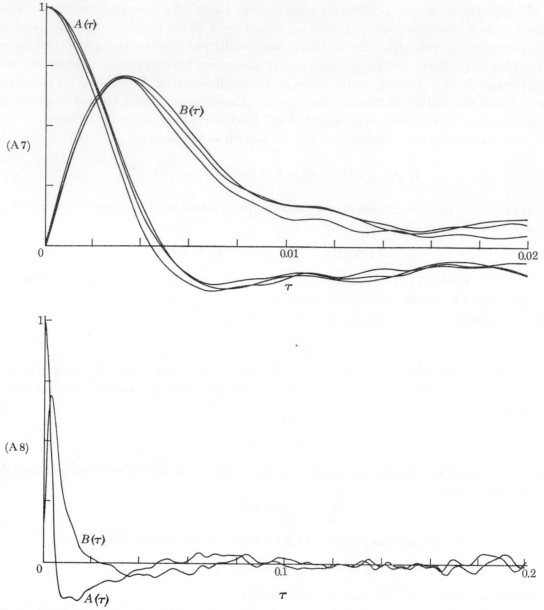

FIGURE A7. Estimates (based on $SEW_{1,2,3}$) of the autocorrelation function $A(\tau)$ for each of the ϕ and ψ stochastic processes, and of their cross-covariance function $B(\tau)$.

FIGURE A8. This is similar to figure A7, but the computations are now based on simulation-sets of size 2000, and extend over a wider range (over which estimates based on the small real data-sets would be quite unreliable).

A and B functions for $0 < \tau < 0.02$ for each of $SEW_{1,2,3}$, while figure A8 shows the A and B functions for $0 < \tau < 0.2$ for very large samples generated by the simulator (5.3). It will be seen that both functions fall rapidly to zero; the peak of the B function near $\tau = 0.004$ corresponds to a systematic lag between the stochastic processes ϕ and ψ, which are otherwise essentially independent of one another.

From these considerations we learn that

$$U(\tau) = \tfrac{1}{2}\{\phi(\tau)^2 + \psi(\tau)^2\}$$

is a stationary stochastic process each of whose individual ordinates has approximately a negative-exponential distribution with unit parameter. An initial analysis, by what I call the modular quantogram (m.q.g.), consists in making a computer plot of $U(\tau)$ against τ over the appropriate τ range, and looking for peaks which are large in relation to the negative-exponential noise. These must then be assessed more carefully, using a Monte Carlo study of the supremum-functional for the U process (I have not done this), and if there are competing significant peaks they can then be compared on the basis of the $ values. The last step is facilitated in my program by the drawing of constant $ loci (isopithons) on the computer output; all this is done automatically, of course. In my version the isopithons are computed from the approximate formula given above, but it would be a simple matter to compute them exactly if a Bessel function sub-routine is available.

If a significantly high peak on the m.q.g. is found, the β-effect must next be tested for. This is done by testing the value of $\psi(\tau)$ at the peak τ as a standard Gaussian variable; if it lies within the range ± 2 we reject the β-effect as an unnecessary complication, and this is what happened with the Thom data.

We then go right back to the beginning and set $\beta = 0$ (no β maximization is now called for), and then proceed as before. This time we find that

$$\$ = \tfrac{1}{2}\phi(\tau)^2 - N \ln \tau, \quad \text{when} \quad \phi(\tau) > 0,$$
$$= -N \ln \tau, \quad \text{when} \quad \phi(\tau) \leqslant 0.$$

Accordingly we now have to examine the plot of $\phi(\tau)$ against τ over the agreed τ range, that is, we have to look at the c.q.g., and the formula for the isopithons becomes

$$\phi = \sqrt{\{2(\$ + N \ln \tau)\}}$$

for so long as the expression within the square-root is non-negative. Detailed study of the argument shows that when the isopithon hits the τ axis, then it is to be continued downwards in a vertical straight line. (Of course this last detail is of no importance, because we shall never consult the isopithons unless we have to discriminate between significantly high positive peaks of ϕ-value 3.5 or more.)

It will not escape the reader that a peak at $\tau = \tau^*$ must imply a peak of sorts at $\tau = 2\tau^*$, $3\tau^*$, and so on for higher harmonics. I wasted a lot of time (before the likelihood argument was evolved) in working with a linear geometrically discounted combination of peaks and their harmonics, like

$$(1 - \rho) \sum_{m \geqslant 0} \rho^m \phi(m\tau).$$

This is *possible*, because the series can be summed in finite terms and so computer-plotted, but I found it quite futile, and I am now satisfied that there is no good theoretical reason for doing anything of the sort. Professor Whittle pointed out to me that if one used a lumped Gaussian distribution for the ϵ *and neglected all but the central ('unlumped') term* (without which approximation the likelihood argument gets stuck right at the start), one winds up with Broadbent's 1955–6 functional statistic (see formulae at the foot of p. 50 in Broadbent (1955)). This can also be regarded as a linear discounted combination of the c.q.g. and its harmonics, but very curiously the 'weights' in Broadbent's linear combination alternate in sign, which seems not natural to me; also they tend to zero very slowly indeed. This did not matter to him because his series, too, can be summed, and it was in that form that he used it. Those who wish to compare and contrast Broadbent's work with mine should be warned that on his typical diagram it is the *minima*, and not the maxima, which have to be looked for. He used, as I do, a Monte Carlo basis for assessing significance, and proposed a partial ordering in the plane to assist in the comparison of rival significant peaks. This last device I do not need, because the isopithons essentially play that role, if it be needed at all.

The last mathematical topic on which I should like to touch very briefly concerns the asymptotic formula of Cramér (Cramér & Leadbetter 1967) for the limiting distribution of the supremum of a section of a 'well-behaved' Gaussian stationary standardized stochastic process. This formula is very beautiful, but so far as I am aware no one in the world has the slightest idea how accurate it is in finite, that is, non-limiting, circumstances. Our Monte Carlo calculations give us an opportunity to throw a first gleam of light on this matter, and it is obviously an opportunity not to be missed.

If we write $P(s)$ for the probability that the supremum random variable S will have a value *s or less*, then Cramér's limit theorem asserts that

$$\lim_{s \to \infty} P(s) = \exp(-e^{-z}),$$

where his z is defined in such a way that, in our notation,

$$e^{-z} = (\tau_1 - \tau_0)\ \text{r.m.s.}\ (X)\ e^{-\frac{1}{2}s^2},$$

and it is supposed that $(\tau_1 - \tau_0)$ is being allowed to increase, as s increases, at such a rate that z is held fixed. This is not, of course, a very helpful statement when we come to apply the formula! A word of explanation is required about r.m.s. (X); this means, as the notation indicates, the root-mean-square value of the circle diameters for the sample (or more strictly, for the coarse-grained distribution from which they are drawn). In Cramér's formula what appears is in fact the absolute value of the second derivative of the autocorrelation function at the origin, which is equal to the expectation of $(2\pi X)^2$. It is interesting to note that the sample-size N nowhere appears in the Cramér formula, and that the only trace of the identity of the data-set used is in the value of r.m.s. (X). For this I have used the value 65.433 ft which was computed from (5.3); that for S_2 would have been 65.4 ft, while that the SEW_2 would have been 72.2 ft. With these explanations the reader will be able to adapt the arguments and charts which follow to suit other data-sets and other values of τ_0 and τ_1.

Now while we do not expect Cramér's formula to be valid in our particular circumstances, because we are not (*and never will be*) in the limit situation, we may none the less reasonably expect that its general analytic form may indicate a helpful way of graduating the information gained empirically from our Monte Carlo experiments using the simulator (5.3), for which r.m.s. $(X) = 65.433$ ft. To be specific, we can usefully look at the experimental data we have accumulated in this way and see if there is any sign of a linear relation between the transformed variables

$$x = 2000\ (2\pi)^{-\frac{1}{2}}\ e^{-\frac{1}{2}s^2},$$

and

$$y = 20 \ln (1/P(s)),$$

where we estimate $P(s)$ by

$$(\text{number of empirical } S \text{ values} \leqslant s)/600.$$

If Cramér's formula were to hold exactly, then we should have the relation

$$y = \{\sqrt{(2\pi)} \times (65.433/100) \times (\tau_1 - \tau_0)\}\, x,$$

or in numerical terms,

$$y = 0.820\, x.$$

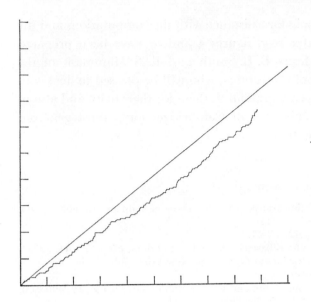

FIGURE A 9. The (y, x) plot corresponding to the simulated sample of 600 S values, compared with the straight line which represents the Cramér limit. On the horizontal scale, $x = 2000$ times the 'standardized Gaussian density' at location s, while $y = 20 \ln (1/P(s))$, where $P(s) = \mathrm{pr}(S \leq s)$, and natural logarithms are used. The diagram relates to the situation $\tau_1 - \tau_0 = 0.5$, and r.m.s.$(X) = 65.433$. In other situations the ordinates should be increased in proportion to $(\tau_1 - \tau_0)$ r.m.s. (X).

FIGURE A 10. This is a tenfold enlargement of the bottom left-hand corner of figure A 9. The scales are as before, but significance levels $100(1-P)$ are now indicated.

Figure A 9 shows that the actual plot of y against x falls below this line, but approaches it very closely near $x = y = 0$ (which is the 'limit situation'). The diagram covers a range 0.00–8.86 for x, and a range 0.00 to 6.57 for y; thus at the top right-hand end of the stepped curve we have $P(s) = \exp(-0.3285) = 0.72$, so that the diagram in fact represents the upper 28% of the S distribution.

Figure A 10 shows the situation near $x = y = 0$ enlarged tenfold. In this diagram the top right-hand end of the stepped curve corresponds to $P(s) = \exp(-0.02703) = 0.9733$, so that the enlarged diagram represents the upper 2.7% of the S distribution. It is this last chart which is most likely to be useful in practice. If the stepped curve is used to estimate significance levels, care must be taken to expand its *ordinates* in proportion to r.m.s. (X) $(\tau_1 - \tau_0)$ (the plotted ordinates correspond to 65.433 (0.59–0.09) = 32.7165 for this parameter). Thus if we wanted to work with the doubled range, $\tau_0 = 0.09$, $\tau_1 = 1.09$, we should *double* the ordinates of the stepped curve, thus effectively *squaring* the estimated value of $P(s)$ (as is obviously reasonable).

Finally, note carefully that the 'significance level' is not $P(s)$, but $1 - P(s)$. The following table will be found useful in connexion with figure A 10.

TABLE A 5. SIGNIFICANCE LEVELS AND THE (x, y) DIAGRAM

$1 - P(s)$ (%)	0.1	0.5	1.0	1.5	2.0
y	0.020	0.100	0.201	0.302	0.404

It will be seen that with our parameter-values, the Cramér limit formula is (perhaps by chance) astonishingly accurate for significance levels of 1% or better.

In conclusion I wish to thank Miss Mary Brooks for assistance with the computations and the preparation of the diagrams. Some of the latter (e.g. figures 4 and 5) have been prepared directly from computer-plotter output, by Messrs E. L. Smith and R. S. Hammans of the Department of Physical Chemistry, University of Cambridge, who will be pleased to deal with any enquiries relating to their technique. I am very grateful to them for their help, and also to the Director of the Computer Laboratory, University of Cambridge, for a most generous allocation of 'space' and 'time' to this investigation.

REFERENCES (Kendall)

Boneva, L. I., Kendall, D. G. & Stefanov, I. 1971 Spline transformations: three new diagnostic aids for the statistical data-analyst. *J. R. statist. Soc.* B **33**, 1–70.

Broadbent, S. R. 1955 Quantum hypotheses. *Biometrika* **42**, 45–57.

Broadbent, S. R. 1956 Examination of a quantum hypothesis based on a single set of data. *Biometrika* **43**, 32–44.

Cramér, H. & Leadbetter, M. R. 1967 *Stationary and related stochastic processes*. New York: John Wiley.

Edwards, A. W. F. 1972 *Likelihood*. Cambridge University Press.

von Mises, R. 1918 Über die 'Ganzzahligkeit' der Atomgewichte und verwandte Fragen. *Phys Z.* **19**, 490–500.

Stephens, M. A. 1963 Random walk on a circle. *Biometrika* **50**, 385–390.

Thom, A. 1955 A statistical examination of the megalithic sites in Britain. *J. R. statist. Soc.* A **118**, 275–295.

Thom, A. 1967 *Megalithic sites in Britain*. Oxford: Clarendon Press.

Contributions to the discussion on ancient astronomy: the unwritten evidence

A. H. A. Hogg (*Royal Commission on Ancient and Historical Monuments in Wales and Monmouthshire, Aberystwyth*) referred to Professor Thom's hypotheses. These were not inseparable; some could be accepted while rejecting others. There was, for example, nothing inherently unlikely in the existence of markers related to astronomical events which could be observed during a single adult life, but it was very hard to believe that in an illiterate society numerical information could be stored up and transmitted to successive generations. The credibility of a theory could properly be taken into consideration, as well as the statistical evidence. For example, a very significant statistical correlation appeared to exist between the occurrence of severe earthquakes and the position of Uranus (Tomaschek 1959), but few people would accept this as other than accidental.

Turning to particular points, he said that, even after Professor Kendall's demonstration that a quantum of 5.44 ft apparently existed, he still felt some doubts. To accept the existence of an approximate quantum was not difficult; what seemed incredible was the extreme accuracy with which it was established, a few hundredths of an inch – and which was supposed to have been maintained over several centuries. He asked what standard deviation for the quantum was indicated by the new investigation† and whether the appearance of a quantum could perhaps arise out of the occurrence of a relatively few circles set out using some standard unit, mixed with a larger number where no standard was used. Referring to 'flattened circles' and 'eggs': the apparent ratio of 3:1 for the ratio of circumference to major diameter did not necessarily imply that the builders were aiming at this. Any slight flattening, for whatever reason, would cause the ratio to approach 3:1. Similarly, the wide range of possible geometrical constructions and of Pythagorean or near-Pythagorean triangles, combined with the inevitable uncertainty of setting-out a ring of large stones, would almost always make it possible to fit some 'theoretical' layout to the remains (*Archaeologia Cambrensis* 1968).

Finally, on the subject of 'fans' which were claimed as providing 'charts' for correcting lunar observations, he pointed out that the stones in some of these did in fact depart widely from the theoretical grid. Whereas this might be acceptable for truly megalithic structures owing to the difficulty of adjustment, for those where the stones were small and could easily have been moved to the required positions such differences could not be reconciled with the desire for extreme accuracy credited to the builders.

References

Tomaschek, R. 1959 *Nature, Lond.* **184**, 177 (criticized by E. J. Burr 1960 *Nature, Lond.* **186**, 336) with reply by R. T. E. J. B. accepts that the result appears significant at the 1 % level.
Archaeologia Cambrensis 1968 **117**, 207–210.

L. E. Maistrov (*Kaliningrad, Moscow*). As late as the beginning of this century in North Ossetia there was a place in each village or settlement where the inhabitants met to discuss matters of general interest. Even now there are some old people still alive who used this calendar in their

† See p. 259.

youth. This was located next to the house of worship, the entrance of which was always at the northern side. Near the door stood a bench on which the village elders would hold their discussions. In some villages these benches have been preserved to the present day. This entire area was known as 'Nykhas'. From this bench one can see the Sun disappearing behind the mountain range in the evening. In each village there was a man who would observe the Sun set each day. In the summer the Sun set farther and farther to the right every day. It then reached the farthest point to the right and started to set at points farther and farther to the left. The silhouette of a mountain range is so individual from every view angle that it is not difficult to memorize the place of the sunset at the farthest point to the right. When the sun returns to this point – a year has passed. The sun sets at this point on the longest day (summer solstice). Winter solstice occurred on the day when the sun set at the point farthest to the left. However, these were not the only days which were fixed in relation to the mountain silhouette. All holidays were similarly fixed in this way; in some villages there were many of these – in others less.

Days of equinox were established by fixing the position of the setting Sun on the day half way through the period between the solstices. If the sunset took place at a certain spot before the next movement of the point of sunset from right to left, this occurred on the day of the autumn equinox. In some villages the space between mountain peaks where solstice occurred, was halved and this point was fixed on the mountain profile. The day on which the Sun passed that point, while setting, was the day of the equinox.

About 40–50 days were fixed in this way; these were basically holidays. The method of fixing days was handed down from one generation to the other. This original mountain calendar contained 365 days (on an average). There was no division into weeks and months but the number of days which occurred between the various holidays was thus established.

The mountain calendar had an advantage in that there was no accumulation of mistakes which may easily occur when fixing the point of sunset on a certain place on the mountain profile; in fact, any such mistake was put right. Thus, if any one year included less than 365 days, the following year would include more than 365 days. In this calendar the mountain profile is essentially outlined against the firmament; the position of the centre of the Sun is then fixed in relation to this profile against the firmament.

A. PENNY AND J. E. WOOD. (*Admiralty Underwater Weapons Establishment, Portland, Dorset*). *Astronomical alinements associated with the Dorset cursus*. The purpose of this contribution is to describe the results of an investigation into the Dorset cursus (figure 1), which was mentioned by Professor Atkinson in his paper, and which has previously been described by him in detail (Atkinson 1955). We have attempted to study the cursus, not in isolation, but in its archaeological landscape. It is in an area noted for the number of barrows, both long and round. Chronologically it comes after, or at any rate at the end of, the spate of long-barrow building and before the round barrows. We have taken the date to be 2500 B.C.

The Thickthorn terminal is at the southwest end on a hill. It is the least damaged part of the cursus; unfortunately most of the rest has suffered badly from agriculture. Going northeast, the cursus descends a valley and then rises again to another hill, on which stands a prominent long barrow, Gussage St Michael III. The cursus banks actually pass on either side of this long-barrow, although not quite symmetrically, as there is a wider gap on the southern side. It

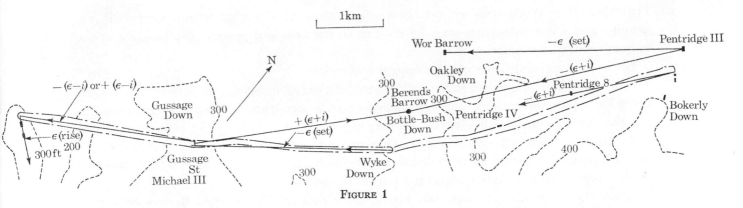

FIGURE 1

goes down another hill, with a distinct kink, and then rises slowly towards Wyke Down, where it is cut by a transverse bank. Continuing northward, the cursus changes direction, bends to incorporate another longbarrow, Pentridge IV, in one of its banks and finally, after one more deflexion, finishes on Bokerly Down just short of Bokerly Ditch. Not far from the last bend, Professor Atkinson discovered a long gap in the northern ditch and a smaller one in the southern ditch. Presumably there were corresponding gaps in the banks at these places, but here the cursus is totally destroyed and can only be followed on aerial photographs.

At the Bokerly Terminal the geometry is quite remarkable. The centre of the end of the cursus is at the intersection of lines along the axes of two very long longbarrows. Furthermore, the projection of one of these passes through yet another longbarrow, Pentridge III, and extended meets yet another nearly $2\frac{1}{2}$ km away.

The purpose of the cursus is obscure. It has been described as a processional way, associated with some funereal cult because of the number of longbarrows in its vicinity.

In this assembly, you will expect me to introduce astronomy, and indeed our results lead us to believe that there are astronomical alinements incorporated in the cursus. Using the conventional terminology we describe the alinements by labelling them with the declination of the Sun or Moon, corresponding to the rising or setting azimuths. Thus if one stands at the centre of the Wyke Down transverse bank and looks towards the Gussage St Michael III longbarrow this represents a $(-\epsilon)$ or midwinter sunset alinement. Last midwinter a few days after the solstice, the sun set slightly to the right of the longbarrow – in 2500 B.C. the Sun would have set in the gap between the east end of the barrow and the eastern bank of the cursus.

There is not time to describe the other seven alinements in any detail. From the Thickthorn terminal to Gussage St Michael III appears to be a lunar $(\epsilon - i)$ sighting; this alinement, incidentally, unlike the others could have been used in both directions. From the Bokerly Terminal to a barrow just outside the cursus is a $-(\epsilon + i)$ alinement. This barrow, Pentridge 8, is catalogued as an oval roundbarrow but ground inspection shows it to be almost certainly a longbarrow. Atkinson's gap in the north bank now appears to be to permit the sighting. The sight-line here is almost along one bank of the cursus. It would not have been possible to sight from the centre of the terminal and retain its meticulous geometrical layout.

Of all the roundbarrows in this area, only one is prominent on the horizon, Berends Barrow. It is tall and conical; we suggest it is not a conventional roundbarrow, but a deliberate lunar foresight. It is carefully placed to give two alinements, one from Gussage St Michael III, one from Pentridge III. When it was dug into by Colt Hoare in 1800 it contained a quantity of charred wood. Possibly it was a beacon to facilitate the lunar sighting. We must record that

Pentridge III to Worbarrow seems to be a solar alinement, somewhat to our surprise. If it is genuine and not accidental it must be the first in the area and precede the building of the cursus by even 1000 years.

Let us now speculate on a possible sequence of events:

In its first phase, the cursus shows both lunar and solar alinements, the three terminals each being backsights and existing longbarrows being used as foresights. The banks serve to emphasize the direction one should observe and may also delineate the area to be kept clear of trees and scrub so as to preserve the sighting. They may also symbolically link the backsights. The incorporation of Pentridge IV in the cursus is presumably to obscure it – for it would otherwise be the only prominent skyline object not part of the system.

Later, the observational scope was extended by constructing Berends Barrow in such a position as to give two alinements. This would not have been possible if the Bokerly terminal was to be one of the backsights, and so an adjacent longbarrow, Pentridge III, was used. Finally, the Neolithic period ended and the Bronze Age roundbarrows were built, not on hill tops as elsewhere, but on low ground, so that they did not confuse the sight-lines. Was this purely out of respect for the earlier function of the cursus, or does it imply a continued interest on into the Early Bronze Age?

REFERENCE

Atkinson, R. J. C. 1955 *Antiquity* **29**, 4–9.

H. L. PORTEOUS (*Department of Pure Mathematics, Liverpool University*) mentioned a hypothetical experiment he had performed to see whether Thom's statistical analysis of the data on which he had based his faith in the megalithic yard was sufficiently sophisticated to distinguish between an accurate quantum and pacing as a method for laying out stone circles. The conclusion drawn from this was that pacing had not been satisfactorily eliminated as a possible method, but further studies would be needed to clarify the position. (cf. Megalithic yard or megalithic myth *J. Hist. Astron.* February 1973).

W. S. REITH (6 *Baronsmead Road, Barnes, London S.W.*13). Professor A. Thom found a linear unit of 2.72 ft (32.64 in), which he called a megalithic yard, as a basic unit of construction of megalithic sites in Britain. Professor D. G. Kendall discussed and found acceptable a quantum of 5.44 ft (2 megalithic yards) as a unit of dimensions of megalithic sites based on Thom's measurements.

It is interesting, I think, to point out that in Assyria of Sargon's time (2300 B.C.) – which is roughly contemporary with the construction of the British megalithic sites – Oppert (1872, 1874) found evidence at Khorsabad of the use of a linear standard, named U (ahu) in the cuneiform script, which was equivalent to 10.8 in. Furthermore, Flinders Petrie (1934) reported a similar unit of 10.9 in used at Ushak, a unit 'eastern foot' of 10.8 in, and 'oscan foot' of 10.85–10.95 in.

One megalithic yard is equivalent to three such 'feet' (3 × 10.9 = 32.7 in), and six such 'feet' are equivalent to the quantum of 5.44 ft discussed by Kendall as an acceptable unit from Thom's measurements of British megalithic sites.

I think it is interesting to point out this equivalence of linear measures used at a similar time of construction of a number of sites in Assyria and in megalithic Britain.

References

Flinders Petrie, W. M. 1934 *Measures and weights* p. 6. London: Methuen & Co.
Oppert, J. 1872 *Asiatique* (6) **20**, 157–177.
Oppert, J. 1874 *Asiatique* (7) **40**, 417–486.
Thom, A. 1967 *Megalithic sites in Britain*, pp. 43, 77–80, 166. Oxford: Clarendon Press.

Concluding remarks

By D. G. King-Hele F.R.S.

It would be very difficult to produce a coherent summary of such a wide-ranging meeting, with so many excellent individual contributions, and I shall not try. Instead I should like to discuss certain features of the meeting that particularly appealed to me; and on this occasion I speak as a semi-mathematician and semi-astronomer, without any specialized knowledge of antiquity.

This meeting was intended to help in building bridges between science and history, or to be more accurate, between scientists and historians; and to convince both groups that bridges improve the quality of life – that each subject is enriched by an injection of the other. There were many examples, of which I will mention just three. First, Dr Needham, whose great work on *Science and civilization in China* is an example to us all, in the very best sense. Then there is Dr Newton's work on the Earth's rotation rate. This depends essentially on ancient manuscripts and similar material. But his results are of great scientific importance in the study of the Earth–Moon system, because the slowing down of the Earth's rotation depends on the orbital behaviour of the Moon, the shape of the oceans (which affects the tides), the goings-on in the Earth's interior, and probably other unknown factors. A decision between the conflicting theories of the Moon's origin, and an understanding of the Earth's internal workings, may well be brought nearer by studies similar to Dr Newton's: he has shown scientists how they must not neglect history. My third example is Professor Lamb's paper on climate and forest cover in the ancient world. Although he needed his expert knowledge in meteorology, over 80 % of the references in his paper are primarily concerned not with meteorology at all, but with fossil coleoptera, Welsh and Irish bogs, sightings of comets, the Nile floods, deep-sea oozes, analysing plant pollen, changes in sea level, the spread of spruce trees, and so on. And his results are also wide-ranging in implication, being important not only for solar physicists trying to decide how and why the Sun's radiation varies, but also for historians, because changes in climate have certainly contributed to the fall of empires in historical times, as well as governing the migrations of prehistoric peoples.

In the past many academic disciplines tended to become intellectual islands, on which were cultivated more and more specialized plants. Each island had a resident professor, and he and his islanders rarely looked outwards. (To judge from Dr Lewis's paper, this metaphor is literally true in Polynesia, where they live on islands and have academic secrecy about navigation.) To return from Polynesia to Academia, it is as well to recognize that this divisive insularity is still quite strong, not because anyone is being wicked but because it is so tidy and self-perpetuating, both administratively and intellectually, to have islands and avoid the traumatic experience of emigration. But a major advance in science most often comes when a bright scientist crosses a bridge to another island and applies what may be old skills to throw light on a new area. The same, I think, will happen more and more in archaeology. Many of the bridges already exist, including some new ones built in the course of this meeting, and the young archaeologists who find the right bridges to cross will arrive in pastures worth ploughing – or more likely in muddy fields!

The second feature of the meeting I should like to discuss is the question of an astronomical culture in non-literate ancient societies. Many people find this idea difficult to credit, but I do not find any difficulty, for three reasons. First, there is the obvious but often forgotten point that these stone-age people probably had very nearly as much brain power as we have (it seems best to avoid using the word 'intelligent' now that it has become so culture-linked and emotive). Human evolution goes back 2 or 3 million years, and it would be surprising if we had made great advances in brain power during the past 2 or 3 thousand years – 0.1 % of our total span. My second reason comes from the impression that we are today rather drunk with words: language was certainly a great invention, but words have perhaps attained too great a dominance, and this tends to blind us to more instinctive forms of culture. The founders of the Royal Society were conscious of this danger when they chose the motto *Nullius in verba*. My third and strongest reason for accepting the idea of an astronomical culture comes from personal experience, in that I find it natural myself to notice the setting points of the Sun and Moon, on the western horizon. For people living an outdoor life with fairly clear skies on a treeless plain or near a coast with hilly islands to the west, I can understand that some of them might feel a compulsion to set up markers – exercising their nascent scientific curiosity, you might say.

As the years went by, they might easily become fascinated with this 'cosmic game', and elaborate it to the limits of their intellectual power by defining the extreme limits of the sunset and moonset points, and perhaps detecting the small perturbation in lunar declination. This implies of course that they had leisure and security, or, as Professor Atkinson remarked in referring to Silbury Hill, a small percentage of their national income was a surplus, which later, in Roman times, would have had to go on defence expenditure. This reminds us that the supreme militarist, Caesar, has much to answer for: his *Gallic Wars*, though badly written, has brain-washed generations of schoolboy Latinists – including archaeologists? – into thinking that the Romans were 'superior' to the barbarians, just because they were more literate (and more efficient at slaughter). This tacit equating of literacy with intelligence now seems in need of drastic revision. It looks as though, more than a thousand years earlier, these 'barbarians' had probably hit upon an azimuthal astronomical sighting method which was far more accurate than the techniques developed in literate societies, such as the method of heliacal risings favoured by Babylonians and Egyptians, and beyond the intellectual grasp of a mere militarist like Caesar.

'Numeracy' is not needed when measuring setting points on the horizon. The arrangement of the stones is the system of numeration, as Professor Renfrew remarked.

The further steps required for eclipse prediction, though possible, seem to me less likely. The time scales are long, and it is not clear how the information would have been transmitted. Also the neolithic peoples of northwest Europe may have been relatively free of the religious imperatives which provoked the obsessive interest in eclipses in many early literate societies.

Another of Professor Thom's ideas, his concept of the megalithic yard of 1.66 m, is now more securely based as a result of Professor Kendall's expert quantum-hunting.

As this meeting went on, I felt that it somehow took on a life of its own and grew into a kind of commemoration, a tribute, to all those nameless but ingenious proto-scientists of antiquity: the Europeans who placed stones to mark astronomical sight-lines; the Maya who achieved an accuracy of 1 part in 2 million in timing Venus; the Polynesian navigators; the megalith builders who lifted slabs weighing 50 tonnes without smashing them; and those unknown Babylonians and Egyptians who still rule our daily lives as we divide the hours and minutes into 60 parts, and divide our years into months and weeks.

Finally, I should like to express the sincere thanks of the Royal Society to those who have come to this meeting, often from far away and at some personal inconvenience, and have presented so many excellent papers or contributed to the discussion. I am sure the Organizing Committee would wish me to voice our gratitude to Dr Roy Hodson, who has from the very beginning taken on much of the burden of organization; and also to the Royal Society's staff and especially Miss Ritchie, for providing such a smooth run-up to the finishing line. Everyone who attended will, I am sure, depart with wider horizons; I also hope everyone found the meeting as enjoyable as I did.

CONCLUDING REMARKS BY S. PIGGOTT F.B.A.

Like Mr King-Hele, I find it quite impossible to summarize even a part of our two-day meeting. He has touched on certain points of particular interest to him, and some of these were features which I found equally outstanding, so here I need mention only briefly our areas of coincidence or overlap. Engagingly, he introduced himself as a 'semi-mathematician and semi-astronomer without any specialized knowledge of antiquity': I can only say of myself that the British Academy could not have chosen among its prehistorians one less numerate than I.

But our hope in planning this meeting, as King-Hele has said, was that we might find common ground between disciplines too long thought to be wholly disparate; to see whether on closer and dispassionate inspection that yawning crevasse between the Two Cultures turned out to be only a crack in the snow. Dr Newton, lucid (as was Professor Kendall later) even to *me*, was of course deeply involved in history, and I shall come back to his reference to Archilochos; Dr Needham is as great an historian as he is a scientist, and with Babylonia, Egypt and the Maya we were involved in history, if only marginally so by reason of conditional literacy.

King-Hele also spoke of the unconscious tribute we paid to 'the nameless but ingenious proto-scientists of antiquity' and it was on these that the most controversial part of our programme centred. This afternoon we paid tribute not only to them, but also to the long, patient, accurate and modestly pursued work of Professor Alexander Thom, on which he has based a thesis which if accepted demands the recognition of considerable mathematical skills among the non-literate societies of northwestern Europe from the fourth to the second millennia B.C.; mute inglorious Newtons who somehow managed to command the labour and organization necessary to construct stone circles or alinements from the Bay of Biscay to the Arctic Ocean. Here, as subsequent discussion showed, however cogent his reasoning may be on purely mathematical grounds, many archaeologists, including myself, would feel that a great number of difficulties have not yet been faced in an evaluation of this hypothesis.

One such problem was touched on by both Professor Atkinson and Professor Lamb – not only the likely heavy incidence of cloudy skies in the north, but even more certainly, the heavy forest cover of all Europe in the temperate botanical climax of the Atlantic and Sub-Boreal phases. A glance for instance at McVean and Ratcliffe's maps in their *Plant communities of the Scottish Highlands* (1962) shows the density of natural woodland over the areas in which so many of the monuments under discussion lie, and the same goes for other regions of their occurrence: indeed Professor Thom's own slides forcibly demonstrated the difficulty of modern survey in the secondary woodland growths of parts of Brittany. And clear skies – how often were they to be seen? Even in the Aegean Homer, in that fine sustained simile at the close of Book VIII of the *Iliad*, where the Achaeans see at night the lights of beleagured Troy shining like stars, adds as a proviso that it would be an exceptional sight, as when 'all the stars are seen, to make glad the heart of the shepherd'. All stars at night, astronomers' delight, and especially when seen from Callanish.

There are other problems too, integral to archaeology and history. Few of us would not like to think of the term 'megalithic' in European prehistory as having anything more than a literal meaning, 'of big stones', and in no way a cultural trait uniting different communities widely separated in space and time; chambered tombs, stone circles, cairns and standing stones need have no relationship to one another, and if we are not careful, we will find alinements as meaningless as the 'old straight tracks' of Alfred Watkins. The astronomy cannot be pursued *in vacuo*, but only with a full knowledge of the archaeology involved, as one hopes this meeting has demonstrated. There is always the danger of seeing ourselves in the past, of becoming victims of the fallacy whereby 'ideas are imported from present-day experience, and ancient man is anachronistically saddled with views he would have found at best strangely unfamiliar', as Ian Richmond put it, or of the unconscious tendency 'to project the axioms, habits of thought and norms of the present day into the past', in Henri Frankfort's phrase. God-like, we try to make ancient man in our own image, and the preferred image varies with the changes of taste and preference of our society. We desire to find admired qualities in the past, and mathematical and scientific qualities are admired today. If ecstasy and shamanism were more highly regarded than these, this is what we might be looking for – and doubtless finding – in prehistory. Observer-imposed categories are dangerous things. Professor Aaboe told us the entrancing story of Neugebauer finding an unsuspected calendrical significance in the teeth of his pocket comb: there is a well-known principle in logic known as Occam's razor, and may we not also need to apply the concept of Neugebauer's comb from time to time in our inquiries?

But perhaps revolutionary concepts are indeed upon us, and we must be prepared for all sorts of surprises. When Dr Newton quoted the Archicholos poem about a solar eclipse I remembered it started –

> There's nothing now
> We can't expect to happen

and, after enumerating several wild improbabilities, ended

> I wouldn't be surprised
> I wouldn't be surprised.

Nor would I.

And finally, and on behalf of the British Academy, I must echo King-Hele in thanking the Royal Society for all they have done to make this meeting possible, and to express the great debt we all owe to the work of Dr Roy Hodson in so effectively carrying the burden of so much complex organization.